The Most Successful Way of JEE Preparation

DPP

Daily Practice Problems

JEE Main & Advanced

Heat & Thermodynamics

PHYSICS Vol-4

The Most Successful Way of JEE Preparation

DPP

Daily Practice Problems

JEE Main & Advanced

Heat & Thermodynamics

PHYSICS Vol-4

DEEPAK PALIWAL

arihant

ARIHANT PRAKASHAN (SERIES), MEERUT

DPP

Daily Practice Problems

arihant

ARIHANT PRAKASHAN (SERIES), MEERUT

卐 Administrative & Production Offices

Corporate Office 'Ramchhaya' 4577/15, Agarwal Road, Darya Ganj, New Delhi -110002
Tele: 011- 47630600, 43518550; Fax: 011- 23280316

Head Office Kalindi, TP Nagar, Meerut (UP) - 250002
Tel: 0121-2401479, 2512970, 4004199; Fax: 0121-2401648

卐 Sales & Support Offices

Agra, Ahmedabad, Bengaluru, Bareilly, Chennai, Delhi, Guwahati, Hyderabad, Jaipur, Jhansi, Kolkata, Lucknow, Meerut, Nagpur & Pune

卐 **ISBN :** 978-93-13193-34-0

Published by Arihant Publications (I) Ltd.

Production Team

Publishing Managers	Mahendra Singh Rawat & Keshav Mohan	*Inner Designer*	Mazher Chaudhary
Project Head	Sorabh Chaudhary	*Page Layouting*	Sunil Sharma, Hemant
Cover Designer	Shanu Mahad	*Proof Reading*	Ashish

For further information about the products from Arihant, log on to www.arihantbooks.com or email to info@arihantbooks.com

/arihantpub /@arihantpub Arihant Publications /arihantpub

ABOUT DPP

Daily Practice Problems

Indian Institutes of Technology better known as IITs have always been the dream destination of each science group students who want to pursue their career as Engineers. These are the institutes which have got very high value not even in India but in the whole world. To get the admission in such institutes one need to face entrance Examinations; JEE Main & JEE Advanced, out of which JEE Main is considered as the preliminary exam & JEE Advanced is the final exam to get into the IITs. Approximately per year 12 lakh students enroll themselves for JEE Main and only the top 2.25 lakh in JEE Main merit can attempt for JEE Advanced, the fact clearly express the degree of toughness of JEE Main & JEE Advanced.

To have success in such an exam one needs to be very planned & systematic in its preparation. A very popular wise saying 'Rome was not built in a day' suggests that to achieve a big target like IITs we need to work continuously. For continuous work, one need to have proper planning that how much work he/she has to do on regular basis, i.e., on daily basis. Each student preparing for JEE has many books but unfortunately none of them gives the students a plan for daily studies, which is the most effective success key for JEE.

Keeping the daily practice needs of the students in mind we have come up with DPP Series (Daily Practice Problems) having approximately 26 books. The primary focus of this series is on achieving success through practice & proper planning. For that we have divided the whole syllabus of JEE Main & JEE Advanced; Physics, Chemistry & Mathematics, into different volumes (around 8 volumes of Mathematics & 9 volumes each of Physics & Chemistry), each volume has Daily Practice Problems Sheets having specific questions on various Topics of the individual chapter of the volume ensuring the complete Practice & Assessment of the topic.

In this way we have tried to give you an effective daily planning for JEE keeping in mind the daily hours that a student can devote to his practice, you are just required to practice one sheet of each subject on daily basis. This habit of daily practice will keep you away from exam stress and thoughts like; how much I have prepared/am I preparing properly or not/ do I need to study more on daily basis or what I am doing is ok, etc., which regularly tease your mind and can give you panic at any stage of your preparation.

Some special features that we have tried to incorporate in our Daily Practice Problems (DPP) series are discussed below

1. Micro Level coverage of all the topics of a Chapter/ Unit given in JEE Main & Advanced Syllabus.
2. Each DPP has various types of questions; Subjective Questions, Single/ Multiple Correct Options, Statements type, Comprehension based, Integer Value Answer type, Matching, etc., making you practice & ready for all the question formats of JEE.
3. New & innovative questions in each DPP challenging & improving your problem solving skills.
4. Along with Topical Coverage, Revisal DPPs for JEE Main & JEE Advanced with each Chapter, covering multidisciplinary questions proving comprehensive practice.
5. JEE Main & JEE Advanced Archive (collection of last 13 years' exams questions) with each chapter.
6. Complete Solutions for each DPP.

We hope that DPP series will prove to be very beneficial for the students preparing for JEE, by giving them daily practice in a systematic & planned manner. The suggestions/criticism from the students, teachers & parents about DPP series will be welcomed whole heartedly and we promise to update this series accordingly from time to time.

DPP
Daily Practice Problems

DAILY TASKS...

1 HEAT, TEMPERATURE AND HEAT TRANSFER

2 KINETIC THEORY OF GASES AND THERMODYNAMICS

1. Heat, Temperature and Heat Transfer

DPP
DAILY PRACTICE PROBLEMS

DPP-1 Heat, Temperature and Expansion of Solids

Subjective Questions

Working Space

Direction (Q. Nos. 1-4) *These questions are subjective in nature, need to be solved completely on notebook.*

1 A platinum wire is used as a resistance thermometer. The wire's resistance was found to be 10 Ω and 16 Ω at ice point and steam point respectively and 30 Ω at boiling point of sulphur at 445°C. Find the resistance of wire at 500°C, it is found that resistance of platinum wire varies with temperature as, $R = R_0 (1 + \alpha t + \beta t^2)$, where R_0 is resistance at 0°C.

2 In a constant volume gas thermometer, following are the pressure values recorded at boiling points of water and sulphur.

Substance	Pressure of Hg (in mm) at BP			
Water	50	100	200	300
Sulphur	96.4	193	387	582

What is the boiling point of sulphur at 1 atm pressure, obtained from the above readings ?

3 i. A pendulum of a wall clock is calibrated at 17°C. How much time is gained or lost by clock in a year if the clock is kept at 28°C? (Given, $\alpha = 19 \times 10^{-6}\,{}^\circ C^{-1}$)

ii. A solid aluminium cylinder is rotating about its own axis in frictionless bearings with a constant angular velocity ω ($\alpha_{Al} = 25 \times 10^{-6}\,{}^\circ C^{-1}$). What is the percentage change in angular speed of cylinder per degree rise of temperature ?

4 Temperature of an aluminium rod is increased from 18°C to 35°C. What stress must be applied to keep the aluminium rod to same length? (Given, $\alpha_{Al} = 25 \times 10^{-6}\,({}^\circ C^{-1})$, $Y_{Al} = 70 \times 10^9\ Nm^{-2}$)

Only One Option Correct Type

Direction (Q. Nos. 5-15) *This section contains 11 multiple choice questions. Each question has four choices (a), (b), (c) and (d), out of which ONLY ONE is correct.*

5 Which of the following is a constant for all temperature scales?

a. UFP
b. LFP
c. UFP − LFP
d. $\dfrac{X - \text{LFP}}{\text{UFP} - \text{LFP}}$, where X = any temperature

6 A new scale N of temperature is divided in such a way that the freezing point of water is 100°N and boiling point is 400°N. The Celsius and new temperature scale readings will be same at

a. −40°C
b. 40°C
c. −50°C
d. 50°C

Working Space

7 The graph between two temperature scales A and B is shown in the figure. Between lower fixed point and upper fixed point, there are 150 equal divisions on scale A and 100 on scale B. Which of the given relation(s) is/are correct?

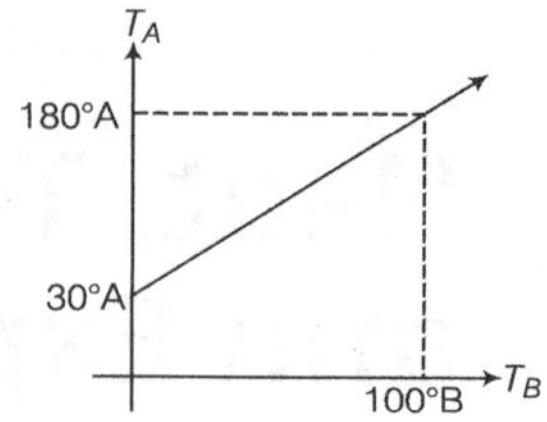

a. $\frac{t_A - 180}{150} = \frac{t_B}{150}$
b. $\frac{t_A - 30}{150} = \frac{t_B}{100}$
c. $\frac{t_B - 180}{150} = \frac{t_A}{100}$
d. $\frac{t_B - 40}{100} = \frac{t_A}{180}$

8 Heat is associated with

a. kinetic energy of random motion of molecules
b. kinetic energy of orderly motion of molecules
c. total kinetic energy of random and orderly motion of molecules
d. kinetic energy of random or orderly motion of molecules

9 A bimetallic strip is made up with aluminium and steel ($\alpha_{Al} > \alpha_{steel}$). On heating, the strip will

a. remain straight
b. get twisted
c. bend with aluminium on concave side
d. bend with steel on concave side

10 As temperature is increased, the time period of a simple pendulum

a. increases as its effective length increases even though its centre of mass still remains at centre of bob
b. remains same as its centre of mass still remains at centre of bob
c. increases as its effective length increases due to shifting of its centre of mass below the centre of bob
d. decreases as its effective length remains same but the centre of mass shifts above the centre of the bob

11 When guests arrive, Shagun put gulab jamuns and pizzas into OTG (oven - toaster - grill) and after few minutes, she touched one of the items of the tray to see whether it is hot or not and then serve them to guests. Choose the correct option.

a. Small gulab jamuns are more hotter than larger ones
b. Large gulab jamuns are more hotter than larger ones
c. Small pizzas are more hotter than larger ones
d. Larger pizzas are more hotter than smaller ones

12 Increase in moment of inertia of a uniform rod of length l (α = coefficient of linear expansion) about its perpendicular bisector when its temperature is slightly increased by Δt, is

a. $l\alpha\Delta T$
b. $\frac{2}{3} l\alpha\Delta T$
c. $2l\alpha\Delta T$
d. $\frac{1}{2} l\alpha\Delta T$

13 A scale is prepared whose length does not change with temperature. For this purpose metallic strips of brass and iron of different lengths are used. If maximum length which can be measured by scale is 10 cm, then lengths of iron and brass strips are respectively. (Given, $\alpha_{iron} = 1.2 \times 10^{-5} K^{-1}$ and $\alpha_{brass} = 1.8 \times 10^{-5} K^{-1}$).

a. 30 cm, 20 cm
b. 20 cm, 30 cm
c. 40 cm, 50 cm
d. 50 cm, 40 cm

14 A thermometer has wrong calibrations. It reads melting point of ice as $-10°C$ and reads 60°C at a temperature of 50°C. Temperature of boiling point noted by this thermometer will be

a. 120°C
b. 110° C
c. 90° C
d. 130° C

Working Space

15 The volume of a metal block changes by 0.12%, when it is heated through 20°C. Coefficient of linear expansion for material of block will be

a. $6\times10^{-5}\,°C^{-1}$ b. $2\times10^{-5}\,°C^{-1}$

c. $4\times10^{-5}\,°C^{-1}$ d. $3\times10^{-5}\,°C^{-1}$

One or More than One Options Correct Type

Direction *(Q. Nos. 16-20) This section contains 5 multiple choice questions. Each question has four choices (a), (b), (c) and (d), out of which ONE or MORE THAN ONE are correct.*

16 When a solid is heated (without phase change), which of the following occurs?

a. Intermolecular distance increases

b. Potential energy of molecules increases

c. Kinetic energy of molecules increases

d. Amplitude of vibrations of molecules increases

17 An aluminium bush is to be fitted over a brass rod. At 20°C, inner diameter of aluminium ring is 5.00 cm and brass rod has a diameter of 5.05 cm. (Given, $\alpha_{Al} = 24\times10^{-6}\,°C^{-1}$, $\alpha_{brass} = 19\times10^{-6}\,°C^{-1}$).

To make the assembly, which of the following can be performed?

a. Aluminium ring is heated above 437°C

b. Brass rod is cooled

c. Both aluminium ring and brass rod can be heated ($\alpha_{Al} > \alpha_{brass}$)

d. Both aluminium ring and brass rod can be cooled using liquid nitrogen

18 A rod of original length L_1 changes its temperature from T_1 to T_2. Let α is temperature coefficient of linear expansion and L_2 is length of temperature T_2. Choose the correct options.

a. $L_2 = L_1[1+\alpha(T_2-T_1)]$, where $T_2 - T_1$ and α are not large

b. $L_2 = L_1 e^{\alpha(T_2-T_1)}$, where α is constant

c. $L_2 = L_1 e^{\int_{T_1}^{T_2}\alpha dT}$, where α is a function of temperature

d. Formula in option (a) is valid in all conditions

19 A hollow copper tube is filled completely using a steel rod. The rod and tube are joined by two rivet pins as shown in the figure.

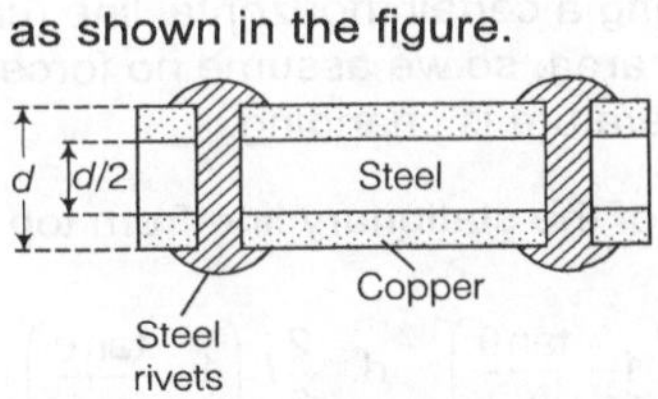

Temperature of assembly is raised by ΔT after setting the rivets. Choose the correct options.

a. Stress in copper (σ_{Cu}) tube $= -\frac{1}{3}$ Stress in steel (σ_{steel}) tube

b. Stress in copper tube = − Stress in steel tube

c. Strain in steel tube $= \alpha_{steel}\,\Delta T$

d. Strain in copper tube $= \frac{\sigma_{Cu}}{Y_{Cu}} + \alpha_{Cu}\,\Delta T$

20 A bimetallic strip is formed using two identical strips of copper and brass. On heating temperature of strip rises by ΔT and strip bends to form an arc of radius of curvature R. Then, R is

a. proportional to ΔT b. inversely proportional to ΔT

c. proportional to $|\alpha_{brass} - \alpha_{copper}|$ d. inversely proportional to $|\alpha_{brass} - \alpha_{copper}|$

Working Space

Statement Type

Direction *(Q. Nos. 21 and 22) This section is based on Statement I and Statement II. Select the correct answer from the codes given below.*

a. Both Statement I and Statement II are correct and Statement II is the correct explanation of Statement I
b. Both Statement I and Statement II are correct but Statement II is not the correct explanation of Statement I
c. Statement I is correct but Statement II is incorrect
d. Statement II is correct but Statement I is incorrect

21 **Statement I** When a metallic sphere is heated, percentage increase in surface area is more than the percentage increase in its volume.

Statement II Coefficient of superficial expansion is $\frac{2}{3}$rd of coefficient of volumetric expansion.

22 **Statement I** Most of the solids expand on heating.

Statement II Potential energy curve is asymmetric about the equilibrium distance between neighbouring atoms.

Comprehension Type

Direction *(Q. Nos. 23-25) This section contains a paragraph, describing theory, experiments, data, etc. Three questions related to the paragraph have been given. Each question has only one correct answer among the four given options (a), (b), (c) and (d).*

Passage

A perfectly plane house roof makes an angle θ with the horizontal surface. A flat square metal plate of side L rests over the roof.

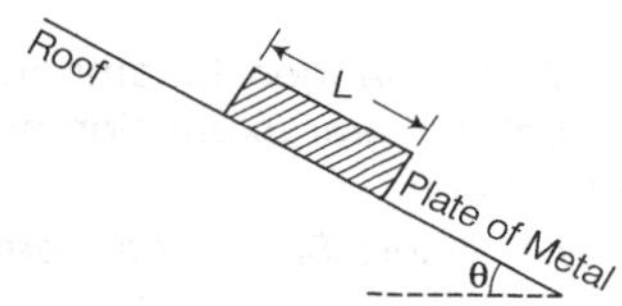

Each day temperature varies between day's high (T_1) and day's low (T_2). Coefficients of thermal expansions of roof and plate are α_1 and α_2 ($\alpha_2 > \alpha_1$) respectively. Coefficient of kinetic friction between plate and roof is μ_k.

Plate and roof are always in contact and both have same temperature at every instance of day.

Because of difference in coefficients of thermal expansions, each bit of plate is moving relative to the roof, except for points along a certain horizontal line (called stationary line). The stationary line occupies no area, so we assume no force of static friction acting on the plate while the temperature is changing.

23 When the temperature is raising, then distance of the stationary line from top edge of plate will be

a. $\frac{L}{2}\left(1-\frac{\tan\theta}{\mu_k}\right)$ b. $L\left(1-\frac{\tan\theta}{\mu_k}\right)$ c. $2L\left(1-\frac{\tan\theta}{\mu_k}\right)$ d. $\frac{2}{3}L\left(1-\frac{\tan\theta}{\mu_k}\right)$

24 When the temperature is falling, then distance of the stationary line from bottom edge of plate will be

a. $\frac{L}{2}\left(1-\frac{\tan\theta}{\mu_k}\right)$ b. $L\left(1-\frac{\tan\theta}{\mu_k}\right)$ c. $2L\left(1-\frac{\tan\theta}{\mu_k}\right)$ d. $\frac{2}{3}L\left(1-\frac{\tan\theta}{\mu_k}\right)$

25 The distance moved by plate relative to roof (down the roof) in each day is

a. $\frac{L(\alpha_2-\alpha_1)\tan\theta}{\mu_k}$ b. $\frac{\frac{2}{3}L(\alpha_2-\alpha_1)\tan\theta}{\mu_k}$

c. $\frac{L(\alpha_2-\alpha_1)(T_1-T_2)\tan\theta}{\mu_k}$ d. $\frac{\frac{2}{3}L(\alpha_2-\alpha_1)(T_1-T_2)\tan\theta}{\mu_k}$

Working Space

Matching List Type

Direction *(Q. No. 26) Choices for the correct combination of elements from Column I and Column II are given as options (a), (b), (c) and (d), out of which one is correct.*

26 Match entries in Column I (type of thermometer) with their correct pairs in Column II (range of temperature).

	Column I		Column II
i.	Liquid filled thermometers	p.	800°C to 6000°C
ii.	Gas thermometers	q.	0°C to 1000°C
iii.	Resistance thermometers	r.	– 100°C to 1200°C
iv.	Thermoelectric thermometers	s.	– 200°C to 500°C
v.	Radiation pyrometers	t.	– 80°C to 350°C

Codes

	i	ii	iii	iv	v
a.	t	s	r	q	p
b.	t	s	r	p	q
c.	q	p	r	s	t
d.	p	q	r	s	t

One Integer Value Correct Type

Direction *(Q. Nos. 27 and 28) This section contains 2 questions. When worked out will result in an integer from 0 to 9 (both inclusive).*

27 Two rods of length $L = 1$ m and coefficient of linear expansion $\alpha = 2.4 \times 10^{-5}\,{}^\circ C^{-1}$ are connected to another rod of length L_1 and coefficient of linear expansion $\alpha_1 = 6 \times 10^{-6}\,{}^\circ C^{-1}$; to form an isosceles triangle. The arrangement is supported on the knife-edge at the mid-point L_1 which is horizontal. The apex of isosceles triangle remains at a constant distance from the knife-edge (at all temperatures) when, $\frac{L_1}{L} = K$. Find the value of K.

28 Two rods one of steel and other of aluminium is equal to cross-sectional area are joined together and the free ends are fixed between two rigid supports as shown in the figure.

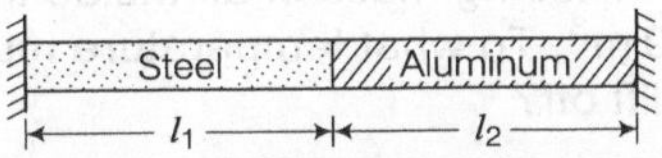

Temperature of surroundings is increased by 100°C. If displacement of joint rods is $(N^2/10)$mm, then find the value of N with the help of following table.

Metal	Coefficient of expansion α (° C^{-1})	Young's modulus Y (Nm^{-2})	Length l (m)
Steel	1.2×10^{-5}	20×10^{10}	20
Aluminium	2.4×10^{-5}	7×10^{10}	7

DPP-2 Expansion of Liquids and Gases

Subjective Questions

Working Space

Direction *(Q. Nos. 1-4) These questions are subjective in nature, need to be solved completely on notebook.*

1 A steel container ($\alpha_{steel} = 11 \times 10^{-6} {}^\circ C^{-1}$) having a volume of 125 cm^3 contains a gas of 202 kPa at 22°C.

i. When container is heated at 195°C, what is the pressure inside the container? Assuming there is no change in volume of the container.

ii. When volume change of steel container is taken in account, what will be the change in the value of pressure obtained previously?

2 A hollow aluminium cylinder 20 cm deep with a capacity of 2 L is filled with turpentine at 20°C. The cylinder is filled upto the top and then it is heated upto 80°C.

i. How much turpentine overflows?

ii. If the cylinder is then cooled back to 20°C. How much level of turpentine surface falls below the rim?

3 Consider a Helium gas enclosure in space of size same as the size of the earth. Helium is initially at a temperature of 50 K and a pressure of 1 atmosphere.

i. Find the mass of gas in the enclosure.

ii. When an extra mass of 8×10^{20} kg of helium is filled into the enclosure, due to heat from the sun, (at same time) the pressure in enclosure is doubled. Find final temperature of enclosure, if volume remains constant.

4 Mass of a hot air balloon and its cargo (without including mass of air inside the balloon) is 200 kg. The volume of balloon is 400 m^3. To what temperature must be the air in balloon be heated so that the balloon lift off?
(Take, $\rho_{air} = 1.25$ kg m^{-3} at 10°C).

Only One Option Correct Type

Direction *(Q. Nos. 5-9) This section contains 5 multiple choice questions. Each question has four choices (a), (b), (c) and (d), out of which ONLY ONE is correct.*

5 A one litre glass beaker is filled upto the top with mercury at 0°C. Coefficient of volumetric expansion of mercury is $1.82 \times 10^{-4} {}^\circ C^{-1}$ and linear expansion of the glass is $0.3 \times 10^{-4} {}^\circ C^{-1}$. How much mercury will flow out, if temperature is raised at 100°C?

a. 21.2 cc
b. 15.2 cc
c. 1.52 cc
d. 2.12 cc

6 Liquid is filled in a flask. When the flask is heated, the level of liquid

a. keeps on increasing
b. keeps on decreasing
c. initially decreases, then increases
d. remains same

Working Space

7 A metal ball immersed in a liquid weights w_1 at $0°C$ and w_2 at $60°C$. The coefficient of cubical expansion of ball is less than that of liquid. Then, the correct option is

a. $w_1 = w_2$ b. $w_1 < w_2$

c. $w_1 > w_2$ d. $w_1 = \frac{w_2}{2}$

8 A metal block floats in mercury at $0°C$ with k_1 fraction of its total volume submerged. When temperature is raised to $60°C$, a fraction k_2 of total volume is found submerged. If the coefficient of volumetric expansion of metal is γ_1 and coefficient of volumetric expansion of mercury is γ_2, then ratio of k_1/k_2 is

a. $\frac{1+60\gamma_1}{1+60\gamma_2}$ b. $\frac{1-60\gamma_1}{1+60\gamma_2}$

c. $\frac{1+60\gamma_1}{1-60\gamma_2}$ d. $\frac{1+60\gamma_2}{1+60\gamma_1}$

9 A piston of area 1 m^2 encloses air column of height 1 m in a piston cylinder arrangement.

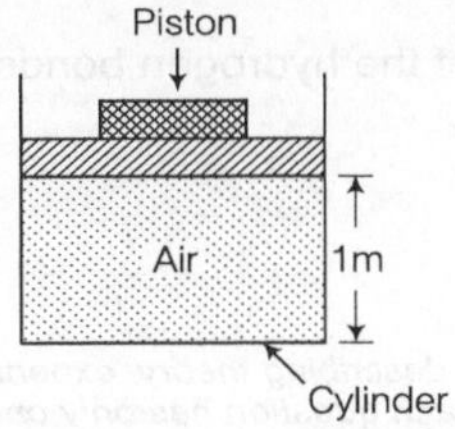

If temperature is raised to 400 K, then the height of piston above the base is

a. $\frac{3}{4}$ m b. 2 m

c. $1\frac{1}{2}$ m d. $\frac{4}{3}$ m

One or More than One Options Correct Type

***Direction** (Q. Nos. 10-14) This section contains 5 multiple choice questions. Each question has four choices (a), (b), (c) and (d), out of which ONE or MORE THAN ONE are correct.*

10 A closed container having a volume of 0.02 m^3 contains a mixture of neon and argon at $27°C$ and pressure of $1 \times 10^5 \text{ Nm}^{-2}$. The mass of mixture is 28 g. Choose the correct options.

a. Mass of neon in mixture is 4.074 g
b. Mass of argon in mixture is 23.926 g
c. Mass of neon in mixture is 23.926 g
d. Mas of argon in mixture is 4.074 g

11 An ideal gas is expanding such that, pT^3 is a constant. The coefficient of volumetric expansion of the gas is

a. $\frac{4}{T}$ b. $\frac{2}{T}$

c. $\frac{3}{T}$ d. $\frac{1}{T}$

Working Space

12 Iron floats over mercury because density of mercury is more than the density of iron. Coefficient of volumetric expansion of iron and mercury are $35 \times 10^{-6} {}^\circ C^{-1}$ and $180 \times 10^{-6} {}^\circ C^{-1}$. Now consider an iron cube floating in a bowl of mercury at $0^\circ C$. When temperature is increased upto $25^\circ C$, then choose the correct options.

a. Cube submerged more at 25° C
b. Cube displayed more at 25° C
c. Percentage change in the fraction of volume submerged is about 1%
d. Percentage change in the fraction of volume submerged is about 0.4%

13 Density of gasoline at $0^\circ C$ is 6.8×10^2 kgm^{-3} and coefficient of volumetric expansion is $950 \times 10^{-6} {}^\circ C^{-1}$. Choose the correct options.

a. Density at 35° C is 660 kg m^{-3}
b. Density at 35° C is 670 kg m^{-3}
c. Percentage change in density at 35° C is – 3%
d. Percentage change in density at 35° C is – 2%

14 Which of these are true regarding anamolous expansion of water?

a. Water expands below 4° C to 0° C
b. Water contracts below 0° C to -4° C
c. Anamolous expansion of water is due to formation of the hydrogen bonds
d. Water expands when heated beyond 4° C

Comprehension Type

Direction (Q. Nos. 15-20) *This section contains 3 paragraphs, each describing theory, experiments, data, etc. Six questions related to the paragraphs have been given. Each question has only one correct answer among the four given options (a), (b), (c) and (d).*

Passage I

Most of the real gases obey Boyle's law at the room temperature. So, the product $p \times V$ is a constant as long as temperature remains constant. Now, consider a tube closed at both ends consists of some mercury trapped in a column of air as shown in the figure.

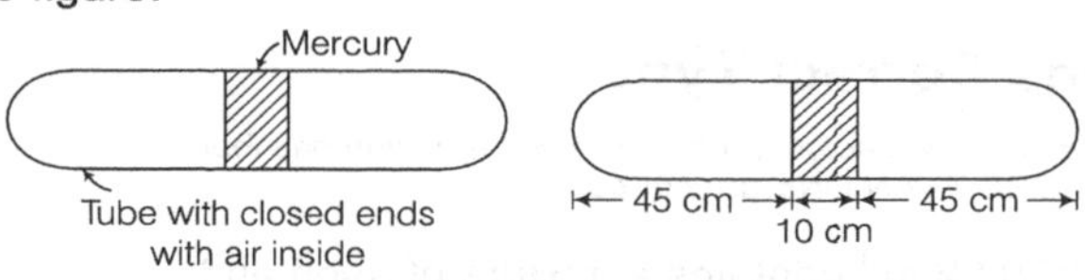

15 If tube is turned to vertical position, what will be the displacement in level of mercury column?

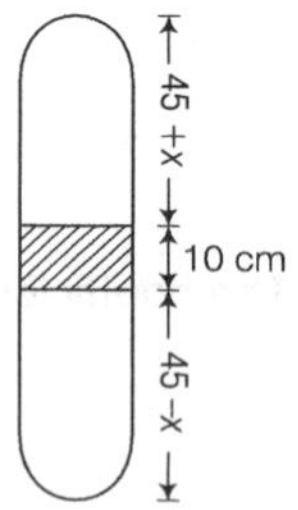

a. 10 cm
b. 5 cm
c. 4 cm
d. 3 cm

Working Space

16 If tube is held at an angle of 60° with vertical, the length of air column above and below mercury column are 46 cm and 44 cm respectively. Initial pressure inside the tube is

a. less than 76 cm of mercury
b. above 110 cm of mercury
c. 5 cm of mercury
d. below 2.5 cm of mercury

Passage II

When a gas is expanded or compressed suddenly, so that no heat is either absorbed or released by the gas, such an expansion (or compression) is called adiabatic expansion (or compression). Equation of state during an adiabatic change is pV^{γ} = constant, where, γ = a constant depending on gas. Also, if gas is considered an ideal gas, then $pV = nRT$ is also true.

17 A monoatomic ideal gas $\left(\gamma = \frac{5}{3}\right)$, is initially at temperature T_1. It is expanded adiabatically to a temperature T_2. If L_1 and L_2 are length of gas columns initially and after expansion, then T_1/T_2 is given by

a. $\left(\frac{L_1}{L_2}\right)^{2/3}$
b. $\frac{L_1}{L_2}$
c. $\frac{L_2}{L_1}$
d. $\left(\frac{L_2}{L_1}\right)^{2/3}$

18 5.6 L of helium gas at STP is adiabatically compressed to 0.7 L. Taking T_1 as initial temperature, final temperature of helium will be $\left(\text{take, } \gamma = \frac{5}{3}\right)$

a. $4T_1$
b. $\frac{T_1}{4}$
c. $16T_1$
d. $\frac{T_1}{16}$

Passage III

A uniform rod of length L and density $2k$ is held in vertical position with its lower end just touching the upper layer of a liquid of density k as shown in the figure.

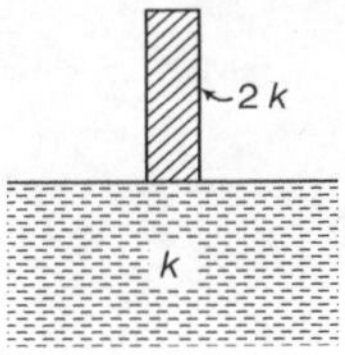

The rod is released at $t = 0$ s.

19 Acceleration of rod as a function of rod's length x going into the fluid, is

a. $g - \frac{gx}{2L}$
b. $g - \frac{gx}{3L}$
c. $\left(g - \frac{gx}{L}\right) \times \frac{1}{2}$
d. $g - \frac{gx}{L}$

20 The rod is dropped inside fluid and its speed is noted, when half of length of rod reached inside fluid. Let it be v_1. Now fluid is heated, so that its volume increases by $\frac{1}{4}$ th of its initial value and rod is again dropped inside hot fluid and speed of rod under similar condition is noted. Let it be v_2. The ratio $\frac{v_2}{v_1}$ will be

a. $\frac{4}{1}$
b. $\frac{\sqrt{2}}{4}$
c. $\frac{6}{\sqrt{35}}$
d. $\frac{8}{\sqrt{35}}$

Matching List Type

Direction (Q. No. 21) *Choices for the correct combination of elements from Column I and Column II are given as options (a), (b), (c) and (d), out of which one is correct.*

21 Potential energy U (for a pair of atoms) *versus* interatomic separation r for two substances A and B are as shown.

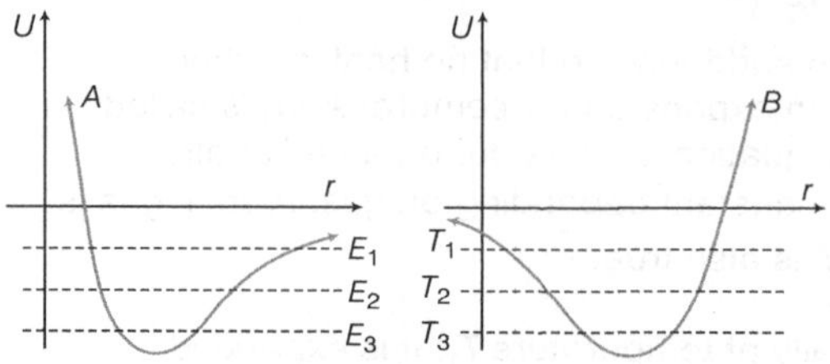

For both substances, graphs of $U_{v/s}\, r$ are on same scale.
where, E_1, E_2 and E_3 are total energy line (for a pair of atoms) and T_1, T_2 and T_3 are the corresponding temperatures. Also, $T_1 > T_2 > T_3$.
Match the following columns and select the correct option from the codes given below.

	Column I		Column II
i.	Expands on heating	p.	Substance A
ii.	Contracts on heating	q.	Substance B
iii.	Liquifies above temperature T_1		
iv.	Liquifies above temperature T_2		

Codes

	i	ii	iii	iv
a.	p	q	p	q
b.	q	p	p	q
c.	p	q	p,q	p
d.	q	p	p	p,q

Statement Type

Direction (Q. No 22) *This section is based on Statement I and Statement II. Select the correct answer from the codes given below.*

a. Both Statement I and Statement II are correct and Statement II is the correct explanation of Statement I
b. Both Statement I and Statement II are correct but Statement II is not the correct explanation of Statement I
c. Statement I is correct but Statement II is incorrect
d. Statement II is correct but Statement I is incorrect

22 **Statement I** A liquid is taken in two flasks, at temperature t_1 and t_2 with volumes V_1 and V_2. When they are mixed, volume of mixtrue formed depends on equilibrium temperature.

Statement II Equilibrium temperature of mixture is

$$t = \frac{m_1 t_1 + m_2 t_2}{m_1 + m_2}$$

Working Space

DPP-3 Calorimetry

Subjective Questions

Working Space

Direction *(Q. Nos. 1-4) These questions are subjective in nature, need to be solved completely on notebook.*

1 A calorimeter contains 10 kg of water and some ice. Variation of temperature with time, when the calorimeter is heated over a slow burner is as follows.

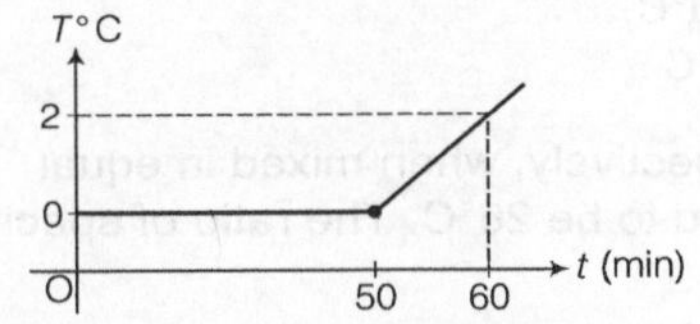

Ignoring heat absorbed by calorimeter, find the amount of ice present.

2 A 670 kg meteorite happens to be composed of aluminium. When it is far from the earth, its temperature is $-15°$C and it moves with a speed of 14 kms^{-1} relative to the earth. As, it crashes into earth, let total energy is equally shared between planet and meteorite. Specific heat of liquid and gaseous aluminium be 1170 J $kg^{-1}C^{-1}$. Find the final temperature of meteorite.

3 An electric kettle has 750 W-240 V heater, is plugged to a 200 V mains. If heat capacity of kettle is 400 JK^{-1} and the initial water temperature is 20°C. Then, how much time is required to boil 500 g of water?

4 The X-ray tube, 10^{18} electrons per second arrive at the metal target with speed of 2×10^6 ms^{-1}. Mass of metal target is 200 g, specific heat of target is 0.5 $Jg^{-1}K^{-1}$. Assuming 98% energy is converted into the heat, how much time is required to raise temperature of target by 5°C?

Only One Option Correct Type

Direction *(Q. Nos. 5-15) This section contains 11 multiple choice questions. Each question has four choices (a), (b), (c) and (d), out of which ONLY ONE is correct.*

5 Water is used as a hot fluid in hot water bags as well as a coolant in cars. Which of the following reasons supports it best?

a. It is easily available
b. It is in liquid form
c. Its boiling point at 100°C
d. It has high specific heat

6 Which of the following has the largest specific heat capacity?

a. Iron
b. Water
c. Mercury
d. Hydrogen

Working Space

7 5 g ice at 0°C is mixed with 1g steam at 100°C. Which of the following statements is correct?

a. Final mixture contains water and ice at 40°C
b. Final mixture contains water at 40°C
c. Final mixture contains ice, water and steam at 40°C
d. Final mixture contains water and steam at 40°C

8 100 g ice at 0°C is mixed with 10 g steam at 100°C. Final mixture contains

a. only water at 0°C
b. only ice at 0°C
c. 90 g water and 20 g ice at 0°C
d. 90 g ice and 20 g water at 0°C

9 30 g ice at 0°C is mixed with 25 g steam at 100°C. The temperature of final mixture is

a. 0°C
b. 100°C
c. 50°C
d. 40°C

10 Two liquids *A* and *B* are at 30°C and 20°C respectively, when mixed in equal masses, the temperature of the mixture is found to be 26°C. The ratio of specific heat of both liquids is

a. 3 : 2
b. 1 : 1
c. 2 : 3
d. 4 : 3

11 A dieter wants to burn 2000 k cal by doing lifting exercise using a 50 kg barbell. He lifts the barbell 2 m each time and gains no energy while lowers down it. Number of times he has to raise barbell, is
(Given, $g = 10\ ms^{-2}$, 4.2 J = 1 cal)

a. 9400
b. 7400
c. 8400
d. 6400

12 Two spheres of same material with ratio of diameters 1 : 2 have their thermal capacities are in the ratio of

a. 1 : 2
b. 1 : 1
c. 1 : 8
d. 1 : 4

13 Specific heat of a substance varies with temperature as,

$$C = 0.2 + 0.14t + 0.023t^2 \ \ (\text{cal g}^{-1}{}^\circ\text{C}^{-1})$$

The heat required to raise the temperature of 2 g of substance from 5°C to 15°C will be

a. 24 cal
b. 56 cal
c. 82 cal
d. 100 cal

14 Work done in converting 1 g of ice at −10°C into steams at 100°C is

a. 3045 J
b. 6056 J
c. 721 J
d. 616 J

15 2 kg of ice at −20°C is mixed with 5 kg of water at 20°C. Final mass of water in the mixture will be

a. 7 kg
b. 6 kg
c. 4 kg
d. 2 kg

Working Space

One or More than One Options Correct Type

Direction *(Q. Nos. 16-20) This section contains 5 multiple choice questions. Each question has four choices (a), (b), (c) and (d), out of which ONE or MORE THAN ONE are correct.*

16 Heat absorbed by a metallic body is in the form of

a. kinetic energy of its molecules
b. potential energy of its molecules
c. kinetic energy and potential energy of its molecules
d. kinetic energy associated with electrons of molecules

17 Refer to plot of temperature and time showing temperature change and state, changes in ice on heating.

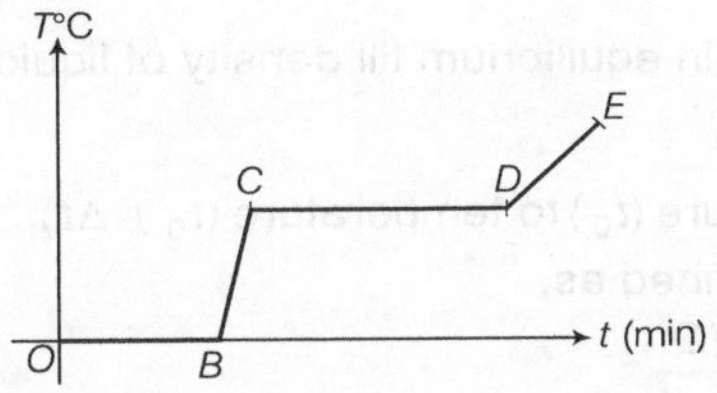

Which of the following are correct ?

a. Region *OB* represents ice and water in thermal equilibrium
b. At *B* water starts boiling
c. At *C* all the water is converted into steam
d. *C* to *D* represents water and steam in equilibrium at boiling point

18 Let p = pressure, L = latent heat of fusion, T = melting point temperature, V_2 = volume of solid state and V_1 = volume of liquid state. Then, which of these are correct regarding melting point of a solid?

a. $\frac{dp}{dT} = \frac{L}{T(V_2 - V_1)}$
b. When $V_2 > V_1$, then $\frac{dp}{dT} > 0$ and melting point increases with increase in pressure
c. When $V_2 < V_1$, then $\frac{dp}{dT} < 0$ and melting point decreases with increase in pressure
d. Melting point of all solids does not vary with pressure

19 An ice cube of mass 0.1 kg at 0°C is placed in an isolated container at 227°C. Specific heat of container is given by $S = A + BT$, where $A = 100$ calkg^{-1}K^{-1} and $B = 2\times10^{-2}$ calkg^{-2}K^{-2}. If final temperature after stabilisation is 27°C, then which of these are correct ? (Take, $L_{water} = 80$cal g^{-1}K^{-1} and $C_{water} = 10^3$ cal kg^{-1}K^{-1})

a. Ice receives a total of 10700 calories of heat
b. Container lost around 10800 calories of heat
c. Container has a mass of around 0.5 kg
d. Water equivalent of container is 300 g

20 Equal masses of three different liquids *A*, *B* and *C* are taken at 10°C, 15°C and 20°C respectively. The steady state temperature when *A* and *B* are mixed, is 13°C and when *B* and *C* are mixed, is 16°C. Which of the following statements is/are correct ?

a. Ratio of specific heats of *A* and *B* is 2/3
b. Ratio of specific heats of *A* and *C* is $\frac{8}{3}$
c. When *A* and *C* are mixed, temperature of mixture is $\left(\frac{140}{11}\right)$°C
d. When all of three *A*, *B* and *C* are mixed, temperature of mixture is 15°C

Statement Type

Direction (Q. Nos. 21 and 22) *This section is based on Statement I and Statement II. Select the correct answer from the codes given below.*

a. Both Statement I and Statement II are correct and Statement II is the correct explanation of Statement I
b. Both Statement I and Statement II are correct but Statement II is not the correct explanation of Statement I
c. Statement I is correct but Statement II is incorrect
d. Statement II is correct but Statement I is incorrect

21 **Statement I** In phase diagram of water, liquid-vapour line ends at 218 atm and 374°C.

Statement II Liquid and vapour phase remains in equilibrium till density of liquid reaches density of vapour.

22 **Statement I** If mass m is heated from temperature (t_0) to temperature $(t_0 + \Delta t)$, then average specific heat of substance is defined as,

$$C = \lim_{\Delta t \to 0} \frac{\Delta Q}{m \Delta t} = \frac{dQ}{m dt}$$

where, ΔQ = heat absorbed by the mass.

Statement II Specific heat of a substance are not a constant quantity.

Matching List Type

Direction (Q. Nos. 23 and 24) *Choices for the correct combination of elements from Column I and Column II are given as options (a), (b), (c) and (d), out of which one is correct.*

23 Match Column I of physical quantity with its correct unit in Column II and mark the correct option from the codes given below.

	Column I		Column II
i.	Specific heat	p.	$J K^{-1}$
ii.	Latent heat	q.	$J kg^{-1} K^{-1}$
iii.	Molar specific heat	r.	$J mol^{-1} K^{-1}$
iv.	Thermal capacity	s.	$J kg^{-1}$

Codes

	i	ii	iii	iv		i	ii	iii	iv
a.	q	s	r	p	b.	p	q	r	s
c.	q	r	s	s	d.	p	s	q	r

24 Match physical quantities in Column I with their dimensional formulae given in Column II and mark the correct option from the codes given below.

	Column I		Column II
i.	Specific heat	p.	$[ML^2T^{-2}K^{-1}]$
ii.	Latent heat	q.	$[M^0 L^2T^{-2}K^{-1}]$
iii.	Molar specific heat	r.	$[ML^2 T^{-2} K^{-1} \mu^{-1}]$
iv.	Thermal capacity	s.	$[M^0 L^2T^{-2}]$

Codes

	i	ii	iii	iv		i	ii	iii	iv
a.	q	s	r	p	b.	q	r	s	p
c.	p	s	r	q	d.	p	r	s	q

Working Space

Working Space

Comprehension Type

Direction *(Q. Nos. 25-27) This section contains a paragraph, describing theory, experiments, data, etc. Three questions related to the paragraph have been given. Each question has only one correct answer among the four given options (a), (b), (c) and (d).*

Passage

A heating curve for 1 g ice at −30°C is converted to steam, at 120°C is shown in figure.

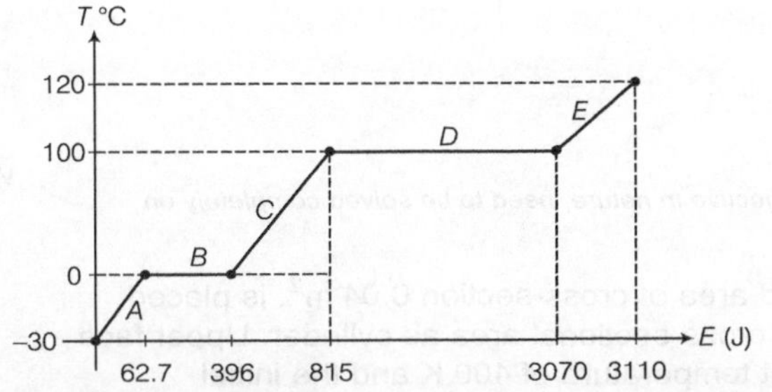

Note that *X*-axis shows energy added in joules and *Y*-axis shows temperature in °C.

25 Ratio of slopes of portion *A*, portion *C* and portion *E* is

a. 2090 : 4190 : 2010

b. 2010 : 4190 : 2090

c. $\frac{1}{2090} : \frac{1}{4190} : \frac{1}{2010}$

d. $\frac{1}{2010} : \frac{1}{4190} : \frac{1}{2090}$

26 Total amount of heat added in part *B* is less than that of part *D* because

a. water is a polar fluid

b. water is a non-polar fluid

c. average distance between two liquid molecules is much less than distance between two gas molecules

d. of hydrogen bonding present between water molecules

27 Ratio of latent heat of vaporisation to latent heat of fusion of water is

a. $\frac{2260}{360}$

b. $\frac{2260}{396}$

c. $\frac{2260}{3070}$

d. $\frac{2260}{333}$

DPP-4 Heat Transfer-I (Conduction of Heat)

Subjective Questions

Working Space

Direction (Q. Nos. 1-4) *These questions are subjective in nature, need to be solved completely on notebook.*

1 A cylindrical block of length 0.4 m and area of cross-section 0.04 m^2, is placed over a disc of mass 0.4 kg and same cross-sectional area as cylinder. Upper face of cylinder is maintained at a constant temperature of 400 K and the initial temperature of disc is 300 K. If coefficient of thermal conductivity of material of the cylinder is 10 $Wm^{-1}K^{-1}$ and specific heat of material of disc is 600 J $kg^{-1}K^{-1}$, how long it will take for the temperature of disc increase to 350 K ? Assume no loss of heat from curved surface of cylinder through the disc.

2 A double pane window consists of two glass sheets, each of area 1 m^2 and thickness 0.01 m separated by a 0.05 m thick air space. In steady state, the room glass interface and glass outdoor interface are at constant temperatures at 27°C and 0°C respectively. Calculate the rate of heat flow through the window pane. Also, find the temperature of other two surfaces. Given, thermal conductivities of glass and air are 0.8 and 0.08 $Wm^{-1}K^{-1}$ respectively.
(Assume air layer is stagnant and no leaks from outside).

3 To prevent overheating of an Integrated Circuit (IC), which is mounted over a device with an electrical input of 1.50 W, a heat sink of aluminium with fins is used. Temperature of heat sink remains constant at 35°C; under steady state conditions. The Integrated Circuit is electrically insulated from heat sink by a rectangular sheet of mica measuring 8.5 mm by 6.25 mm and 0.085 mm thick. The thermal conductivity of mica is equal to 0.075 $Wm^{-1}{}^{\circ}C^{-1}$. What will be the operating temperature for the Integrated Circuit?

4 A pond of water at 0°C is covered with a layer of ice of 4 cm thick.
If air temperature outside stays at $-10^{\circ}C$, how long does it take for the ice thickness increase to 8 cm?

Only One Option Correct Type

Direction (Q. Nos. 5-15) *This section contains 11 multiple choice questions. Each question has four choices (a), (b), (c) and (d), out of which ONLY ONE is correct.*

5 A cylinder of radius R made of a material of thermal conductivity K_1 is surrounded by a cylindrical shell of inner radius R and outer radius $2R$ made of a material of the thermal conductivity K_2. Two ends of combined system are maintained at two different temperatures. Combined effective thermal conductivity of system with its cylindrical surface insulated is

a. $K_1 + K_2$

b. $\dfrac{K_1 + 3K_2}{4}$

c. $\dfrac{K_1K_2}{K_1 + K_2}$

d. $\dfrac{3K_1 + K_2}{4}$

Working Space

6 Three rods of same metal and of same cross-sectional area are joined to form sides of an isosceles triangle ABC and right angled at B. The points A and B are maintained at temperatures T and $(\sqrt{2})T$ respectively. In steady state, the temperature of point C is T_C. Assuming only conduction takes place, T_C / T is

a. $\dfrac{1}{2(\sqrt{2}-1)}$ b. $\dfrac{3}{\sqrt{2}+1}$ c. $\dfrac{1}{\sqrt{3}(\sqrt{2}-1)}$ d. $\dfrac{1}{\sqrt{2}+1}$

7 Three rods of same material and having same cross-section have been joined as shown.

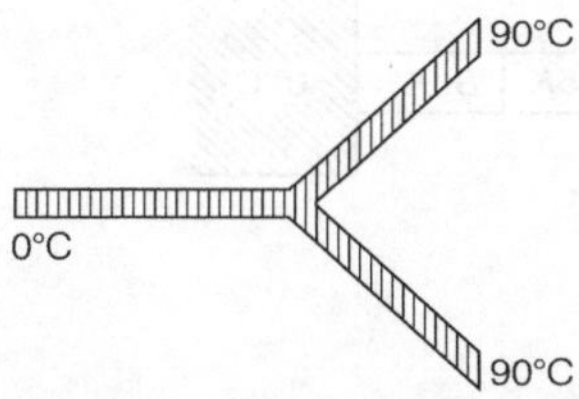

Each rod is of same length with one end kept at a constant temperature as marked.

Temperature of junction of the three rods will be

a. 45°C b. 60°C c. 30°C d. 20°C

8 Coefficient of thermal conductivity for three metals A, B and C are such that, $K_A > K_B > K_C$. If same quantity of heat flow rate per unit area is obtained from each rod by providing a temperature gradient X_A, X_B and X_C in A, B and C respectively, then,

a. $X_A = X_B = X_C$ b. $X_A > X_B > X_C$

c. $X_A < X_B < X_C$ d. $\dfrac{X_A + X_B}{2} = X_C$

9 Six identical conducting rods are joined as shown in the figure.

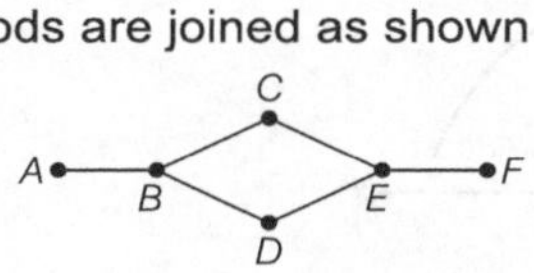

End A and F are maintained at 200°C and 0°C respectively. Temperature of point B in steady state will be

a. 100°C b. 55.56°C

c. 66.67°C d. 80°C

10 Water in a lake starts freezing at $t = 0$ when the outside temperature is $-\theta°$C. The time intervals in which thickness changes from 0 to y, from y to $2y$, from $2y$ to $3y$ and so on, will be in ratio

a. 1 : 2 : 3 b. 1 : 4 : 9

c. 1 : 3 : 5 d. 1 : 9 : 25

11 A room is maintained at 25°C by using a heater of 20Ω and 200 V supply. Room is well insulated from everywhere except for a glass window of area 1 m^2 and glass pane of thickness 0.2 cm. Thermal conductivity of glass is 0.2 cal $m^{-1}\,°C^{-1}\,s^{-1}$.

Assuming heat conduction takes place through window only, outside temperature will be

a. less than 16°C

b. less than 20°C but more than 16°C

c. exactly 20°C

d. above 20°C

Working Space

12 Two identical rods of silver and copper are coated with wax. One end of each is put in a hot resorvoir and other end is kept at room temperature. Ratio of lengths upto which wax will melt, is

a. 6 : 10 b. $\sqrt{10}:3$

c. 100 : 81 d. 10 : 9

13 Ratio of heat current, when rods are in parallel to heat current, with when rods are in series (in given case) will be

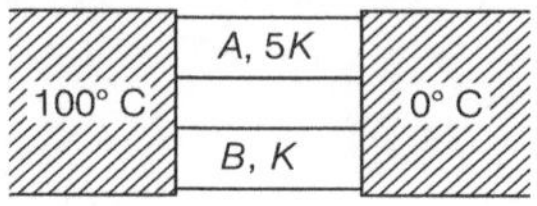

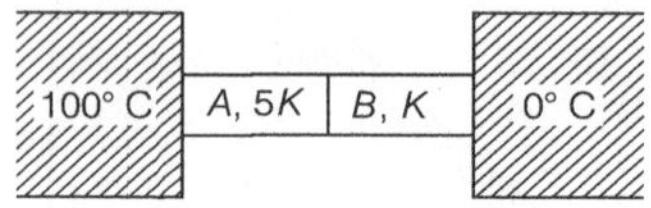

a. $\frac{16}{3}$ b. $\frac{11}{5}$

c. $\frac{6}{5}$ d. $\frac{36}{5}$

14 Radius of a conductor increases from one end to other gradually as shown.

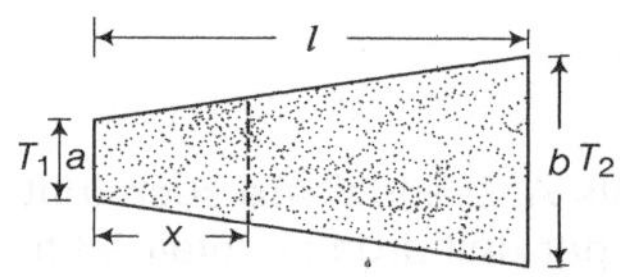

Its ends are maintained at constant temperatures T_1 and T_2 and curved surface is insulated. If H is heat flow rate and x is distance from hot end, then which of the following graphs is correct, when steady state is reached.

a. 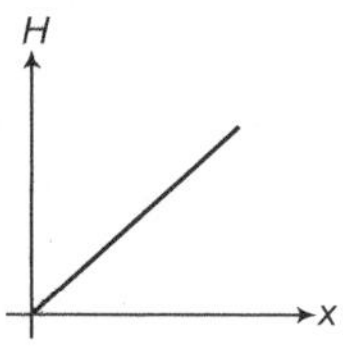

b. 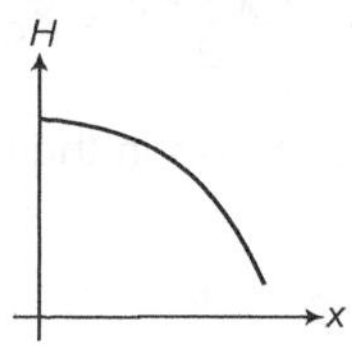

c. 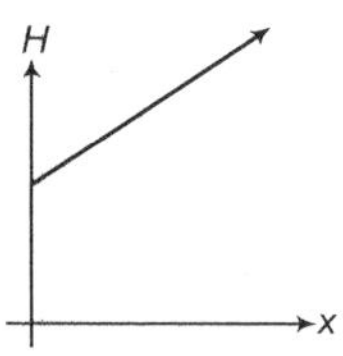

d. 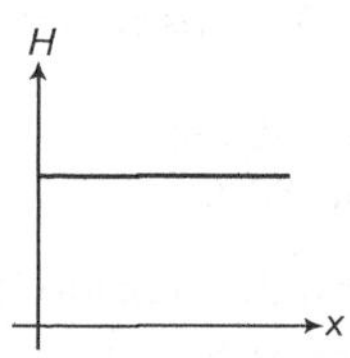

15 Two identical rods AB and CD are connected across the end points of a diameter of a ring as shown in the figure.

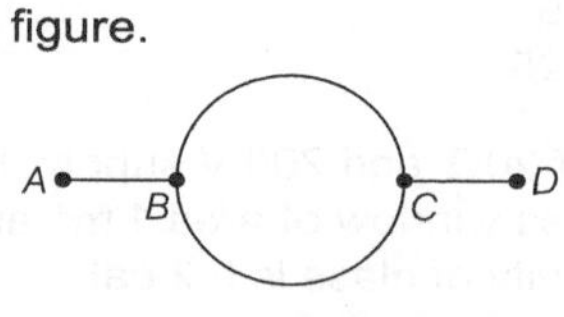

Radius of ring is equal to length of rod AB (or CD) and area of cross-section is same everywhere in the assembly. Rods and ring also have same material. If point A is maintained at 100°C and point D is maintained at 0°C, then temperature of point C will be

a. 62°C b. 37°C

c. 28°C d. 45°C

One or More than One Options Correct Type

Working Space

Direction *(Q. Nos. 16-20) This section contains 5 multiple choice questions. Each question has four choices (a), (b), (c) and (d), out of which ONE or MORE THAN ONE are correct.*

16 Two plates thickness l_1 and l_2 with thermal conductivities K_1 and K_2 respectively are kept in thermal contact as shown in the figure.

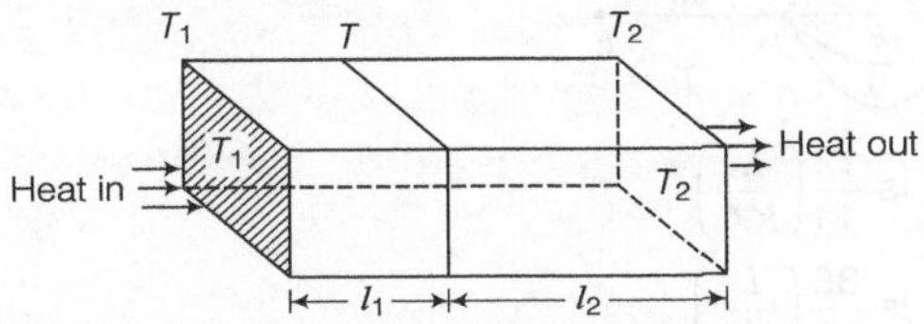

If A is area of each face, then mark the correct options.

a. Thermal resistance of combination is $R = \dfrac{l_1}{AK_1} + \dfrac{l_2}{AK_2}$

b. Heat flow rate through the combination is $H = \dfrac{A(T_1 - T_2)}{\left(\dfrac{l_1}{K_1} + \dfrac{l_2}{K_2}\right)}$

c. Temperature of joint of slabs is $T = \dfrac{T_1 \dfrac{l_2}{K_2} + T_2 \dfrac{l_1}{K_1}}{\left(\dfrac{l_1}{K_1} + \dfrac{l_2}{K_2}\right)}$

d. Equivalent thermal conductivity of combination will be $K_{eq} = \dfrac{l_1 + l_2}{\left(\dfrac{l_1}{K_1} + \dfrac{l_2}{K_2}\right)}$

17 A hollow sphere of glass whose external and internal radii are 11 cm and 9 cm respectively, is completely filled with ice at 0°C is placed in a bath of boiling water (at 100°C). Given, density of ice is 0.9 g cm^{-3}, latent heat of fusion of ice is 80cal g^{-1} and thermal conductivity of glass is $0.002 \text{ cal cm}^{-1} \text{s}^{-1} {}^\circ\text{C}^{-1}$.

Choose the correct options.

a. Rate of melting of ice is nearly $\dfrac{\pi}{2} \text{gs}^{-1}$

b. Rate of melting of ice is nearly $2\pi \text{ gs}^{-1}$

c. Time taken by ice to melt completely is nearly 10 min

d. Time taken by ice to melt completely is more than 10 min

18 Three identical conducting rods are connected as shown in the figure.

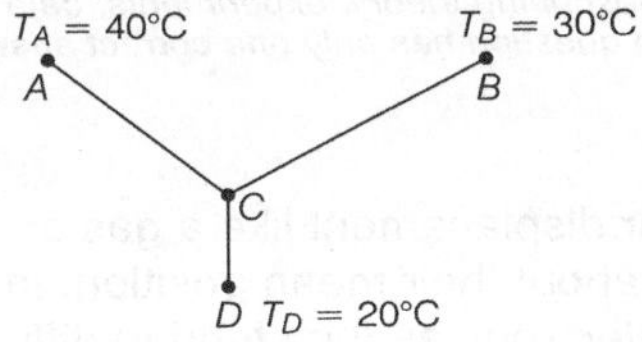

Choose the correct options.

a. Temperature of junction C is 15° C

b. Temperature of junction C is 30° C

c. Heat will flow from B to C

d. Heat will flow from C to D

19 One end of a rod is maintained at 50°C and the other end is kept in ice at 0°C. Rate of fusion of ice is doubled, when

a. temperature of hot end is made 200°C and area of cross-section is doubled

b. temperature of hot end is made 100°C and length of rod is made four times

c. area of cross-section is reduced to half and length is doubled

d. temperature is made 100°C and area of cross-section and length both are doubled

Working Space

20 Ten identical rods of length L, area A made of same material with conductivity K are joined as shown in the figure. Ends A and E are maintained at constant temperatures of 100°C and 0°C. Assuming no radiative and convective heat loss. Choose the correct options.

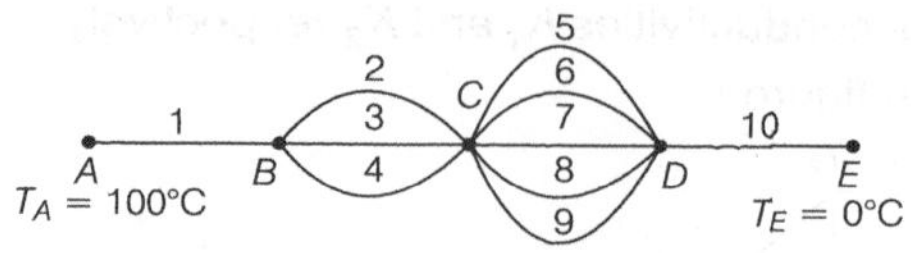

a. Total thermal resistance of combination is $\frac{12}{11}\left(\frac{L}{KA}\right)$

b. Total thermal resistance of combination is $\frac{38}{15}\left(\frac{L}{KA}\right)$

c. Heat flow rate is $\frac{750}{19}\left(\frac{KA}{L}\right)$

d. Heat flow rate is $\frac{225}{19}\left(\frac{KA}{L}\right)$

Statement Type

Direction (Q. Nos. 21 and 22) *This section is based on Statement I and Statement II. Select the correct answer from the codes given below.*

a. Both Statement I and Statement II are correct and Statement II is the correct explanation of Statement I
b. Both Statement I and Statement II are correct but Statement II is not the correct explanation of Statement I
c. Statement I is correct but Statement II is incorrect
d. Statement II is correct but Statement I is incorrect

21 **Statement I** Value of thermal conductivity generally increases with temperature.

Statement II A high temperature favours ionisation and dissociation of molecules/atoms.

22 **Statement I** Temperature gradient across the length of an uninsulated rod (whose ends are kept at different temperatures) decreases along the length of the rod.

Statement II Rate of heat transmitted by conduction is $\frac{dQ}{dt} = KA\left(\frac{\Delta T}{\Delta x}\right)$.

Comprehension Type

Direction (Q. Nos. 23-25) *This section contains a paragraph, describing theory, experiments, data, etc. Three questions related to the paragraph have been given. Each question has only one correct answer among the four given options (a), (b), (c) and (d).*

Passage

In solids, heat is transferred not due to molecular displacement like a gas or liquid, but due to random vibration of molecules about their mean position. In metals, there is an additional effect due to free electrons as they tend to diffuse through metal from hot to cold region, transferring energy by collision with other electrons and with the lattice ions in the older region.

$$\therefore \quad \frac{dQ}{dt} = j_E = -K\frac{dT}{dx}$$

In steady state conduction, j_E remains constant.

23 For a rod of uniform cross-section lying along X-axis with one of its end at $x = 0$ and at T_0 temperature, temperature of any section at a distance x from origin is

a. $T = T_0 + \frac{j_E x}{K}$ b. $T = T_0 - \frac{j_E x}{K}$ c. $T = \frac{T_0 + j_E K}{K}$ d. $T = \frac{T_0 - j_E x}{K}$

Working Space

24 Two slabs with thickness L_1 and L_2 and thermal conductivities K_1 and K_2 are in contact as shown in the figure.

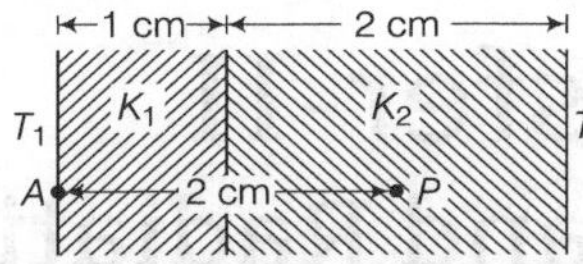

Let $L_1 = 1\text{cm}$, $L_2 = 2\text{cm}$, $K_1 = K$ and $K_2 = 3K$, then temperature at a point P which is such that $AP = 2\text{cm}$ (Take, $T_1 = 0°\text{C}$ and $T_2 = 100°\text{C}$) is

a. 50°C b. 60°C c. 80°C d. 90°C

25 As in metals electrons diffuse to carry energy from hot region to cold region that means. Choose the correct options.

a. $V_{Hot} - V_{Cold} > 0$ b. $V_{Hot} - V_{Cold} < 0$

c. $V_{Hot} = V_{Cold}$ d. Data insufficient

Matching List Type

***Direction** (Q. No. 26) Choice for the correct combination of elements from Column I and Column II are given as options (a), (b), (c) and (d), out of which one is correct.*

26 Match Column I with Column II. Mark the correct option from the codes given below.

	Column I		Column II
i.	Conduction in solids at low temperatures.	p.	Phenon vibration
ii.	Conduction in solids at high temperatures.	q.	Drift of free electrons
iii.	Conduction in gases.	r.	Molecular collision
iv.	Conduction in liquids.		

Codes

	i	ii	iii	iv		i	ii	iii	iv
a.	p	q	r	r	b.	p	p,q	r	q,r
c.	p	p	r	r	d.	q	p,q	r	p,r

One Integer Value Correct Type

***Direction** (Q. Nos. 27 and 28) This section contains 2 questions. When worked out will result in an integer from 0 to 9 (both inclusive).*

27 A hollow sphere of inner radius r_0 and outer radius $2r_0$ is made of a uniform material of constant thermal conductivity K. The temperature with in the ball is maintained at $2T_0$ and outside the ball is T_0. If temperature at distance $\frac{3r_0}{2}$ from centre is $\frac{pT_0}{3}$, then find the value of p.

28 A large tank of water at a constant temperature of 90°C is connected to a small vessel containing 10 kg water at 10°C, through a metal rod of length 2 m, area of cross-section 0.5 m^2 and thermal conductivity of 20 units. If the time taken for the temperature of the water in small vessel at 30°C is $t = N \times 4200 \log(4/3)$, then find the volue of N. (Given, specific heat of water = 4200 J kg^{-1} $°C^{-1}$)

DPP-5 Heat Transfer-II (Convective and Radiative Heat Transfer)

Subjective Questions

Working Space

Direction *(Q. Nos. 1-4) These questions are subjective in nature, need to be solved completely on notebook.*

1 i. Give the basic relationship for heat transfer by convection.

ii. A 6 m long uninsulated steel pipe of an AC (air conditioner) passes across a room with a constant temperature 20°C.

Outer diameter of steel pipe is 30 cm and surface temperature is about 110°C. Convective heat transfer coefficient for air is about 2 J $m^{-2}{}^\circ C^{-1}s^{-1}$. Calculate heat transfer rate from the pipe into the room.

2 Flat wall of a boiler is exposed to the atmosphere. The wall is covered with a layer of insulation 2 cm thick whose thermal conductivity is 0.08 $Wm^{-1}{}^\circ C^{-1}$. Inside temperature of boiler wall is 600°C. The wall loses heat to atmosphere by convection on the surface of insulation. Convective heat transfer coefficient on the insulation surface is 0.2×10^{-3} $W m^{-2}\,{}^\circ C^{-1}$. Calculate the bulk temperature of environment T_b, if temperature of outer surface is not to exceed 500°C.

3 Let your body temperature is 35°C and you are standing in a room maintained at a constant temperature of 20°C. Let emissivity of your skin is 0.9 and surface area of your skin is 1.5 m^2.

What is the mass of ice at 0°C that can be melted using amount of heat lost by your body by radiation only in 1 h?

4 If there is no atmosphere on the surface of the earth, what would be equilibrium temperature of the earth? (Take, value of solar constant $= 1340\, Wm^{-2}$)

Only One Option Correct Type

Direction *(Q. Nos. 5-14) This section contains 10 multiple choice questions. Each question has four choices (a), (b), (c) and (d), out of which ONLY ONE is correct.*

5 Variation of temperature T of two identical bodies A and B with time t, when they loose heat by radiation only is as shown in the figure.

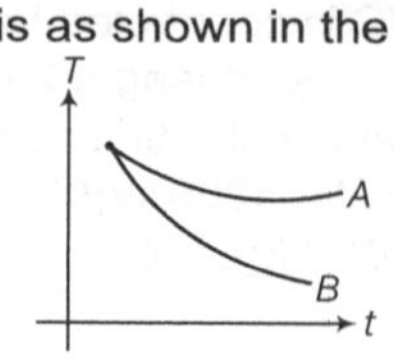

Then, correct relation between emissivity (e) and absorpitivity (a) of the two bodies is

a. $e_A > e_B$ and $a_A < a_B$
b. $e_A < e_B$ and $a_A > a_B$
c. $e_A > e_B$ and $a_A > a_B$
d. $e_A < e_B$ and $a_A < a_B$

Working Space

6 Spectral energy density distribution E_λ of a black body at two different temperatures T_1 and T_2 is shown in the figure.

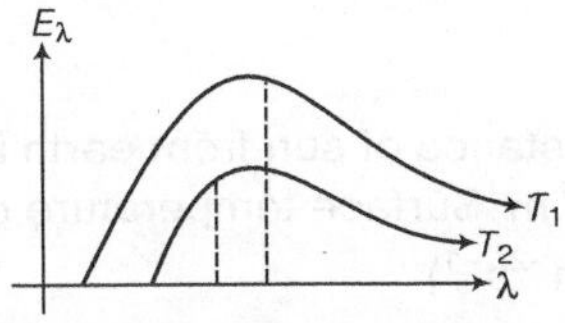

If areas under the curves are in the ratio 16 : 1 and T_2 is 2000 K, then value of temperature T_1 will be

a. 32000 K
b. 16000 K
c. 8000 K
d. 4000 K

7 Energy spectrum of a body shows a maxima around a wavelength of λ_0. Temperature of this body is increased and it is found that energy maxima now occurs around a wavelength $\frac{3\lambda_0}{4}$. Power radiated by the black body now increased by a factor of

a. $\frac{256}{81}$
b. $\frac{64}{27}$
c. $\frac{16}{9}$
d. $\frac{4}{3}$

8 Following are the observations made for a black body at 3000 K :

Amount of energy emitted per second	Range of wavelength of radiation
U_1	490 nm to 500 nm
U_2	990 nm to 1000 nm
U_3	1490 nm to 1500 nm

Wien's constant $= 3 \times 10^6$ nmK. Mark the correct option.

a. $U_1 > U_2$
b. $U_3 > U_2$
c. $U_2 > U_3$
d. $U_1 = U_2 = U_3$

9 Two hollow metallic spheres of the same metal and radii R and $2R$ are at temperatures $2T$ and T respectively. The ratio of total energy radiated per second by the spheres is

a. 1 : 1
b. 4 : 1
c. 16 : 1
d. 8 : 1

10 A solid cuboid with edges 1 cm, 2 cm and 3 cm is suspended in vacuum, its temperature falls from 100°C to 90°C in 100 s.
Another solid cuboid of same material and similar surface finish is kept in same enclosure. If sides of cuboid are 2 cm, 4 cm and 6 cm, then time taken for this cuboid to cool from 100°C to 90°C will be

a. 100 s
b. 150 s
c. 50 s
d. 200 s

11 Ram wants to drink tea, when it is at 50°C. He ordered tea which arrives at a temperature of 80°C. Tea takes one minute to cool from 80°C to 60°C. If room temperature is 30°C, how long he has to wait to drink the tea ?

a. Less than one minute
b. Two more minutes
c. Half a minute
d. Nearly three minutes

Working Space

12 A body, initially at 64°C cools to 52°C in 10 minutes. Temperature of surroundings is 16°C. Temperature of body after 10 more minutes is

a. 48°C
b. 43°C
c. 35°C
d. 50°C

13 Earth receives about 1400 Wm^{-2} from the sun. Distance of sun from earth is about 1.4×10^{11} m and the radius of sun is 7×10^8 m. Surface temperature of sun can be approximated as, (Take, $\sigma = 5.6 \times 10^{-8}$ $Jm^{-2}s^{-1}$)

a. 10^4 K
b. 10^3K
c. $10^{\frac{15}{4}}$ K
d. 5883 K

14 A body P is connected to a large body Q through a conducting rod of length L, cross-sectional area A and thermal conductivity K. This assembly is placed in an atmosphere of temperature T_A and body Q is also maintained at temperature T_A. Let heat capacity of body P is C and it is initially at temperature T_1. If in time t second temperature of body P falls to T_2. Then, choose the correct option.

a. $\log\left(\frac{T_2 - T_A}{T_1 - T_A}\right) = -\left(K_1 + \frac{KA}{LC}\right)t$
b. $\log\left(\frac{T_2 - T_A}{T_1 - T_A}\right) = -\left(K_1 + \frac{KA}{LC} \cdot t\right)$
c. $\log\left(\frac{T_1 - T_A}{T_2 - T_A}\right) = -\left(K_1 + \frac{KA}{LC} \cdot t\right)$
d. $\log\left(\frac{T_1 - T_A}{T_2 - T_A}\right) = -\left(K_1 + \frac{KA}{LC}\right)t$

One or More than One Options Correct Type

Direction (Q. Nos. 15-19) *This section contains 5 multiple choice questions. Each question has four choices (a), (b), (c) and (d), out of which ONE or MORE THAN ONE are correct.*

15 Two bodies have thermal emissivities of 0.01 and 0.81 respectively. Ratio of outer surface area of the bodies is 1. Two bodies emit total radiant power at same rate. Wavelength corresponding to maximum spectral radiancy in the radiation from B is λ, shifted from the wavelength corresponding to maximum spectral radiancy in the radiation from A by 1.00 μm. If the temperature of A is 5802 K, then choose the correct options.

a. Temperature of B is 1934 K
b. $\lambda_B = 1.5\mu m$
c. Temperature of B is 11604 K
d. Temperature of B is 2901 K

16 A room is initially at the outdoor temperature of 25° C. A large fan that consumes 200 W of electricity, when running is turned on. The overall heat transfer coefficient from room is 6 Wm^{-2} and total area of room is 30m^2. Then, choose the correct options.

a. Final temperature of room after a long time will be 26.1°C
b. Temperature of room keep on raising due to work done by the fan
c. Temperature of room falls below 25°C due to cooling produced by fan
d. Fan acts like a heater of power much less than 200 W

17 A person is standing in a large room. Temperature of person is 29°C and temperature of room is 20°C. Outer surface area of person is 1.6 m^2 and the convective heat transfer coefficient is 6 $Wm^{-2}{}^\circ C^{-1}$. Then, choose the correct options.

a. Heat transfer by conduction only is nearly zero
b. Heat loss due to convection only is 86.4 W
c. Heat loss due to radiation cannot exceed 86 W
d. Total heat loss does not exceed 123 W

Working Space

18 Which of these are correct regarding different modes of heat transfer ?

a. In natural convection, we move away from a hot surface, air speed increases whereas temperature decreases

b. Amount of energy transferred by radiation reduces with temperature of medium

c. Conversion of energy from one form to another is always associated with adverse effects on atmosphere

d. In equation $\frac{dQ}{dt} = -KA\frac{dT}{dx}$, negative sign shows temperature decreases with increasing x

19 Choose the correct options.

a. If energy spectrum of a body does not contain wavelengths of 7800 Å to 300000 Å, then it cannot loss heat by radiation mode

b. If an ideal black body is thrown into surface it initially appears black, then it cannot be distinguished from its background

c. Gravity is necessary for convection currents

d. A green glass, when heated in a furnace when it melts glows red whereas a red glass when heated glows green in colour

Comprehension Type

Direction *(Q. Nos. 20-24) This section contains 2 paragraphs, each describing theory, experiments, data, etc. Five questions related to the paragraphs have been given. Each question has only one correct answer among the four given options (a), (b), (c) and (d).*

Passage I

Radiant energy spectrum graph for black body at temperature T_0 is as shown in the figure.

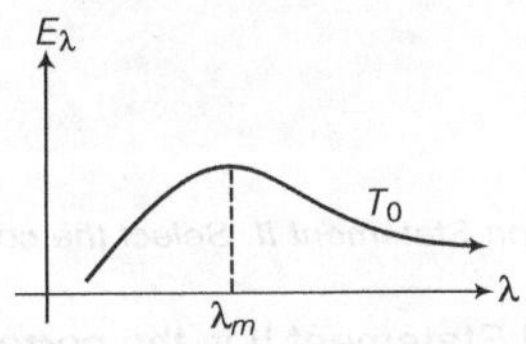

20 If temperature of body is increased to some temperature $T\,(T > T_0)$, then choose the correct option.

a. Intensity of radiation for every wavelength decreases

b. Maximum intensity occurs at a greater wavelength

c. Area under graph decreases

d. Area under graph is proportional to fourth power of temperature

21 Which graph correctly represents the spectral intensity *versus* wavelength for temperatures T and T_0.

a.
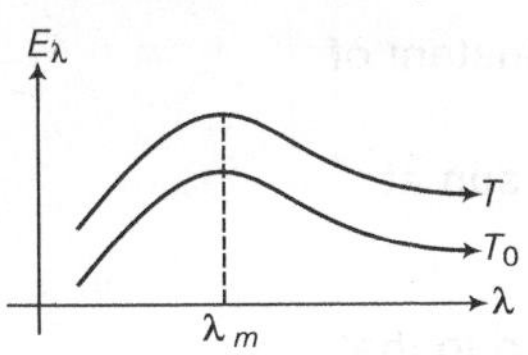

b.
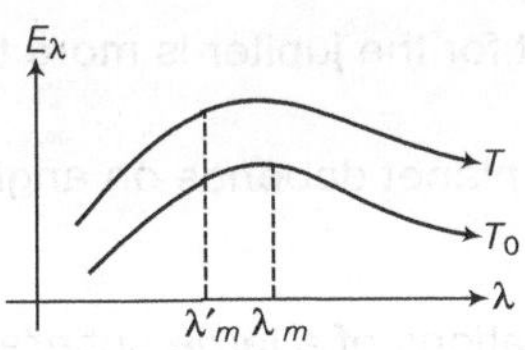

c.
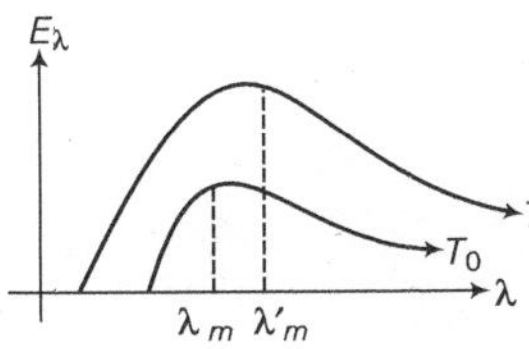

d.
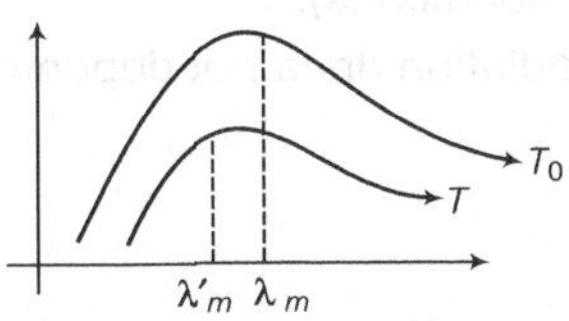

Working Space

Passage II

In an experiment of study of Newton's law of cooling, a ball of metal is heated from inside a heater of 20 W in a room maintained at constant temperature of 20°C.

Following observations are made.

i. Mass of the ball is 1 kg.
ii. Rate of increase of temperature decreases with rise of temperature.
iii. Temperature of ball becomes 50°C after some time and then remains constant.
iv. Specific heat of metal is 500 $Jkg^{-1}°C^{-1}$.

Corresponding to above observations, answer the following questions.

22 Rate of loss of heat, when ball is at 50°C is

a. 10 W
b. 15 W
c. 20 W
d. 25 W

23 Assuming Newton's law of cooling holds for the ball, rate of loss of heat, when the ball is at 30°C is

a. $\frac{10}{3}$ W
b. $\frac{20}{3}$ W
c. $\frac{30}{3}$ W
d. $\frac{40}{3}$ W

24 In 5 min temperature of ball rises from 20°C to 30°C. Total energy lost from the ball in this time interval is

a. 250 J
b. 500 J
c. 750 J
d. 1000 J

Statement Type

Direction (Q. Nos. 25-27) *This section is based on Statement I and Statement II. Select the correct answer from the codes given below.*

a. Both Statement I and Statement II are correct and Statement II is the correct explanation of Statement I
b. Both Statement I and Statement II are correct but Statement II is not the correct explanation of Statement I
c. Statement I is correct but Statement II is incorrect
d. Statement II is correct but Statement I is incorrect

25 **Statement I** A blue star is hotter than a red star.

Statement II A black body at higher temperature radiates more power per unit area.

26 **Statement I** Value of solar constant for the jupiter is more than the solar constant of the earth.

Statement II Solar constant S for a planet depends on angle subtended by sun at the planet as $S \propto \theta^2$.

27 **Statement I** Rate of cooling by radiations of a large sphere is less compared to that of a small sphere (under similar conditions).

Statement II Rate of cooling by radiation does not depend on mass.

Working Space

Matching List Type

Direction (Q. No. 28) Choices for the correct combination of elements from Column I and Column II are given as options (a), (b), (c) and (d), out of which one is correct.

28 Match reactance (r), absorptance (a) and transmittance (t) coefficients in Column II with the type of surface they belong in Column I. Mark the correct option from the codes given below.

	Column I		Column II
i.	Perfect reflecting surface	p.	$r + a + t = 1$
ii.	Perfect absorbing surface	q.	$a = 0$
iii.	Perfect transmitting surface	r.	$r = 0$
iv.	Any practical surface	s.	$t = 0$

Codes

	i	ii	iii	iv		i	ii	iii	iv
a.	p,q,s	p,q,r	p	q,s	b.	p,q	p,r	p,s	t
c.	p,q,s	p,r,s	p,q,r	p	d.	p,q,r	p,q,s	p,q,s	p,q,r,s

One Integer Value Correct Type

Direction (Q. Nos. 29-30) This section contains 2 questions. When worked out will result in an integer from 0 to 9 (both inclusive).

29 A liquid takes 30 s to cool from 95°C to 90°C and 70 s to cool from 55°C. If time taken by the liquid to cool from 50°C to 45°C is $21k$ (in seconds), then find the value of k.

30 A long rod is heated from one end and the bar attains the conditions of steady state.

(Given, emissivity of the material of rod = 0.9, coefficient of thermal conductivity of rod = $100\,\text{Wm}^{-1}\,{}^\circ\text{C}^{-1}$, radius of rod = 10cm = 0.1 m)

This setup is placed in vacuum. If θ is temperature at steady state of any section of rod and

$$\frac{d^2\theta}{dx^2} = 2 \cdot N^2 \cdot \theta \times 10^{-2}$$

where, x is the distance of section from hot end, find the value of N.

SET - 1

Revisal Problems
for JEE Main

Only One Option Correct Type

Direction (Q. Nos. 1-20) *This section contains 20 multiple choice questions. Each question has four choices (a), (b), (c) and (d), out of which ONLY ONE is correct.*

Working Space

1 A bullet strikes an aluminium plate in space initially at rest and gets embeded in it. Initial speed of bullet is 10 ms^{-1}, mass of bullet is 50 g, mass of aluminium plate is 950 g. Specific heat of Al $= 900$ J kg^{-1} K^{-1} and specific heat of bullet $= 150$ J $kg^{-1}K^{-1}$. Heat generated raises temperature of bullet and slab. Difference in amount of heat absorbed by aluminium slab and bullet is

a. about 95% of initial KE
b. about $\frac{7}{19}$ of 95% of initial KE
c. about $\frac{113}{115}$ of 95% of initial KE
d. about $\frac{17}{222}$ of 95% of initial KE

2 An open glass tube is filled upto the top, with 25 cm^3 of mercury at a temperature of 20° C. If this tube is now kept in an enclosure of 100° C, amount of mercury overflow from the tube will be (Take, $\gamma_{mercury} = 1.8 \times 10^{-4} {}^\circ C^{-1}$ and $\gamma_{glass} = 0.27 \times 10^{-4} {}^\circ C^{-1}$)

a. 0.36 cm^3 b. 0.54 cm^3 c. 0.054 cm^3 d. 0.30 cm^3

3 Pressure p at a height H above mean sea level is (Assuming temperature of atmosphere same everywhere. Where, $M =$ molar mass of air, $R =$ universal gas constant, $T =$ temperature of atmosphere, $p_0 =$ atmosphere pressure at mean sea level)

a. $p = p_0$
b. $p = p_0 - H\left(\frac{Mg}{RT}\right)$
c. $p = p_0 e^{\left(\frac{Mg}{RT}\right)H}$
d. $p = p_0 e^{-\left(\frac{Mg}{RT}\right)H}$

4 A solid whose volume does not change with temperature floats in a liquid. At temperature of 20° C nearly half part of solid is submerged in the fluid. At temperature of 80° C nearly three-fourth part of solid is submerged in the liquid. The coefficient of volume expansion of the liquid is equal to

a. 10^{-11} °C^{-1} b. 10^{-5} °C^{-1} c. 10^{-2} °C^{-1} d. 10^{-4} °C^{-1}

5 Two glass bulbs of equal volume are connected by a narrow tube. Initially bulbs are at 0° C and contains air at pressure of 1 atm. One of the bulbs is then placed in melting ice and other in a water both of constant temperature is 62° C.

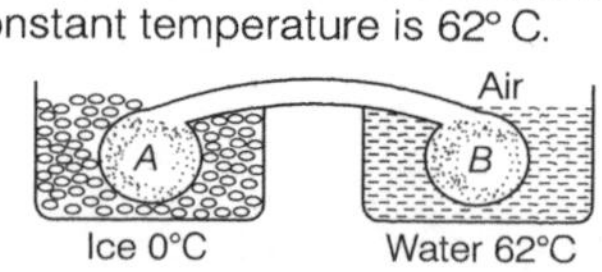

After a long time, value of pressure in the bulbs is

a. 76 cm of Hg
b. 100 cm of Hg
c. 83.83 cm of Hg
d. 79.25 cm of Hg

6 A thin rod of negligible mass and cross-section area 4×10^{-6} m^2 suspended vertically from one end, has a length of 0.5 m at 100° C. Rod is cooled to 0° C but prevented from contracting by attaching a mass at the lower end. Given, Young's modulus for rod $Y = 10^{11}$ Nm^{-2} and coefficient of linear expansion $= 10^{-5}K^{-1}$. Energy stored in the rod will be

a. 500 J b. 250 J c. 1000 J d. 350 J

7 A thermometer of mass 55 g and specific heat 0.2 cal g^{-1} °C^{-1}, reads 15° C in air. When it is dipped in 300 g water, it reads 44.4° C. Temperature of water was

a. 40°C b. 45.5°C c. 50°C d. 100°C

Working Space

8 An isosceles triangle is formed using a thin rod of length l_1 and coefficient of linear expansion α_1 as the base and two thin rods of length l_2 and coefficient of linear expansion α_2 as two sides. If the distance between the apex and the mid-point of the base remains unchanged as the temperature changes, then correct relation is

a. $\frac{l_1}{l_2} = \sqrt{\frac{\alpha_2}{\alpha_1}}$ b. $\frac{l_1}{l_2} = \sqrt{\frac{\alpha_1}{\alpha_2}}$ c. $\frac{l_1}{l_2} = 2\sqrt{\frac{\alpha_2}{\alpha_1}}$ d. $\frac{l_1}{l_2} = 2\sqrt{\frac{\alpha_1}{\alpha_2}}$

9 A uniform solid brass sphere is rotating with angular speed of ω_0 about a diameter. If its temperature is increased by 100° C, then its new angular speed ($\alpha_{brass} = 2 \times 10^{-5}\ °C^{-1}$) will be

a. $\frac{\omega_0}{1-0.002}$ b. $\frac{\omega_0}{1+0.002}$ c. $\frac{\omega_0}{1-0.004}$ d. $\frac{\omega_0}{1+0.004}$

10 Two halves of a wooden cylinder are held together by a metal ring with coefficient of linear expansion α and Young's modulus of elasticity Y. Area of cross-section of ring is A. Initially radius of ring r is slightly less than radius of cylinder R. To fit ring over the cylinder, ring is heated till it just slips over the cylinder and then its presses the cylinder together.

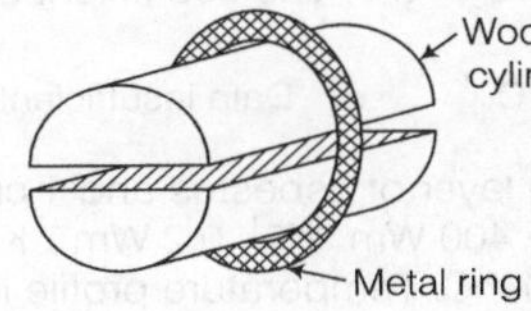

Force applied by one part of wooden cylinder over the ring is

a. $AY\alpha\Delta T$ b. $2AY\alpha\Delta T$
c. $2\pi AY\alpha\Delta T$ d. $\pi AY\alpha\Delta T$

11 A metal piece of mass 1 kg at constant atmospheric pressure and temperature 10° C is given 1000 J of heat. Fraction of total heat that is used up net in raising the temperature of metal piece will be (Given, specific heat of metal = 500 J kg^{-1} °C^{-1}, coefficient of linear expansion of metal $= 3 \times 10^{-5}\ °C^{-1}$, density of metals at 10° C = 1000 kg m^{-3}, $p_{atm} = 10^5$ Nm^{-2}).

a. $\frac{9}{5} \times 10^{-4}$ b. $\frac{5}{9} \times 10^{-4}$ c. $\frac{5}{9} \times 10^{-5}$ d. $\frac{9}{5} \times 10^{-5}$

12 A metal cube with coefficient of linear expansion $10^{-5}\ °C^{-1}$ is floating in a liquid. It is found that on increasing the temperature, cube remains submerged upto same depth in the liquid. Coefficient of volumetric expansion of liquid will be

a. $10^{-5}\ °C^{-1}$ b. $2 \times 10^{-5}\ °C^{-1}$ c. $3 \times 10^{-5}\ °C^{-1}$ d. $5 \times 10^{-6}\ °C^{-1}$

13 In given glass tube of uniform cross-section everywhere outer liquid columns are open to the atmosphere. Temperature of limbs and height of liquid columns is equilibrium is marked as,

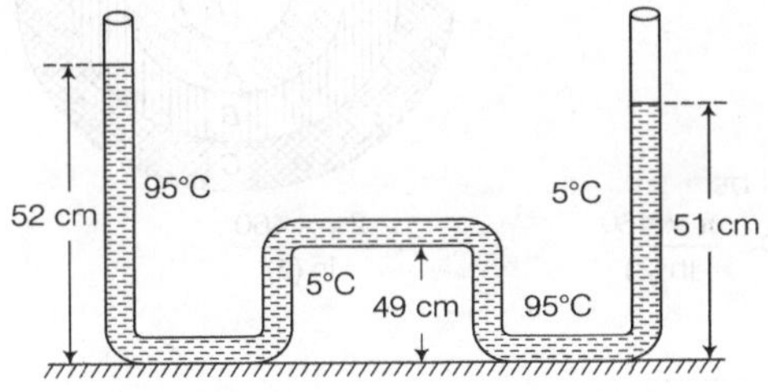

Coefficient of linear expansion of the fluid will be

a. $\frac{1}{8995}\ °C^{-1}$ b. $\frac{3}{8995}\ °C^{-1}$ c. $\frac{1}{9500}\ °C^{-1}$ d. $\frac{1}{26985}\ °C^{-1}$

14 A piece of metal weights 46 g in air. When it is immersed in a liquid of specific gravity 1.24 at 27° C it weights 30 g. At temperature of 42° C it weights 30.5 g. Specific gravity of the liquid at 42° C is 1.20. Coefficient of linear expansion of the metal will be

a. $3.316 \times 10^{-5}\ °C^{-1}$ b. $2.316 \times 10^{-5}\ °C^{-1}$
c. $4.316 \times 10^{-5}\ °C^{-1}$ d. $1.316 \times 10^{-5}\ °C^{-1}$

15 Heating curves for two solids A and B are given as below :

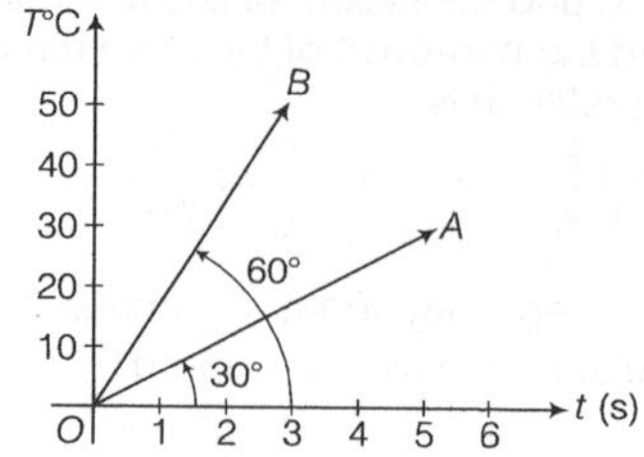

Ratio of time taken by A to time taken by B to heat upto 30° C under same rate of heat supplied is

a. $\frac{1}{2}$ b. $\frac{3}{1}$ c. $\frac{1}{3}$ d. $\frac{2}{1}$

16 For a non-isentropic solid, coefficient of linear expansion in one direction is $1.2 \times 10^{-5}\,°C^{-1}$ and in every other direction (perpendicular to this) is $2.4 \times 10^{-5}\,°C^{-1}$. The coefficient of volumetric expansion for this solid will be

a. no unique value b. $6.0 \times 10^{-5}\,°C^{-1}$ c. $1.8 \times 10^{-5}\,°C^{-1}$ d. Data insufficient

17 A composite wall consists of 2 cm of copper plate, 1.5 cm layer of asbestos and 1 cm layer of fiber glass. Thermal conductivities of materials are 400 $Wm^{-1}K^{-1}$, 0.2 $Wm^{-1}K^{-1}$ and 1.0 $Wm^{-1}K^{-1}$. Temperature difference across the wall is 500° C. Temperature profile in steady state most closely resembeled by

a. 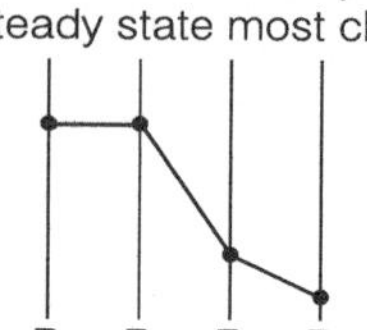

b. 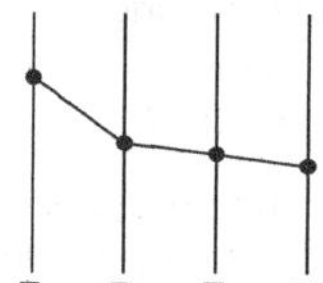

c. 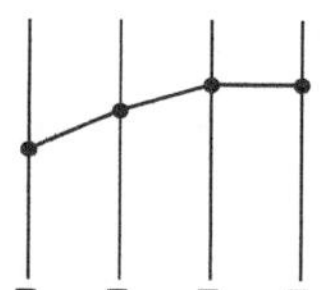

d. 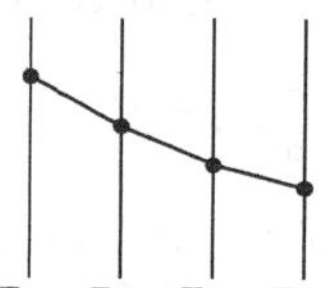

18 An insulated cylindrical tube of an air-conditioner's condenser contains a hot fluid. Temperature of fluid is 500° C and outside temperature is 40° C. Hot fluid tube is very thin and is covered with 3-layers of different insulating materials. Cross-section of the tube is as shown in the figure.

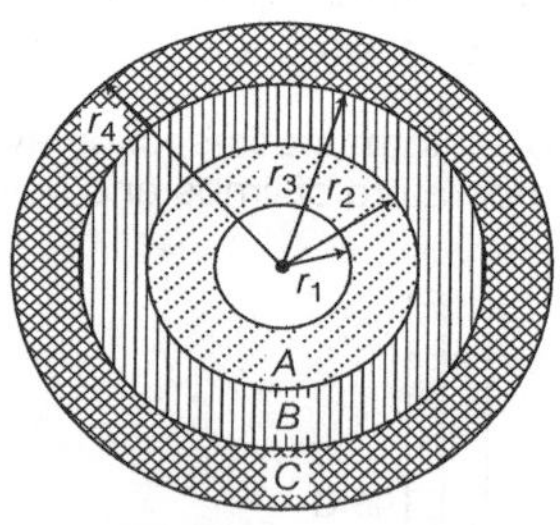

Given, $r_1 = 1$ cm, $r_2 = 2$ cm, $r_3 = 8$ cm

$r_4 = 64$ cm, $K_A = 1\ Wm^{-1}\,°C^{-1}$

$K_B = 2\ Wm^{-1}\,°C^{-1}$, $K_C = 3\ Wm^{-1}\,°C^{-1}$

Heat loss per unit length (in watts) of tube will be

a. $\frac{460}{\ln(2)}$ b. $\frac{460}{\pi \ln(2)}$ c. $\frac{\pi \times 460}{\ln(2)}$ d. $\frac{2\pi \times 460}{\ln(2)}$

19 Consider a composite wall.

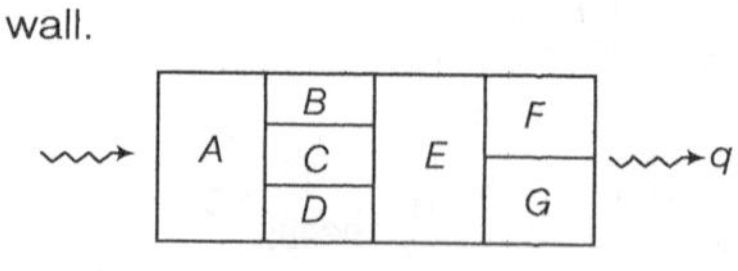

If resistances of regions A, B, C, D, E, F and G are 0.1, 0.2, 0.4, 0.8, 0.9, 1.0 and 2.0 (in $m^2\,°CW^{-1}$) respectively. Then, total thermal resistance of the equivalent wall in ($m^2\,°CW^{-1}$) will be

a. 1.67 b. 1.56
c. 1.78 d. 1.89

Working Space

Working Space

20 A flat rectangular surface of dimension 15 cm ×2 cm is placed perpendicular to the direction of radiations (electromagnetic waves) falling over it. Surface absorbs half of radiation and reflects remaining half. Total momentum transferred to the surface in 1 s is 3×10^{-4} kg - ms^{-1}.
Emissive power of the source of radiation will be

a. 10^8 Wm^{-2} b. 10^9 Wm^{-2}
c. 10^{12} Wm^{-2} d. 10^6 Wm^{-2}

Statement Type

Direction *(Q. Nos. 21-25)* ***This section is based on Statement I and Statement II. Select the correct answer from the codes given below.***

a. Both Statement I and Statement II are correct and Statement II is the correct explanation of Statement I
b. Both Statement I and Statement II are correct but Statement II is not the correct explanation of Statement I
c. Statement I is correct but Statement II is incorrect
d. Statement II is correct but Statement I is incorrect

21 **Statement I** A more precise definition of kelvin scale is

$$T = 273.16\left(\frac{Q}{Q_{TP}}\right)_{\text{rev. cycle}}$$

Q_{TP} and Q are heat transfer between cycle and the reservoir at 273.16 K and at temperature T.

This definition leads to fact that a negative kelvin temperature is not possible.

Statement II Energy rejected by a heat transfer cannot be negative.

22 **Statement I** Normal boiling point of organic compound with given phase diagram is 40° C.

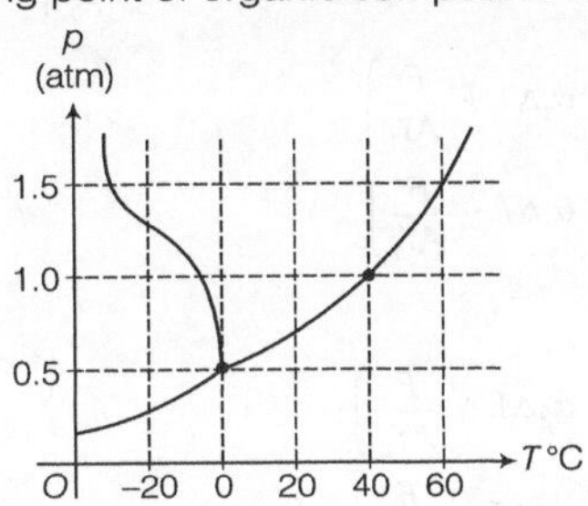

Statement II Compounds with stronger dipole moments tend to have higher boiling point.

23 **Statement I** For a rod *AB*, in steady state, if temperature at a distance of 2 cm from end *A* is 80° C and from end *B* is 20° C, then temperature difference between ends *A* and *B* is 100° C.

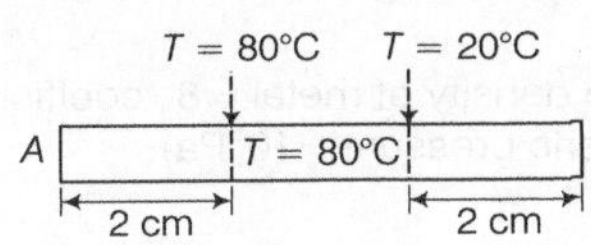

Statement II In steady state, heat received by any section of rod *AB* equals to heat lost by the section.

24 **Statement I** If a body at T° C is placed in an enclosure of T° C and the enclosure is evacuated, then temperature of body remains constant.

Statement II Heat is conducted between two bodies, only when they are at different temperatures.

25 **Statement I** Natural convection is not possible in an orbiting satellite.

Statement II For an orbiting satellite, gravitational pull is zero.

SET - 2

Revisal Problems for JEE Advanced

Comprehension Type

Working Space

Direction (Q. Nos. 1-11) *This section contains 4 paragraphs, each describing theory, experiments, data, etc. Eleven questions related to the paragraphs have been given. Each question has only one correct answer among the four given options (a), (b), (c) and (d).*

Passage I

Two rods P and Q of different metals having the same area A and the same length L are placed between two rigid walls as shown in the figure.

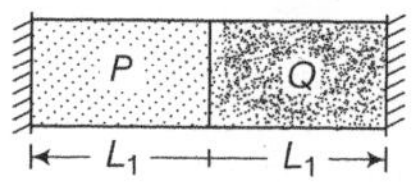

Coefficients of linear expansion of P and Q are α_1 and α_2 and their Young's modulus are Y_1 and Y_2, respectively. Temperature of rods is increased by $T°$ C.

1 Force exerted by one rod on other is

a. $\dfrac{TA(\alpha_1+\alpha_2)}{\dfrac{1}{Y_1}+\dfrac{1}{Y_2}}$ b. $TAY_1Y_2(\alpha_1+\alpha_2)$

c. $TA(Y_1+Y_2)\alpha_1\alpha_2$ d. None of these

2 The new length of the rod P is

a. $L\left(1+\alpha_1\Delta T+\dfrac{F}{AY_1}\right)$ b. $L\left(1-\alpha_1\Delta T+\dfrac{F}{AY_1}\right)$

c. $L\left(1+\alpha_1\Delta T-\dfrac{F}{AY_1}\right)$ d. $L\left(1-\alpha_1\Delta T-\dfrac{F}{AY_1}\right)$

3 The new length of the rod Q is

a. $L\left(1+\alpha_2\,\Delta T+\dfrac{F}{AY_2}\right)$ b. $L\left(1-\alpha_2\Delta T+\dfrac{F}{AY_2}\right)$

c. $L\left(1-\alpha_2\Delta T-\dfrac{F}{AY_2}\right)$ d. $L\left(1+\alpha_2\Delta T-\dfrac{F}{AY_2}\right)$

Passage II

20000 J of heat energy is supplied to a metal block of mass 500 g at atmospheric pressure. The initial temperature of the block is 30° C.

(Given, specific heat of metal $=400\text{ J kg}^{-1}\text{°C}^{-1}$, relative density of metal $=8$, coefficient of volumetric expansion of metal $=8\times10^{-5}\text{°C}^{-1}$, atmospheric pressure $=10^5$Pa).

4 Final temperature of block is

a. 120°C b. 130°C

c. 140°C d. 150°C

5 Work done by the block on the surroundings is

a. 0.05 J b. 0.1 J

c. 1.0 J d. 10 J

6 Change in internal energy of block is

a. zero

b. 20000 J

c. less than 20000 J

d. more than zero but less than 2 J

Passage III

We define, heat flux as heat conducted per unit area per unit time.

$$\therefore \quad \text{Heat flux, } H = \frac{q}{A} = \frac{-KA\frac{dT}{dx}}{A} = -K\frac{dT}{dx}$$

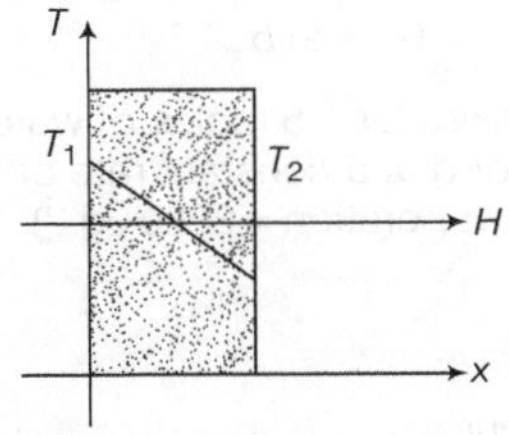

Heat flux is a directional quantity in one-dimensional flow of heat as heat flow is always perpendicular to a surface of constant temperature called an isotherm. Heat flux vector is defined as,

$$\mathbf{H} = -K\Delta T = -K\left(\hat{\mathbf{i}}\frac{\partial T}{\partial x} + \hat{\mathbf{j}}\frac{\partial T}{\partial y} + \hat{\mathbf{k}}\frac{\partial T}{\partial z}\right)$$

where, $\frac{\partial T}{\partial x}, \frac{\partial T}{\partial y}$ and $\frac{\partial T}{\partial z}$ are partial derivatives of temperature function.

(To find $\frac{\partial T}{\partial x}$, treat y and z as constants ... etc.).

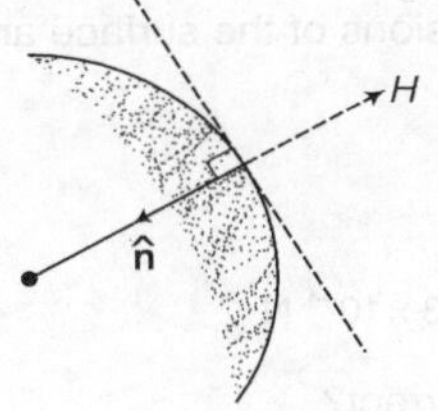

Now, consider a metal ball with temperature at any point of it is proportional to square of distance from centre. Centre of ball is at origin. Conductivity of ball is 80.0 $Wm^{-1}K^{-1}$. Temperature at point (1, 1, 1) m is initially 400 K. Radius of ball is 4 m.

7 Heat flux across the sphere of radius 2 m is

a. 4267 Wm^{-2} into the sphere
b. 4267 Wm^{-2} out of the sphere
c. 6400 Wm^{-2} into the sphere
d. 6400 Wm^{-2} out of the sphere

8 Rate of heat flow across the surface of sphere of radius 2 m will be

a. 42667 W into the sphere
b. 42667 W out of the sphere
c. 6400 W into the sphere
d. 6400 W out of the sphere

Passage IV

Heating curve for a substance as shown in the figure.

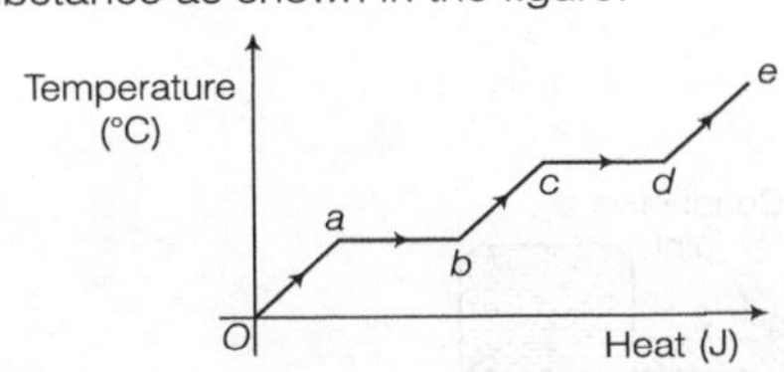

9 Match entries of Column I with the process they shows in Column II and mark the correct option from the codes given below.

	Column I		Column II
i.	*a* to *b*	p.	Condensation
ii.	*a* to *d* directly without reaching *b* or *c*	q.	Deposition
iii.	*d* to *a* directly without reaching *c* or *b*	r.	Sublimation
iv.	From *d* to *c*	s.	Melting

Codes

	i	ii	iii	iv
a.	t	r	q	p
b.	p	q	r	s
c.	p	r	s	t
d.	s	t	q	p

Working Space

10 Intermolecular bonds are formed during the process

a. 0 to *a* b. *a* to *b* c. *b* to c d. *b* to *a*

11 How much time it takes for 1.5 moles of water at 100° C to be completely converted into steam, if heat is added at a constant rate of 17 Js^{-1}.
(Take, enthalpy of evaporation = 2250 Jg^{-1})

a. Around 1 h
b. Around 5 min
c. Around 3 min
d. Around half an hour

One or More than One Options Correct Type

Direction **(Q. Nos. 12-16)** ***This section contains 5 multiple choice questions. Each question has four choices (a), (b), (c) and (d), out of which ONE or MORE THAN ONE are correct.***

12 Solar constant is about 10^3 Wm^{-2}. Let a perfectly absorbing surface is placed perpendicularly into the path of solar radiation. Let dimensions of the surface are 8 m ×20 m. Choose the correct options.

a. Average power reaching the surface is about 1.60×10^5 W
b. Maximum power reaching the surface is about 1.6×10^5 W
c. Force on the surface due to radiation pressure is zero
d. Force on the surface due to radiation pressure is about 5.33×10^{-4} N

13 Solar constant is about 1000 Wm^{-2}. Which of these are correct?

a. 1000 Wm^{-2} reaches surface of the earth
b. Nearly half of this, i.e. 500 Wm^{-2} reaches earth's surface
c. About 300 Wm^{-2} is reflected back into the space
d. About 200 Wm^{-2} is absorbed in the atmosphere

14 Which of these options are correct regarding emission of radiation by matter?

a. For gases and semi-transparent solids, emission is a volumetric phenomenon
b. In most solids radiations is a surface phenomena
c. In most liquids radiations is a volumetric phenomena
d. In all fluids radiation is a volumetric phenomena

15 Five identical rods numbered 1 to 5 each with conductivity K, length l and area A are connected as shown in the figure and the ends of combination are maintained at 100° C and 0° C, respectively.

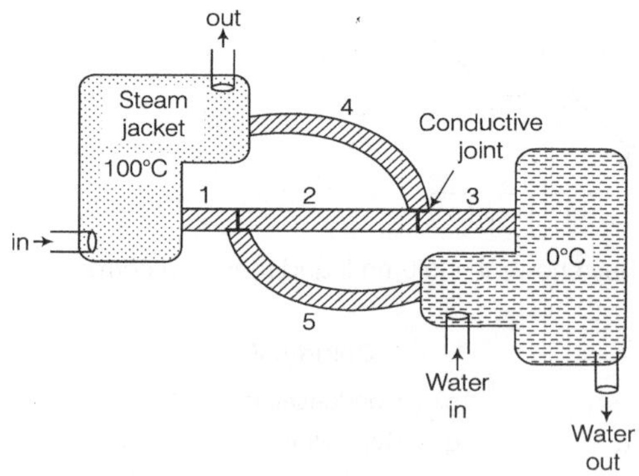

Choose the correct options.

a. Equivalent thermal resistance of rod combination is l/KA
b. Equivalent thermal resistance of combination of rods is $3l/KA$
c. No heat flows through rods 4 and 5
d. No heat flows through rod 2

Working Space

16 A 100 cm long cylindrical flask with inner and outer diameters 2 cm and 4 cm respectively, is completely filled with ice ($L_{ice} = 80$ calg^{-1}).

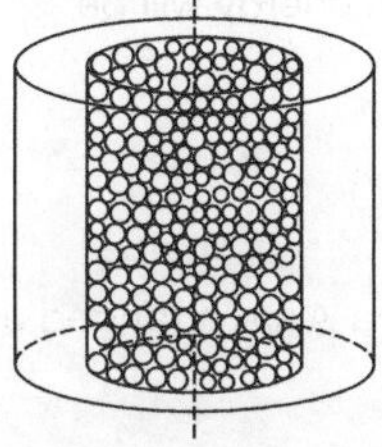

Temperature outside is 40° C and is constant. Thermal conductivity of flask is 0.693 $Wm^{-1°}C^{-1}$. Then, choose the correct option(s).

a. Rate of heat in flow to flask is 80π Js^{-1}

b. Rate of melting of ice is $\frac{\pi}{4200}$ kg s^{-1}

c. Rate of melting of ice is $100\ \pi$ kg s^{-1}

d. Rate of heat in flow to flask is 40π Js^{-1}

Only One Option Correct Type

Direction *(Q. Nos. 17-21) This section contains 5 multiple choice questions. Each question has four choices (a), (b), (c) and (d), out of which ONLY ONE is correct.*

17 An uninsulated steam pipe passes through a room in which the air and walls are at 25° C. The outside diameter of the pipe is 70 mm and its surface temperature and emissivity are 200° C and 0.8, respectively. Coefficient of convective heat transfer from the surface to the air is 15 $Wm^{-2}K^{-1}$. Rate of heat loss per unit length of pipe is given by,

a. 421 Wm^{-1}
b. 444 Wm^{-1}
c. 577 Wm^{-1}
d. 998 Wm^{-1}

18 Phase diagram for iodine is

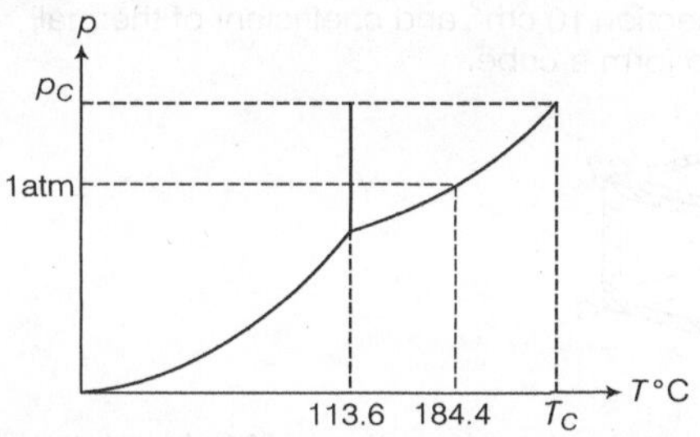

Choose the correct options.

a. Triple point temperature is more than normal boiling point of iodine
b. Normal boiling point of iodine is not fixed
c. Normal boiling point of iodine is 113.6°C
d. Normal melting point of iodine is 113.6°C

19 A piece of wood is floating in benzene.
(Given, density of benzene at 0° C $= 9 \times 10^2$ kg m^{-3},
density of wood at 0° C $= 8.8 \times 10^2$ kg m^{-3},
coefficient of cubical expansion of benzene $= 1.2 \times 10^{-3}$ K^{-1} and
coefficient of cubical expansion of wood $= 1.5 \times 10^{-4}$ K^{-1})
The temperature at which wood piece will sink in benzene is

a. 20.7°C
b. 21.7°C
c. 22.7°C
d. 23.7°C

Working Space

20 Power radiated by a black body is P_0 and the wavelength corresponding to maximum energy is λ_m. On changing temperature of black body, power radiated is increased to $\frac{256}{81}P_0$. The shift in wavelength corresponding to the maximum energy will be

a. $+\frac{\lambda_m}{4}$ b. $+\frac{\lambda_m}{2}$

c. $-\frac{\lambda_m}{4}$ d. $-\frac{\lambda_m}{2}$

21 Rods *AC*, *CB* and *CD* of same length l and cross-sectional area *A* are arranged as shown in the figure.

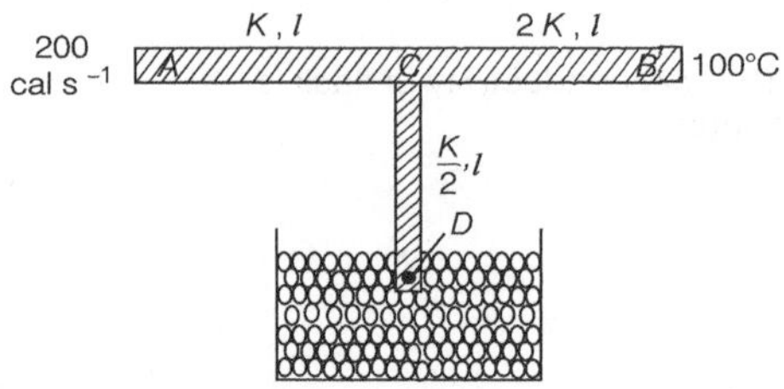

Heat is supplied through end *A* and the rate of 200 cal s^{-1} and end *B* is kept at 100° C. End *D* is dipped in a beaker containing ice at 0° C (latent heat of ice = 80 cal g^{-1}). Thermal conductivites of *AC*, *CB* and *CD* and *K*, 2*K* and *K*/2 respectively with $K = 100$ cal $m^{-1}s^{-1}{}^\circ C^{-1}$. Area of cross-section *A* is 10 cm^2. Length of each rod $l = 1$ m. Time in which 440 g of ice melts is

a. 800 s b. 700 s

c. 400 s d. 600 s

Matching List Type

***Direction** (Q. No. 22) Choices for the correct combination of elements from Column I and Column II are given as options (a), (b), (c) and (d), out of which one is correct.*

22 Twelve rods each of length 1 m, area of cross-section 10 cm^2 and coefficient of thermal conductivity $K = 100$ cal $m^{-1}s^{-1}{}^\circ C^{-1}$ are joined to form a cube.

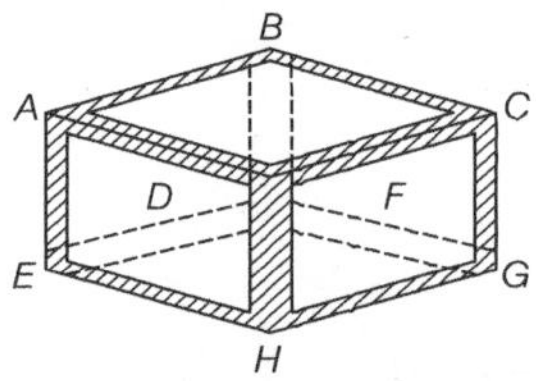

Temperature of hot junction is 100° C and temperature of cold junction is 0° C. Match the Column I with the Column II and mark the correct code from options given below. Make use of the electrical thermal analogy.

	Column I		Column II
i.	Hot junction is *A* and cold junction is *D*	p.	Thermal resistance = $\frac{0.3}{4}$ ($cal^{-1}s^\circ C$)
ii.	Hot junction is *A* and cold junction is *H*	q.	Thermal resistacne = $\frac{0.5}{6}$ ($cal^{-1}s^\circ C$)
iii.	Hot junction is *A* and cold junction is *G*	r.	Thermal resistance = $\frac{0.7}{12}$ (cal^{-1} $s^\circ C$)

Codes

	i	ii	iii			i	ii	iii
a.	q	q	r		b.	p	q	r
c.	r	p	q		d.	r	q	p

Working Space

One Integer Value Correct Type

Direction *(Q. Nos. 23 and 24) This section contains 2 questions. When worked out will result in an integer from 0 to 9 (both inclusive).*

23 A rod of length L, area of cross-section A is rotating in a horizontal plane about a vertical axis through one of its end. The density of rod is ρ and it is rotating at a constant angular speed ω rads^{-1}. To keep rod at constant length, we reduce its temperature by amount $\Delta T = \dfrac{\rho\omega^2 L^2}{kY\alpha}$. Find of the value of k where, Y = Young's modulus and α = coefficient of linear expansion for material of rod.

24 A small surface of area $A_1 = 10^{-3} m^2$ is emitting radiation with intensity of 7000 W per square metre per unit steradian solid angle, in perpendicular direction. A_1 is known to emit radiation diffusely.

Radiation emitted by A_1 is intercepted by three surfaces A_2, A_3 and A_4, ($A_2 = A_3 = A_4 = 10^{-3}$ m^{-2}) each of these lies at a distance of 0.5 m from A_1

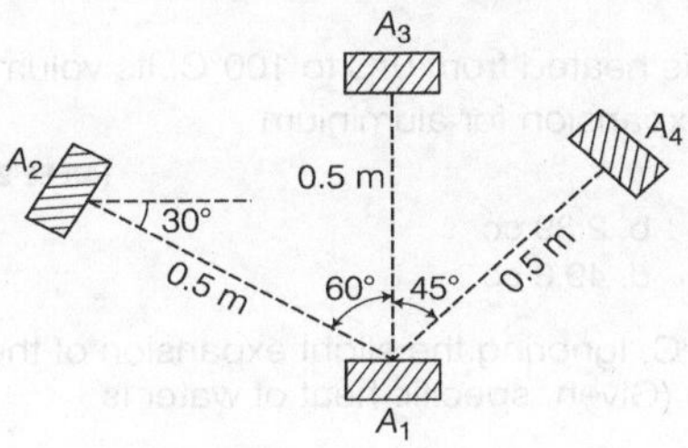

Assuming complete absorption and assuming plates are in space free from any other field or gravitation. The ratio of squares of initial accelerations of A_2 and A_4 is $k/8$. Find the value of k.

JEE Main & AIEEE Archive

(Compilation of Last 13 Years' Questions)

1 Consider a spherical shell of radius R at temperature T. The black body radiation inside it can be considered as an ideal gas of photons with internal energy per unit volume $u = \frac{U}{V} \propto T^4$ and pressure $p = \frac{1}{3}\left(\frac{U}{V}\right)$. If the shell now undergoes an adiabatic expansion the realation between T and R is **(JEE Main 2015)**

a. $T \propto e^{-R}$ b. $T \propto e^{-3R}$

c. $T \propto \frac{1}{R}$ d. $T \propto \frac{1}{R^3}$

2 An aluminium sphere of 20 cm diameter is heated from 0°C to 100°C. Its volume changes by (Given, coefficient of linear expansion for aluminium $\alpha_{Al} = 23 \times 10^{-6}/°C$) **(AIEEE 2011)**

a. 28.9 cc b. 2.89 cc

c. 9.28 cc d. 49.8 cc

3 100 g of water is heated from 30°C to 50°C. Ignoring the slight expansion of the water, the change in its internal energy is (Given, specific heat of water is 4184 J/kg/K) **(AIEEE 2011)**

a. 8.4 kJ b. 84 kJ

c. 2.1 kJ d. 4.2 kJ

4 A long metallic bar is carrying heat from one of its ends to the other end under steady state. The variation of temperature θ along the length x of the bar from its hot end is best described by which of the following figure? **(AIEEE 2009)**

a.
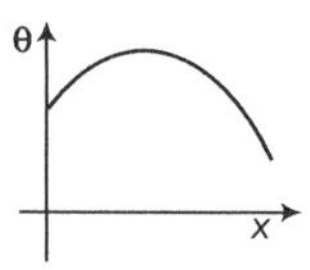

b.
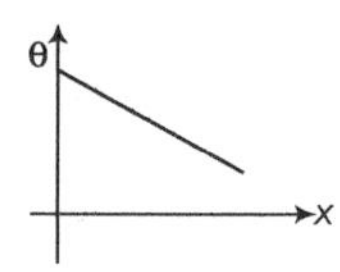

c.
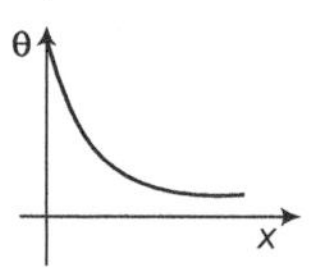

d.
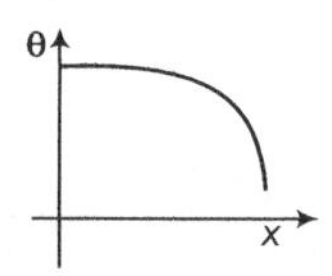

JEE Advanced & IIT JEE Archive

(Compilation of Last 10 Years' Questions)

1 Two spherical stars A and B emit black body radiation. The radius of A is 400 times that of B and A emits 10^4 times the power emitted from B. The ratio $\left(\frac{\lambda_A}{\lambda_B}\right)$ of their wavelengths λ_A and λ_B at which the peaks occur in their respective radiation curves is **(2015 Adv., Integer Type)**

2 Two rectangular blocks, having identical dimensions, can be arranged either in configuration I or in configuration II as shown in the figure. One of the blocks has thermal conductivity K and the other $2K$. The temperature difference between the ends along the X-axis is the same in both the configurations. It takes 9s to transport a certain amount of heat from the hot end to the cold end in the configuration I. The time to transport the same amount of heat in the configuration II is **(2013 Adv., Only One Option Correct Type)**

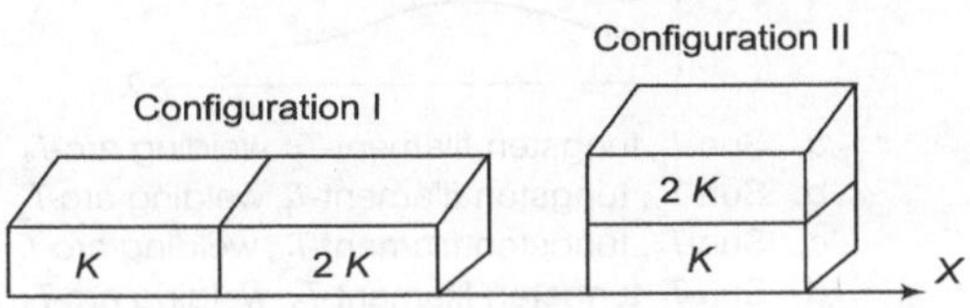

a. 2.0 s b. 3.0 s
c. 4.5 s d. 6.0 s

3 Two moles of ideal helium gas are in a rubber balloon at 30°C. The balloon is fully expandable and can be assumed to require no energy in its expansion. The temperature of the gas in the balloon is slowly changed to 35°C. The amount of heat required in raising the temperature is nearly (Take, $R = 8.31$ J/mol-K) **(2012, Only One Option Correct Type)**

a. 62 J b. 104 J
c. 124 J d. 208 J

4 Three very large plates of same area are kept parallel and close to each other. They are considered as ideal black surfaces and have very high thermal conductivity. The first and third plates are maintained at temperatures $2T$ and $3T$ respectively. The temperature of the middle (i.e. second) plate under steady state condition is **(2012, Only One Option Correct Type)**

a. $\left(\frac{65}{2}\right)^{\frac{1}{4}} T$ b. $\left(\frac{97}{4}\right)^{\frac{1}{4}} T$ c. $\left(\frac{97}{2}\right)^{\frac{1}{4}} T$ d. $(97)^{\frac{1}{4}} T$

5 A composite block is made of slabs A, B, C, D and E of different thermal conductivities (given in terms of a constant K) and sizes (given in terms of length L) as shown in the figure. All slabs are of same width. Heat Q flows only from left to right through the blocks. Then, in steady state **(2011, One or More than One Options Correct Type)**

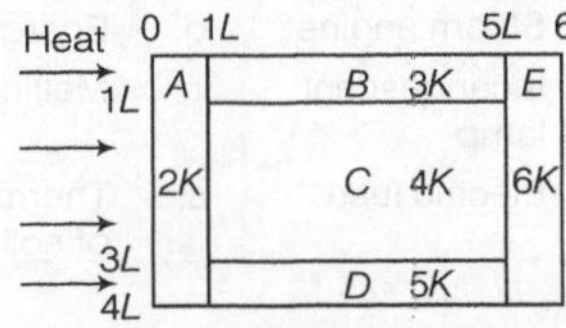

a. heat flow through slabs A and E are same
b. heat flow through slab E is maximum
c. temperature difference across slab E is smallest
d. heat flow through C = heat flow through B + heat flow through D

6 Steel wire of length L at 40° C is suspended from the ceiling and then a mass m is hung from its free end. The wire is cooled down from 40° C to 30° C to regain its original length L. The coefficient of linear thermal expansion of the steel is 10^{-5}/° C, Young's modulus of steel is 10^{11} Nm^{-2} and radius of the wire is 1 mm. Assume that, $L >>$ diameter of the wire. Then, the value of m (in kg) is nearly **(2011, Integer Type)**

7 Two spherical bodies A (radius 6 cm) and B (radius 18 cm) are at temperatures T_1 and T_2, respectively. The maximum intensity in the emission spectrum of A is at 500 nm and in that of B is at 1500 nm. Considering them to be black bodies, what will be the ratio of the rate of total energy radiated by A to that of B? **(2010, Integer Type)**

8 A piece of ice (heat capacity = 2100 Jkg^{-1}° C^{-1} and latent heat = 3.36×10^5 J kg^{-1}) of mass m gram is at –5°C at atmospheric pressure. It is given 420 J of heat so that the ice starts melting. Finally when the ice-water mixture is in equilibrium, it is found that 1 g of ice has melted. Assuming there is no other heat exchange in the process, the value of m is **(2010, Integer Type)**

9 A metal rod AB of length $10x$ has its one end A in ice at 0°C and the other end B in water at 100°C. If a point P on the rod is maintained at 400°C, then it is found that equal amounts of water and ice evaporate and melt per unit time.

The latent heat of evaporation of water is 540 $cal g^{-1}$ and latent heat of melting of ice is 80 cal g^{-1}. If the point P is at a distance of λx from the ice end A, find the value of λ. (Neglect any heat loss to the surrounding.) **(2009, Integer Type)**

10 Column I gives some devices and Column II gives some processes on which the functioning of these devices depend. Match the devices in Column I with the processes in Column II.
(2007, Matching Type)

	Column I		Column II
i.	Bimetallic strip	p.	Radiation from a hot body
ii.	Steam engine	q.	Energy conversion
iii.	Incandescent lamp	r.	Melting
iv.	Electric fuse	s.	Thermal expansion of solids

Codes

	i	ii	iii	iv
a.	r	r,s	p,r	q,r
b.	p	q,s	q	r,s
c.	s	q	p,q	q,r
d.	p,q	q	r,s	p,s

11 A black body of temperature T is inside a chamber of temperature T_0. Now, the closed chamber is slightly opened to Sun such that temperature of black body (T) and chamber (T_0) remains constant.
(2006, One or More than One Options Correct Type)

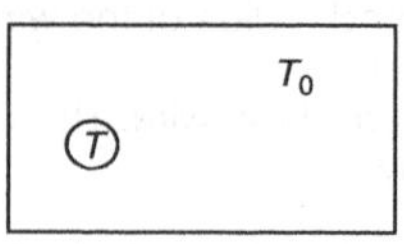

Which of these statements is/are correct?
a. Black body will absorb more radiation
b. Black body will absorb less radiation
c. Black body emit more energy
d. Black body emit energy equal to energy absorbed by it

12 In an insulated vessel, 0.05 kg steam at 373 K and 0.45 kg of ice at 253 K are mixed. Find the final temperature of the mixture (in kelvin).
(2006, Subjective Type)
(Given, $L_{fusion} = 80$ cal g^{-1} = 336 J g^{-1},
$L_{vaporisation} = 540$ cal g^{-1} = 2268 J g^{-1},
$S_{ice} = 2100$ J kg^{-1}, $K = 0.5$ cal $g^{-1} K^{-1}$
and $S_{water} = 4200$ J kg^{-1}, $K = 1$ cal $g^{-1} K^{-1}$)

13 Water of volume 2 L in a container is heated with a coil of 1 kW at 27°C. The lid of the container is opened and energy dissipates at rate of 160 Js^{-1}. In how much time temperature will rise from 27°C to 77°C?
(Take, specific heat of water = 4.2 $kJkg^{-1}$)
(2005, Only One Option Correct Type)
a. 8 min 20 s
b. 6 min 2 s
c. 7 min
d. 14 min

14 In which of the following process, convection does not take place primarily?
(2005, Only One Option Correct Type)
a. Sea and land breeze
b. Boiling of water
c. Warming of glass of bulb due to filament
d. Heating air around a furnace

15 Variation of radiant energy emitted by Sun, filament of tungsten lamp and welding arc as a function of its wavelength is shown in the figure. Which of the following options is the correct match? **(2005, Only One Option Correct Type)**

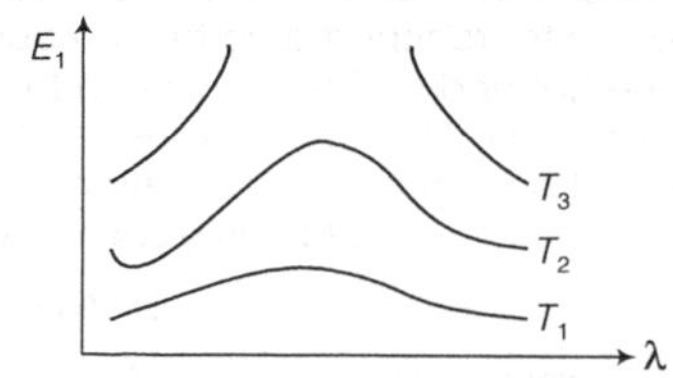

a. Sun-T_1, tungsten filament-T_2, welding arc-T_3
b. Sun-T_2, tungsten filament-T_1, welding arc-T_3
c. Sun-T_3, tungsten filament-T_2, welding arc-T_1
d. Sun-T_1, tungsten filament-T_3, welding arc-T_2

16 Calorie is defined as, the amount of heat required to raise temperature of 1 g of water by 1°C and it is defined under which of the following conditions ? **(2005, Only One Option Correct Type)**
a. From 14.5°C to 15.5°C at 760 mm of Hg
b. From 98.5°C to 99.5°C at 760 mm of Hg
c. From 13.5°C to 14.5°C at 76 mm of Hg
d. From 3.5°C to 4.5°C at 76 mm of Hg

17 A metal of mass 1 kg at constant atmospheric pressure and at initial temperature 20°C is given a heat of 20000J. Find the following :
i. Change in temperature, **(2005, Subjective Type)**
ii. Work done and
iii. Change in internal energy.
(Given, specific heat = 400 J/kg/°C, coefficient of cubical expansion, $\gamma = 9 \times 10^{-5}/°C$, density $\rho = 9000$ kg m^{-3}, atmospheric pressure $= 10^5 N m^{-2}$)

2. Kinetic Theory of Gases and Thermodynamics

DPP
DAILY PRACTICE PROBLEMS

DPP-1 Kinetic Theory of Gases-I (Real and Ideal Gases)

Subjective Questions

Working Space

Direction *(Q. Nos. 1-4) These questions are subjective in nature, need to be solved completely on notebook.*

1 Twelve molecules having speeds (in cms^{-1}), 6.0, 2.0, 4.0, 6.0, 6.0, 4.0, 1.0, 8.0, 5.0, 3.0, 7.0 and 8.0 respectively, are kept in a box of size 1 cm × 1 cm × 1cm. Calculate :

i. Mean speed
ii. Rms speed
iii. Number of back and forth round trips per second
iv. Most probable speed

2 i. Calculate rms speed of helium atoms near the surface of Sun.

ii. If rms speed of helium atoms is to made double of (i), to what temperature, it must be raised ?

iii. If pressure of helium sample is tripled keeping its volume at constant, by what factor does v_{rms} change?

3 i. Calculate molecular energy of one gram of hydrogen gas at 323 K. Given, molecular weight of hydrogen is 2×10^{-3} kg and $R = 8.3\,\text{Jmol}^{-1}\text{K}^{-1}$.

ii. Consider a gas sample at 2500 K, whose atoms can occupy only two energy levels, separated by 0.216 eV. Determine the number of atoms in the higher energy level.

iii. One cubic metre of atomic hydrogen at 0°C and atmospheric pressure contains about 2.70×10^{25} atoms. Energy value of a hydrogen atom in a stationary state is $\frac{-13.6}{n^2}$ (eV). Use Boltzmann's factor to find the number of atoms in the first excited state at 0°C.

4 A rectangular block is moving with a speed of 10 ms^{-1} in a gas at a temperature of 300 K and density 4×10^{-3} kg m^{-3}. The block moves along a straight line. Find drag force on the block. The mass of one molecule of the gas is about 4.14×10^{-23} kg and area of cross-section of block is 1 m^2.

Only One Option Correct Type

Direction *(Q. Nos. 5-15) This section contains 11 multiple choice questions. Each question has four choices (a), (b), (c) and (d), out of which ONLY ONE is correct.*

5 The rms speed of a gas molecule at 27°C and 1.0 atm pressure is 100 ms^{-1}. The rms speed of same molecule at 127°C and 2.0 atm pressure will be

a. 100 ms^{-1}
b. 200 ms^{-1}
c. $\frac{100}{\sqrt{3}}$ ms^{-1}
d. $\frac{200}{\sqrt{3}}$ ms^{-1}

Working Space

6 A gas mixture contains three type of molecules A, B and C with masses $m_A > m_B > m_C$. Then which of the following options is correct?
(K indicates kinetic energy and V indicates rms speed)

a. $K_A = K_B = K_C, v_A = v_B = v_C$
b. $K_A > K_B > K_C, v_A > v_B > v_C$
c. $K_A = K_B = K_C, v_A < v_B < v_C$
d. $K_A < K_B < K_C; v_A = v_B = v_C$

7 A sample of 0.5 g of hydrogen gas in a cube of side 3 cm is initially at NTP. The gas in the cube is compressed to a final pressure of 100 atm. Choose the correct option (s).

a. Ideal gas law is applicable initially but not in final state
b. Ideal gas law is applicable initially as well as in final state
c. Ideal gas law is not applicable in any state
d. Ideal gas law is applicable in final state

8 A planet contains a single type of gas in its atmosphere having molecular mass, $m = 3.0 \times 10^{-26}$ kg. Distribution of speeds in atmosphere is

Speed (ms^{-1})	Percentage of molecules
200	10
400	20
600	40
800	20
1000	10

If escape velocity of the planet is 900 ms^{-1}, assuming other conditions are stable. The possible estimated reduction in temperature of the planet in few years will be

a. 100 K
b. 96 K
c. 296 K
d. 48 K

9 A cylinder contains two type of gases A and B. Number of molecules of gas A in cylinder is N, each having a mass m. Number of molecules of gas B in cylinder is $2N$, each having a mass $2m$. The mean square of x-component of velocity of A is v_{Ax}^2 and mean square of x-component of velocity of B is v_{Bx}^2. The ratio of $v_{Ax} : v_{Bx}$ is

a. 1 : 1
b. 2 : 1
c. $1 : \sqrt{2}$
d. $\sqrt{2} : 1$

10 The temperature at which the rms speed of oxygen molecule will be sufficient for escaping from the earth is

a. 100° C
b. 6000° C
c. 1.61×10^5 K
d. 0.82×10^5 K

11 In an interval of 30s, 200 hailstones strike a glass window of area 0.5 m^2 at an angle of 30° to the surface of window. Each hailstone has a mass of 5 g and moves with a speed of 6 ms^{-1}. Assuming elastic collision, pressure over the window pane is

a. 1 Pa
b. $\frac{2}{3}$ Pa
c. $\frac{1}{5}$ Pa
d. $\frac{2}{5}$ Pa

12 A sealed cubical container with volume V contains 20 particles with each mass m and having various speeds. Two particles have speed v, three particles have speed $2v$, five particles have speed $3v$, four particles have speed $4v$, three particles have speed $5v$, two particles have speed $6v$ and one has speed $7v$. Pressure exerted by these particles on the walls of container is p. Then, $p \cdot V$ is

a. $106\, mv^2$
b. $\frac{8}{3} mv^2$
c. $51\, mv^2$
d. $20\, mv^2$

Working Space

13 Pressure of CO_2 in a container is given by

$$p = \frac{RT}{2V - b} - \frac{a}{4V^2} \text{ then,}$$

mass of the gas in the container is

a. 11 g
b. 22 g
c. 33 g
d. 44 g

14 A container of volume 30 L contains an ideal gas at a temperature of 0°C. A part of gas is expelled from the container, till the pressure changes by 0.5 atm. Temperature is kept constant. Density of gas initially is 1.3 gL^{-1}. Mass of the released gas is

a. 100 g
b. 20 g
c. 50 g
d. 40 g

15 For a gas, equation of state is $\left(p + \frac{aT^2}{V}\right)V^c = RT + b$, where *a*, *b*, *c* and *R* are constants. Isotherms for this gas have equation

$$p = AV^m - BV^n$$

where, *A* and *B* depends only on temperature.
Choose the correct option(s).

a. $m = -c$ and $n = -1$
b. $m = c$ and $n = 1$
c. $m = -c$ and $n = 1$
d. $m = c$ and $n = -1$

One or More than One Options Correct Type

Direction *(Q. Nos. 16-20) This section contains 5 multiple choice questions. Each question has four choices (a), (b), (c) and (d), out of which ONE or MORE THAN ONE are correct.*

16 Let $\bar{v}$, v_{rms} and v_p denote the mean speed, root mean square speed and most probable speed of the molecules in an ideal monatomic gas at absolute temperature *T*. Let *m* is the mass of a molecule.Then, choose the correct options.

a. Average KE of a molecule is $\frac{3}{4}mv_p^2$
b. $\bar{v} : v_{rms} : v_p :: \sqrt{3} : \sqrt{2.5} : \sqrt{2}$
c. $v_p < \bar{v} < v_{rms}$
d. KE of 1g of a gas $= \frac{3}{2}\frac{k}{m} \cdot T$

17 Assuming ideal gas behaviour, choose the correct options.

a. The coefficient of volume expansion at constant pressure is same for all ideal gases
b. Average translational KE per molecule of gas is $3kT$
c. In a gaseous mixture, the average translational KE of the molecules of each component of velocity is same
d. The mean free path of molecules increases with decrease in pressure

18 Which of the following are true regarding distribution of speeds of molecules in an ideal gas ?

a. v_p : the most probable speed is the root of equation $\frac{d}{dv}f(v) = 0$, where $f(v)$ is Maxwell distribution of speeds
b. $\int_0^\infty f(v) \cdot dv$ = total number of molecules in the gas sample
c. Maxwell's speed distribution curve shifts to right at higher temperatures
d. As, temperature increases maximum value of $f(v)$ on Maxwell's distribution curve decreases

Working Space

19 In the atmosphere of the earth, number of molecules per unit volume varies with altitude as per the law :

$$n_V(h) = n_0 e^{-mgh/k_B T}$$

where, $n_V(h)$ = number of molecules per unit volume at height h above the earth's surface.

$$n_0 = \frac{N}{V} \quad \text{at} \quad h = 0$$

m = mass of 1 molecule, k_B = Boltzmann's constant

T = absolute temperature

Using this law, choose the correct options.

a. Atmosphere is denser at lower altitudes because of gravity
b. Gravitational potential energy of a gas molecule increases with altitude
c. Average gravitational PE of a molecule depends only on temperature
d. Average height of a molecule increases as temperature increases

20 A hollow cube made up of hard plastic is filled with ionised hydrogen and face *ABCD* is given a positive charge.

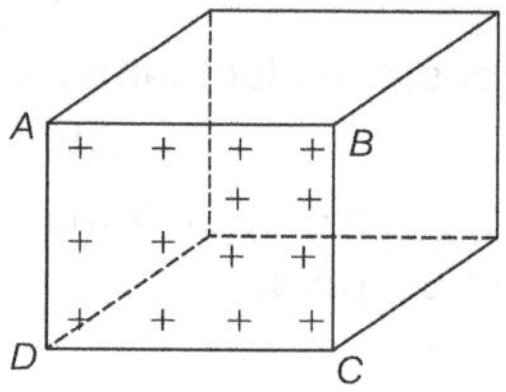

The usual kinetic theory expression for pressure

a. will be valid
b. will be valid, if face *ABCD* is not charged
c. will not be valid, since ions would experience forces other then due to collisions with the walls
d. will not be valid because isotropy is lost

Comprehension Type

Direction (*Q. Nos. 21-23*) *This section contains a paragraph, describing theory, experiments, data, etc. Three questions related to the paragraphs have been given. Each question has only one correct answer among the four given options (a), (b), (c) and (d).*

Passage

Ideal gas equation $pV = kNT$ is a good approximation for real gases only, when effect of intermolecular forces and of molecular sizes are negligible. Such condition occurs only at high temperatures and at low pressures.

In 1873 JD van der Waals, proposed an equation which took into account the effect of intermolecular forces and size of molecules in a simple way.

The van der Waals, equation can be written as,

$$\left[p + a\left(\frac{N}{V}\right)^2\right](V - Nb) = kNT$$

constants a and b are determined experimentally for a gas. Now, answer the following questions.

21 The van der Waals' equation may be cast into the viral form (in powers of N/V) in the following way.

a. $p = \frac{kNT}{V}\left[1 + b\frac{N}{V} + b^2\left(\frac{N}{V}\right)^2 + \cdots\right]$

b. $p = \frac{kNT}{V}\left[1 + \frac{a}{kT}\cdot\frac{N}{V} + b^2\left(\frac{N}{V}\right)^2 + \cdots\right]$

c. $p = \frac{kNT}{V}\left[1 + \left(b + \frac{a}{kT}\right)\frac{N}{V} + b^2\left(\frac{N}{V}\right)^2 + \cdots\right]$

d. $p = \frac{kNT}{V}\left[1 + \left(b - \frac{a}{kT}\right)\frac{N}{V} + b^2\left(\frac{N}{V}\right)^2 + \cdots\right]$

Working Space

22 Boyle's temperature, the lowest temperature at which a real gas closely resembles an ideal gas is given by

a. $T_B = \frac{a}{kb}$ b. $T_B = \frac{b}{ka}$

c. $T_B = \frac{kb}{a}$ d. $T_B = \frac{ka}{b}$

23 Which of these is van der Waals' equation in terms of critical pressure p_c, critical volume V_c and critical temperature T_c?

a. $\left(\frac{p}{p_c} + \frac{V_c^2}{V^2}\right)\left(\frac{V}{V_c} - 1\right) = \frac{8}{3}\left(\frac{T}{T_c}\right)$ b. $\left(\frac{p}{p_c} + \left(\frac{V_c}{V}\right)^2\right)\left(\frac{V}{V_c} + \frac{1}{3}\right) = \frac{8}{3}\left(\frac{T}{T_c}\right)$

c. $\left(\frac{p}{p_c} + \frac{3V_c^2}{V^2}\right)\left(\frac{V}{V_c} + \frac{1}{3}\right) = \frac{8}{3}\left(\frac{T}{T_c}\right)$ d. $\left(\frac{p}{p_c} + \frac{3V_c^2}{V^2}\right)\left(\frac{V}{V_c} - \frac{1}{3}\right) = \frac{8}{3}\left(\frac{T}{T_c}\right)$

Statement Type

Direction *(Q. Nos. 24-27) This section is based on Statement I and Statement II. Select the correct answer from the codes given below.*

a. Both Statement I and Statement II are correct and Statement II is the correct explanation of Statement I
b. Both Statement I and Statement II are correct and Statement II is not the correct explanation of Statement I
c. Statement I is correct but Statement II is incorrect
d. Statement II is correct but Statement I is incorrect

24 **Statement I** If a vessel contains equal amounts (by mass) of helium and argon, then the average speed of helium atom is same as that of argon atom.

Statement II After some time of filling, average kinetic energy of helium atoms is same as that of argon atoms.

25 **Statement I** If the pressure of a fixed amount of gas increases, then the temperature of gas must be increase.

Statement II $pV = nRT$.

26 **Statement I** Nitrogen is the most common constituent of atmosphere while helium is found only in traces.

Statement II Helium molecules have speeds exceeding the earth's escape speed.

27 **Statement I** Translational kinetic energy of 1g molecule of any ideal gas at 0°C is 3.4 kJ.

Statement II For an ideal gas, kinetic energy of 1 molecule is $\frac{3}{2} k_B T$.

DPP-2 Kinetic Theory of Gases-II (Gas Laws)

Subjective Questions

Working Space

Direction (Q. Nos. 1-4) These questions are subjective in nature, need to be solved completely on notebook.

1 Density of CO_2 gas at STP is 1.977 gL^{-1}. Find the volume per molecule. Interpret the results you find.

2 A gas mixture at 100°C and 0.8 atm pressure contains 50% of Helium and 50% Xenon by weight. What are the partial pressures of each component of the mixture?

3 A hollow 40 cm long cylinder of negligible mass rests over a frictionless horizontal surface. The cylinder contains nitrogen in one section and oxygen in other section. The nitrogen and oxygen volumes are separated by a vertical non-porous membrane. Pressure of the nitrogen is twice that of oxygen.
How far will the cylinder move, if the membrane breaks?

4 i. Show that 1 pascal $= 1 \frac{J}{m^3}$.

ii. Find energy density in space of translational KE of an ideal gas.

Only One Option Correct Type

Direction (Q. Nos. 5-14) This section contains 10 multiple choice questions. Each question has four choices (a), (b), (c) and (d), out of which ONLY ONE is correct.

5 One litre of oxygen at a pressure of 2 atm is compressed to a volume of 800 cm^3. What pressure is required? (keep temperature constant).

a. 2 atm b. 3 atm c. 2.5 atm d. 4 atm

6 Total number of molecules per unit volume (cm^3) at 27°C and a pressure of 10^{-10} mm of Hg in a sample of ideal gas is

a. 6.023×10^{23} b. 3.2×10^{23} c. 3.2×10^{6} d. 6.023×10^{6}

7 A person with body temperature 33°C inhales about 500 mL of air at a temperature of 27°C. Volume exhaled by the person will be

a. more than 500 mL b. exactly 500 mL
c. more than or equal to 500 mL d. less than 500 mL

8 The quantity $\frac{pV}{k_B T}$ represents the

a. mass of gas b. energy of gas
c. number of moles of gas d. number of molecules of the gas

Working Space

9 A gas obeys $pV^{3/2}$ = constant. If this gas is compressed quickly to half of its initial volume, then final temperature of the gas is

a. $\sqrt{2}T$ b. $2T$ c. $2\sqrt{2}T$ d. $4T$

10 An ideal gas undergoes a process whose process equation is

$$p = \frac{a}{\left[1+\left(\frac{V}{b}\right)^2\right]}$$

where, a and b are constants. When volume of gas $V = b$ the temperature of the gas will be

a. $\frac{ab}{2R}$ b. $\frac{ab}{R}$ c. ab d. zero

11 Two containers A and B are such that,

$$V_A = 0.10 \text{ m}^3,\ p_A = 1.40 \text{ MPa}$$
$$V_B = 0.15 \text{ m}^3,\ p_B = 0.7 \text{ MPa}$$

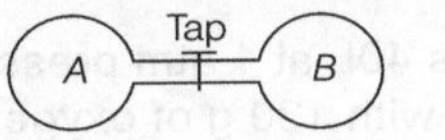

After opening the tap, final pressure in the containers is (Assume temperature of A and B are same)

a. 0.7 MPa b. 0.98 MPa c. 0.14 MPa d. 2.10 MPa

12 The gauge pressure in a car tyre is 305 kPa, when its temperature is 15°C. After running at high speed, the tyre gauge reads 360 kPa. If atmospheric pressure is 101 kPa, then final temperature of tyre is

a. 45°C b. 35°C c. 68°C d. 54°C

13 A tank is designed to store 18 kg of N_2 (Nitrogen) gas at a pressure of 4.5 atm. If H_2 (Hydrogen) gas is filled in the tank at 3.5 atm, then amount of gas in the tank will be

a. 18 kg b. 5 kg c. 1 kg d. 7 kg

14 Let us assume that, ideal gases behaves as follows.

i. At constant temperature, pressure is inversely proportional to the square of volume.
ii. At constant pressure the volume varies directly with the $\frac{2}{3}$ power of the temperature.
iii. At 273 K and 1 atm pressure 1.00 mole of an ideal gas occupies 22.4 L.

Then, the ideal gas equation will be

a. $pV = nRT$ b. $pV^2 = n^2RT$
c. $pV^2 = n^2RT^2$ d. $pV^2 = n^2RT^{4/3}$

One or More than One Options Correct Type

Direction (Q. Nos. 15-19) This section contains 5 multiple choice questions. Each question has four choices (a), (b), (c) and (d), out of which ONE or MORE THAN ONE are correct.

15 If a room of size 5 m × 3 m × 2.5 m at STP,

a. number of moles present is around 1700
b. number of moles present is around 2500
c. mass of air is about 68 kg
d. mass of air is about 50 kg

Working Space

16 A light container having a diatomic gas enclosed with in, is moving with velocity *v*. Mass of gas is *m*, number of moles is *n*. Then, choose the correct options.

a. KE of gas with respect to COM of system is $\frac{3}{2}nRT$

b. KE of gas with respect to COM of system is $\frac{5}{2}nRT$

c. KE of gas with respect to ground is $\frac{5}{2}nRT$

d. KE of gas with respect to ground is greater than $\frac{5}{2}nRT$

17 A vessel of volume 2×10^{-2} m^3 contains a mixture of H_2 and He at 47°C and at 4.15×10^5 Nm^{-2} pressure. Mass of mixture is 10^{-2} kg. Then,

a. mass of H_2 is 2.5×10^{-3} kg

b. mass of He is 7.5×10^{-3} kg

c. mass of H_2 is 7.5×10^{-3} kg

d. mass of He is 2.5×10^{-3} kg

18 A mixture of ethane (C_2H_6), ethene (C_2H_4) occupies 40L at 1 atm pressure and 400 K temperature. This mixture reacts completely with 130 g of oxygen (O_2) to produce carbon dioxide (CO_2) and water (H_2O).

Choose the correct options.

a. Mole fraction of ethane is 0.67

b. Mole fraction of ethene is 0.33

c. Mole fraction of ethane is 0.33

d. Mole fraction of ethene is 0.67

19 A 4 :1 molar mixture of Helium and Methane is contained in a vessel at 20 atm pressure. Due at a hole, the gas mixture leaks out. Choose the correct options.

a. Partial pressure of methane is same as the partial pressure of helium initially

b. Initially partial pressure of methane is 4 atm

c. Initial rate of diffusion of He and CH_4 are in ratio 1 : 8

d. Initial rate of diffusion of He and CH_4 are in ratio 8 : 1

Matching List Type

Direction (Q. No. 20) *Choices for the correct combination of elements from Column I and Column II are given as options (a), (b), (c) and (d), out of which one is correct.*

20 *A* and *B* are two identical containers. *A* contains 5 moles of H_2 and *B* contains 10 moles of O_2 at same temperature. Assuming real gas behaviour, match the following columns.

	Column I		Column II
i.	In container *A*	p.	Pressure is more
ii.	In container *B*	q.	RMS speed is more
iii.	As, number of moles is more	r.	Average thermal energies are same
iv.	Due to a smaller molecular mass	s.	Internal energy is more

Codes

	i	ii	iii	iv
a.	q,r	p,r,s	p,s	q
b.	q	p	r	s
c.	q,r	p,r	s	q
d.	r	p	s	q

Working Space

Comprehension Type

***Direction** (Q. Nos. 21-23) This section contains a paragraph, describing theory, experiments, data, etc. Three questions related to the paragraphs have been given. Each question has only one correct answer among the four given options (a), (b), (c) and (d).*

Passage

To estimate size of a molecule, an experiment is done as explained.

I. A pot containing 10L of water is taken.

II. It is kept in the open, it is found that 90% of the water is evaporated at the end of day.

III. It is topped up every morning up to the original mark for filling.

21 Number of molecules originally in the pot is

a. 6.023×10^{23} b. 6.023×10^{27}

c. 3.34×10^{25} d. 3.34×10^{26}

22 In how much time there is not a single molecule of the first batch is remaining? (of the water molecules in the pot on the vary first day)

a. One week b. One year

c. One month d. One decade

23 Let the water topping experiment is performing for a very long time, so that pot water initially kept on day, one part is of the hydrosphere (mass of hydrosphere $= 1.32 \times 10^{21}$ kg). How many water molecules are originally in the pot likely to present today?

a. 6.023×10^{23} b. 3.34×10^{26}

c. 2.53×10^{6} d. None of these

DPP-3 Kinetic Theory of Gases-III (Degree of Freedom and Energy of Gas)

Subjective Questions

Working Space

Direction (Q. Nos. 1-4) *These questions are subjective in nature, need to be solved completely on notebook.*

1 Calculate the average KE of translation of molecules of an ideal gas at 0° C.

i. For 1 molecule

ii. For 1 mole

iii. If molar mass of a gas is 4, find molecular translational kinetic energy of 1*g* of the gas.

2 Calculate

i. The average KE of translation of an oxygen molecule at 27° C.

ii. The total KE of an oxygen molecule at 27° C.

iii. What fraction of total energy is translational and what fraction of total energy is rotational?

3 First excited state of a hydrogen atom is 10.2 eV above its lowest state.

i. At what temperature hydrogen atom will be exicited to this state?

ii. Why atoms de-excite by emitting radiation rather than by thermal methods?

4 i. Give a relation between atomicity and ratio of specific heats of a gas. Also, explain its implications.

ii. Give a formula to easily obtain number of degree of freedom of a gas.

iii. Give a formula relating ratio of specific heats and degree of freedom.

Only One Option Correct Type

Direction (Q. Nos. 5-14) *This section contains 10 multiple choice questions. Each question has four choices (a), (b), (c) and (d), out of which ONLY ONE is correct.*

5 Degree of freedom for a solid molecule (or of a molecule in a solid) is

a. 3 b. 4 c. 5 d. 6

6 Energy of all molecules of a monatomic gas having a volume *V* and pressure *p* is $\frac{3}{2}pV$. Ratio of total translational kinetic energy of all molecules to total energy of all molecules of a diatomic gas at same pressure and volume is

a. $\frac{3}{2}$ b. $\frac{5}{2}$

c. $\frac{5}{3}$ d. $\frac{3}{5}$

Working Space

7 Temperature of argon gas kept in a vessel is increased from room temperature by 5°C. Then, choose the correct options.

a. 50% of heat is absorbed in form of translational energy of molecules and 50% is lost
b. 50% of heat is absorbed as translational energy and 50% of heat is absorbed as rotational energy of molecules
c. 100% of heat is absorbed as translational energy of molecules
d. 100% of heat is absorbed as rotational energy of molecules

8 At STP, speed of sound in a gas sample is found exactly 330 ms^{-1}. Density of gas sample is 1.3 gm^{-3}. Number of degree of freedom for a gas molecule will be

a. 3 b. 4
c. 5 d. 6

9 Certain ideal gas molecules behave like rigid spheres of 3×10^{-10} m radius. The mean free path for the gas will be

a. 3×10^{-10} m b. 3×10^{-8} m
c. 2.4×10^{-10} m d. 2.4×10^{-8} m

10 At what pressure will the mean free path be 50 cm for spherical molecules of an ideal gas of radius is 3×10^{-10} m, at the temperature 20° C?

a. 1 atm b. 4.1 MPa
c. 5.1 MPa d. 3.1 MPa

11 A nitrogen molecule has velocity vector perpendicular to the surface of the earth. If temperature is taken 27° C and if the molecule go straight without colliding with other molecules, about how much higher molecule can rise above the surface of the earth, if magnitude of velocity vector is same as that of rms velocity.

a. 10 km b. 20 km
c. 18 km d. 14 km

12 Equal amount of heat is given to a diatomic gas sample and a monatomic gas sample. Sample masses, initial temperatures and pressures are same. Then, choose the correct option.

a. Temperature of monatomic sample is found more than that of diatomic sample
b. Temperature of diatomic sample is found more than that of monatomic sample
c. Temperature of both monatomic and diatomic samples will be same
d. Data given is insufficient to conclude

13 Mean free path of oxygen molecules at 0°C and pressure p is $\lambda_1 = 9.5\times10^{-8}$ m. Keeping temperature constant pressure is reduced 100 times, then new mean free path is

a. 9.5×10^{-8} m b. 9.5×10^{-10} m
c. 9.5×10^{-6} m d. 9.5×10^{-12} m

14 Value of mean free path for nitrogen at NTP is around is 6.85×10^{-8} m. RMS speed of nitrogen molecule at NTP is 4.53×10^{2} ms^{-1}. The frequencey of collision for nitrogen molecules at NTP will be

a. $6.6\times10^{9}\times\sqrt{\frac{2}{3}}(s^{-1})$
b. $6.6\times10^{9}\times\sqrt{\frac{3}{2}}(s^{-1})$
c. $6.6\times10^{9}\times\sqrt{\frac{3\pi}{8}}(s^{-1})$
d. $6.6\times10^{9}\times\sqrt{2}(s^{-1})$

One or More than One Options Correct Type

Working Space

Direction (Q. Nos. 15-19) *This section contains 5 multiple choice questions. Each question has four choices (a), (b), (c) and (d), out of which ONE or MORE THAN ONE are correct.*

15 Which of these are true regarding a diatomic gas?

a. $(C_V)_{\text{diatomic gas}} = \frac{5}{3}(C_V)_{\text{monatomic gas}}$

b. A diatomic gas has only two degrees of freedom for rotation as moment of inertia of gas molecule through an axis and atoms is very small

c. At higher temperatures, a diatomic molecule have one more degree of freedom due to vibration of atoms

d. Range of C_V for a diatomic gas is $\left[\frac{3}{2}R, \frac{7}{2}R\right]$

16 Which of these are true regarding crystalline solids?

a. Principle of equipartition of energy cannot be applicable for solids

b. Molar specific heat for any solid can have any value ranging from 0 to $3R$

c. At high temperatures nearly all solids have same specific heat value

d. In any crystalline solid, heat is absorbed in form of KE and PE and ratio of KE and PE is always one

17 A cubical box of 10.35 m side length having a pressure about $\sqrt{2} \times 10^{-3}$ Pa at 27°C. Box contains an ideal gas with molecular radius $\frac{1}{\sqrt{\pi}}$(Å). Which of these are correct?

a. Collision with wall are not elastic at such low pressure

b. Mean free path length is more than 50 m

c. Mean free path length is more than mean interatomic spacing

d. On an average atom/molecule of gas makes 5 collisions with wall before hitting another molecule

18 At ordinary temperature and pressure, the observed value of C_V for SO_2, (sulphur dioxide) gas is 7.5 cal $mol^{-1} K^{-1}$.

Which of these are correct regarding above observation?

a. Theoretical value of C_V is about 6 cal $mol^{-1} K^{-1}$

b. SO_2 is not triatomic

c. Few SO_2 molecules have degree of freedom more than 6

d. Atoms in few SO_2 molecules are capable of vibration at ordinary temperature

19 Translational energy of 20 g of oxygen is 2202 J. Then, choose the correct options.

a. Rotational energy associated with 20 g oxygen is nearly 1468 J

b. Total energy of 20 g oxygen is 3600 J

c. An oxygen molecule has 7 degree of freedom

d. Temperature of 20 g oxygen (under consideration) is 10°C

Statement Type

Direction (Q. Nos. 20-22) *This section is based on Statement I and Statement II. Select the correct answer from the codes given below.*

a. Both Statement I and Statement II are correct and Statement II is the correct explanation of Statement I

b. Both Statement I and Statement II are correct and Statement II is not the correct explanation of Statement I

c. Statement I is correct but Statement II is incorrect

d. Statement II is correct but Statement I is incorrect

Working Space

20 **Statement I** Maximum frequency of sound waves, which can be transmitted through a column of gas decreases with increase in pressure of the gas.

Statement II Sound waves through a gas column are longitudinal in nature.

21 **Statement I** Hydrogen gas can have only discreet values of C_V. C_V for H_2 can be $\frac{3}{2}R$, $\frac{5}{2}R$ and $\frac{7}{2}R$ only.

Statement II Energy is quantised.

22 **Statement I** In polyatomic gases, atoms with in the molecules may also vibrate with respect to each other. In such cases, molecule will have one more degree of freedom.

Statement II A degree of freedom is the way (mode) in which a system (or molecule) can possess energy.

Comprehension Type

Direction *(Q. Nos. 23-25) This section contains a paragraph, describing theory, experiments, data, etc. Three questions related to the paragraphs have been given. Each question has only one correct answer among the four given options (a), (b), (c) and (d).*

Passage

Consider a crade model for rotating diatomic chlorine molecule.

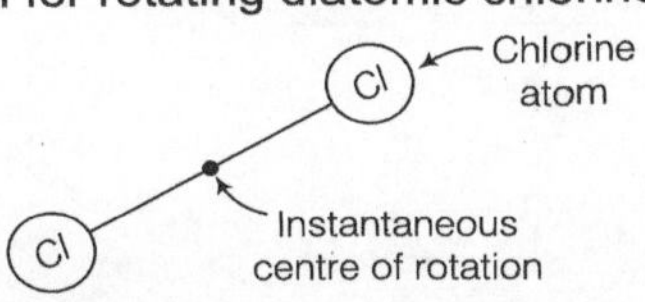

Distance between atomic centres of Cl atoms $= 2 \times 10^{-10}$ m.

Rotational speed, $\omega = 2 \times 10^{12}$ rads^{-1}.

Molar mass of chlorine is 70 g per mole.

Temperature is not very high.

23 What is the rotational KE of 1 molecule of Cl_2?

a. 2.33×10^{-21} J
b. 2.33×10^{-11} J
c. 2.33×10^{-10} J
d. 2.33×10^{-9} J

24 Total energy of 1 molecule of Cl_2 (on an average) is

a. 0.932×10^{-21} J
b. 5.8×10^{-21} J
c. 1.398×10^{-21} J
d. 3.8×10^{-21} J

25 Temperature of sample of Cl_2 given is around

a. 300 K
b. 105 K
c. – 105°C
d. 0°C

One Integer Value Correct Type

Direction *(Q. Nos. 26 and 27) This section contains 2 questions. When worked out will result in an integer from 0 to 9 (both inclusive).*

26 At pressure p and tempeature 200 K, mean free path of a gas is 2.5×10^{-7} m. If a pressure of $\frac{p}{4}$ and tempeature of 800 K, mean free path is $N \times 10^{-6}$ m, then find the value of N.

27 Sound waves in a gas can propogate only if the gas molecules collide with each other on the time scale of the sound waves period. The highest possible frequency for a sound wave in a gas and is approximately equal to the inverse of average collision time (or mean relaxation time) for molecules of that gas. For a gas containing molecules of mass m, radius r at pressure p and temperature T, then the maximum frequency, if $f_{max} = (N^2) \cdot pr^2 \sqrt{\frac{\pi}{mkT}}$. Find the value of N.

Working Space

Matching List Type

Direction (Q. No. 28) *Choices for the correct combination of elements from Column I and Column II are given as options (a), (b), (c) and (d), out of which one is correct.*

28 Match Column I with value of KE of 1 mole of gas given in Column II and mark the correct option from the codes given below.

	Column I		Column II
i.	Monatomic	p.	$\frac{7}{2}RT$
ii.	Diatomic	q.	$3RT$
iii.	Triatomic non-linear	r.	$\frac{5}{2}RT$
iv.	Triatomic linear	s.	$\frac{3}{2}RT$

Codes

	i	ii	iii	iv
a.	p	q	r	s
b.	p	q	s	r
c.	s	r	q	p
d.	s	r	p	q

DPP-4 Specific Heat of Solids, Liquids and Gases

Only One Option Correct Type

Working Space

Direction *(Q. Nos. 1-15) This section contains 15 multiple choice questions. Each question has four choices (a), (b), (c) and (d), out of which ONLY ONE is correct.*

1 If m mass of a substance absorbs dQ amount of heat and its temperature shows a raise of dT, then specific heat of a substance is

a. $\frac{mdQ}{dT}$ b. $\frac{mdT}{dQ}$

c. $\frac{dQ}{mdT}$ d. $\frac{dT}{mdQ}$

2 A real gas can have

a. only one value of specific heat which changes with temperature
b. only two values of specific heat which changes with temperature
c. only two values of specific heat which remains constant with temperature
d. infinite number of specific heats

3 Internal energy of an ideal gas is a function of temperature only. If a gas is taken from temperature T_1 to T_2, its internal energy changes from U_1 to U_2. For m mass of gas,

a. $U_2 - U_1 = mC_V(T_1 - T_2)$ b. $U_2 - U_1 = mC_p(T_2 - T_1)$
c. $U_2 - U_1 = mC_p(T_1 - T_2)$ d. $U_2 - U_1 = mC_V(T_2 - T_1)$

4 m gram of a gas is heated from T_1 temperature to T_2 temperature. Then, which of these is incorrect (consider MKS units)?

a. $C_p - C_V = \frac{R}{J}$ b. $C_V = \frac{R}{J(\gamma - 1)}$

c. $C_p = \frac{\gamma R}{(\gamma - 1)J}$ d. $C_V = \frac{R}{\gamma - 1}$

5 Enthalpy of a gas is a property of the gas. It is not given by

a. $h = u + p\Delta V$ b. $h = C_p\Delta T$ c. $h = C_V\Delta T + R\Delta T$ d. $h = u + p\Delta V - \Delta Q$

6 0.3 kg of nitrogen gas at 100 kPa and 40°C is contained in a cylinder with a piston.

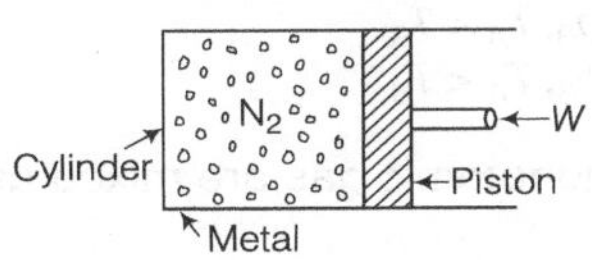

Piston is moved compressing nitrogen, till pressure becomes 1 MPa and temperature becomes 160°C. Work done in the process is 30 kJ. Heat transferred from nitrogen to the surrounding is

a. 33 kJ b. –3 kJ c. 30 kJ d. 63 kJ

Working Space

7 At constant volume, heat absorbed by 1 mole of a gas is $\frac{3R}{2}$ for each kelvin rise of temperature. Then, gas must be

a. monatomic
b. diatomic
c. triatomic
d. polyatomic

8 For a diatomic gas, difference between the two specific heats is 4150 J/kg-K. If 2.5 kg of this gas is supplied with 20750 J of heat at constant volume, then raise of temperature as shown by gas is

a. $\frac{4}{5}$K
b. $\frac{5}{4}$K
c. $\frac{3}{2}$K
d. $\frac{2}{3}$K

9 Specific heat of argon at constant volume is 0.075 $\frac{\text{kcal}}{\text{kg-K}}$ and its gram molecular specific heat is 2.98 $\frac{\text{cal}}{\text{mole-K}}$. Mass of argon atom (in grams) is

a. 6.60×10^{-23}
b. 3.30×10^{-23}
c. 2.20×10^{-23}
d. 13.20×10^{-23}

10 When a certain mass of an ideal diatomic gas is supplied with 105 cal of heat, then what amount of heat, which raises its temperature?

a. 105 cal
b. 75 cal
c. 27cal
d. 60 cal

11 A gas is heated at constant pressure. Fraction of heat that can be used to do external work is ($\gamma = C_p/C_V$).

a. $\frac{1}{\gamma}$
b. $1-\frac{1}{\gamma}$
c. $\gamma - 1$
d. $1-\frac{1}{\gamma^2}$

12 Let degree of freedom for a gas has an average value 6. This gas does 25 J of piston work at constant pressure. Heat absorbed by the gas must be

a. 75 J
b. 100 J
c. 150 J
d. 125 J

13 A vessel contains a gas of molecular mass M and is moving with a constant speed v. When the vessel is suddenly stopped, rise in temperature shown by gas ($\gamma = C_p/C_V$) will be

a. $\frac{Mv^2}{2R(\gamma+1)}$
b. $\frac{Mv^2(\gamma-1)}{2R}$
c. $\frac{Mv^2}{2R(\gamma+1)}$
d. $\frac{Mv^2}{2R(\gamma-1)}$

14 For a gas in a state A, $C_p - C_V = R$ and in an other state B, $C_p - C_V = 1.05R$. Then, choose the correct option (s).

a. $p_A = p_B, T_A = T_B$
b. $p_A = p_B, T_A < T_B$
c. $p_A < p_B, T_A > T_B$
d. $p_A > p_B, T_A < T_B$

15 If two moles of a diatomic gas and one mole of monatomic gas are mixed and this mixture is supplied with 190 J of heat.
Then, amount of heat which is used to do external work is

a. $\frac{19}{13}$
b. $\frac{13}{19}$
c. $\frac{6}{19}$
d. $\frac{19}{6}$

Working Space

One or More than One Options Correct Type

Direction (Q. Nos. 16-20) This section contains 5 multiple choice questions. Each question has four choices (a), (b), (c) and (d), out of which ONE or MORE THAN ONE are correct.

16 Which of these are correct about a metal at high temperatures?

a. Atoms in lattice have both kinetic and potential energies
b. Atoms in lattice have only potential energies
c. Total energy of an atom in lattice in $3kT$
d. Total energy of an atom in lattice is $3RT$

17 A certain pure metal has a specific heat of 230 $Jkg^{-1}K^{-1}$ at high temperatures. Which of these are correct?

a. Molar specific heat of metal is 24942 J/mol-K
b. Molar specific heat of metal remains constant for a slight temperature change
c. Molecular weight of metal is 108 kg mol-K
d. All atoms in metal vibrate at same frequency

18 When 0.44 kg of air at 180°C expands adiabatically to three times its original volume and during the process, there is a fall in temperature to 15°C. Work done during the process is 52.5 kg. Then,

a. specific heat at constant volume is 0.722 kJ/kg-K
b. specific heat at constant volume is 0.722 kJ/(K.mol) (K)
c. specific heat at constant pressure is 1.018 kJ/kg-K
d. specific heat at constant pressure is 1.018 kJ/K mol. K

19 For a process, specific heat is given by $C_n = C_V\left(\frac{\gamma - n}{1 - n}\right)$ for $n = 1$, which of these are correct?

a. Work done in the process is equal to heat added to the gas
b. Specific heat in the process is not defined
c. Process is isotropic
d. Process is irreversible

20 An ideal monatomic gas is heated from temperature T_1 to T_2 at constant volume. The gas is expanded back to its initial temperature as per the law pV^n = constant. It is found that entropy change in both process are equal. Then,

a. the net entropy change is zero
b. index of expansion is 4/3
c. index of expansion is 5/3
d. specific heat of expansion of gas is $\frac{3}{2}R$

Matching List Type

Direction (Q. No. 21) Choices for the correct combination of elements from Column I and Column II are given as options (a), (b), (c) and (d), out of which one is correct

21 Specific heat of a gas does not have a unique value, it depends on the process undergoes by the gas. Match the process (Column I) with the value of specific heat (Column II) and mark the correct option from given codes.

	Column I		Column II
i.	Gas is compressed suddenly and no heat is supplied from outside.	p.	$C < 0$
ii.	Gas is heated and also allowed to expand.	q.	$C > 0$
iii.	Heat is supplied to gas but expansion is restricted to a slower rate than its natural rate.	r.	$C = \infty$
iv.	Heat is supplied but gas is expanded forcefully.	s.	$C = 0$

Codes

	i	ii	iii	iv		i	ii	iii	iv
a.	s	r	q	p	b.	p	q	r	s
c.	s	q	r	p	d.	p	r	q	s

DPP-5 Thermodynamics-I (Work, Heat and Internal Energy)

Only One Option Correct Type

Working Space

Direction (*Q. Nos. 1-10*) *This section contains 10 multiple choice questions. Each question has four choices (a), (b), (c) and (d), out of which ONLY ONE is correct.*

1 Which of the following quantities can be called as thermodynamics property?

I. $\int p \cdot dV$

II. $\int V \cdot dp$

III. $\int p \cdot dV + \int V \cdot dp$

a. Both I and II
b. Both II and III
c. I, II and III
d. Only III

2 A cylinder contains compressed helium, is used to inflate in an elastic flexible balloon originally folded completely flat, to a volume of 0.6 m^3. If barometer reads 760 mm of Hg, then the work done by the atmosphere on the balloon is

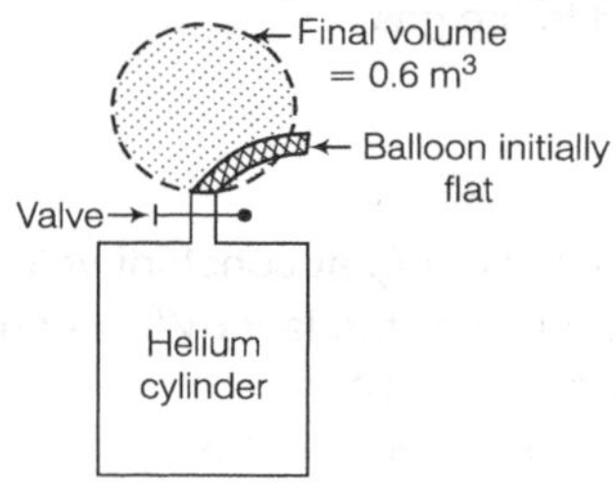

a. 60.8 kJ
b. – 60.8 kJ
c. 0
d. 60.8 J

3 In an experimental arrangement, a piston and cylinder arrangement has a stirring device. Piston is frictionless and is held down by the atmospheric pressure (101.3 kPa). Stirring device makes 9500 revolutions with an average torque of $\frac{1}{2\pi}$ N-m churning the fluid. The piston of diameter $\frac{1}{\sqrt{\pi}}$ m^2 moves out by 10 cm due to churning of fluid. Net work transfer for the system is

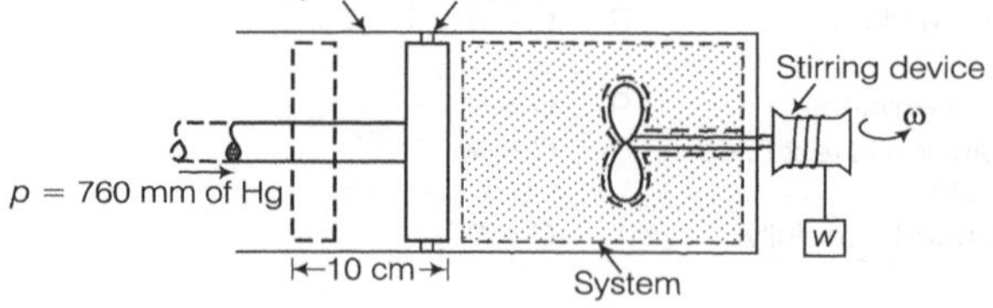

a. 2.53 kJ
b. – 2.53 kJ
c. 9497.47 kJ
d. – 9497.47 kJ

Working Space

4 Properties of a closed system change as per the relation

$$pV = 3.0$$

where, p = pressure in Nm^{-2} and V = volume in m^3. Work done on the system, when pressure increases from $1.5 \times 10^5\ Nm^{-2}$ to $7.5 \times 10^5\ atm^{-2}$, is : (Given, $\log_e 5 = 1.61$)

a. 4.83 J
b. – 4.83 J
c. 3.00 J
d. – 3.00 J

5 An ideal gas is taken from state A to state B as shown on pV diagram by 3-different processes 1, 2 and 3.

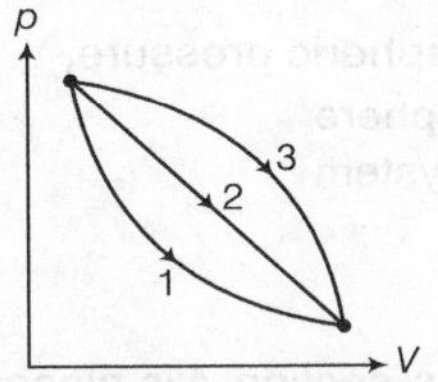

If Q = heat absorbed and U = change in internal energy. Then, choose the correct options.

a. $Q_1 = Q_2 = Q_3$, $U_1 < U_2 < U_3$
b. $Q_1 > Q_2 > Q_3$, $U_1 = U_2 = U_3$
c. $Q_1 < Q_2 < Q_3$, $U_1 = U_2 = U_3$
d. $\frac{Q_1}{U_1} = \frac{Q_2}{U_2} = \frac{Q_3}{U_3}$

6 If a system is provided with 150 J of heat and system does a work of 110 J, then change in internal energy of the system is

a. 260 J
b. – 260 J
c. 40 J
d. – 40 J

7 Volume of a van der Waals' gas changes from V_1 to V_2 at constant temperature. The work done is the process is

a. $nRT \ln\left(\frac{V_2 - nb}{V_1 - nb}\right) + an^2\left(\frac{V_1 - V_2}{V_1 V_2}\right)$
b. $nRT \ln\left(\frac{V_2 - na}{V_1 - na}\right) + bn^2\left(\frac{V_1 - V_2}{V_1 V_2}\right)$
c. $nRT \ln\left(\frac{V_2 - na}{V_1 - na}\right) + n^2\left(\frac{V_1 - V_2}{V_1 V_2}\right)$
d. $nRT \ln\left(\frac{V_1 - nb}{V_2 - nb}\right) + an^2\left(\frac{V_1 V_2}{V_1 - V_2}\right)$

8 A gas cylinder with a movable frictionless piston and conducting walls contains n moles of a gas at tempeature T.

If the piston is slowly displaced by an external agent to make the volume double of its initial value, then work done by external agent is

a. nRT
b. $nRT(1 + \ln 2)$
c. $nRT(1 - \ln 2)$
d. $2nRT$

9 In the following indicator diagram, ratio of work done in the process 1 with that of the process 2 is

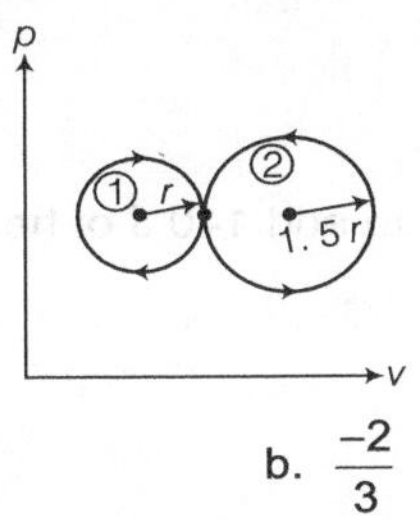

a. $\frac{3}{2}$
b. $\frac{-2}{3}$
c. $\frac{4}{9}$
d. $\frac{-4}{9}$

Working Space

10 1 cm^3 of water absorbs 540 cal of heat to become steam with a volume of 1671 cm^3. If atmospheric pressure is 1.013×10^5 Nm^{-2} and 1 cal = 4.19 J, the energy spent in this process in overcoming molecular attraction is

a. 540 cal
b. 40 cal
c. 500 cal
d. zero

One or More than One Options Correct Type

Direction (Q. Nos. 11-15) *This section contains 5 multiple choice questions. Each question has four choices (a), (b), (c) and (d), out of which ONE or MORE THAN ONE are correct.*

11 During melting of a slab of ice at 273 K and at atmospheric pressure,

a. positive work is done on ice-water system by atmosphere
b. positive work is done on atmosphere by ice-water system
c. internal energy of the ice-water system increases
d. internal energy of atmsophere increases

12 A non-conducting piston of mass m and area of cross-section A is placed on a non-conducting cylinder.

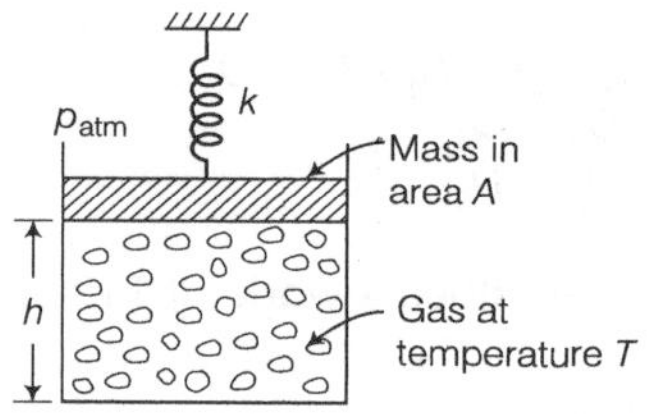

Initial height of piston is h and spring is relaxed with a spring constant k. Then, choose the correct option(s).

a. Number of moles present, $n = \frac{\left(p_{atm} - \frac{mg}{A}\right)Ah}{RT}$

b. Number of moles present, $n = \frac{\left(p_{atm} + \frac{mg}{A}\right)Ah}{RT}$

c. Work done in displacing piston by heating gas $= mgx + p_{atm} Ax + \frac{1}{2}kx^2$

(x = displacement of piston)

d. Work done in displacing piston by heating gas $= mgx + p_{atm} Ax - \frac{1}{2}kx^2$

13 One mole of helium is kept in a cylinder of area 8.5 cm^2. Cylinder is closed by a light frictionless piston. If a heat of 42 J given to gas causes a rise of temperature through 2°C, then (take, p_{atm} = 100 kPa).

a. $\Delta U = 30$ J
b. $\Delta W = 17.1$ J
c. displacement of piston = 20 cm
d. displacement of piston = 10 cm

14 An ideal diatomic gas is heated at constant pressure to add 140 J of heat to the gas. Then, choose the correct option(s).

a. Change in internal energy of gas is 100 J
b. Work done by the gas is 40 J
c. Work done by the gas is zero
d. Change in internal energy of gas is 140 J

Working Space

15 n-moles of a diatomic gas undergoes a cyclic process ABC. Temperature at A is T_0. Choose the correct alternatives.

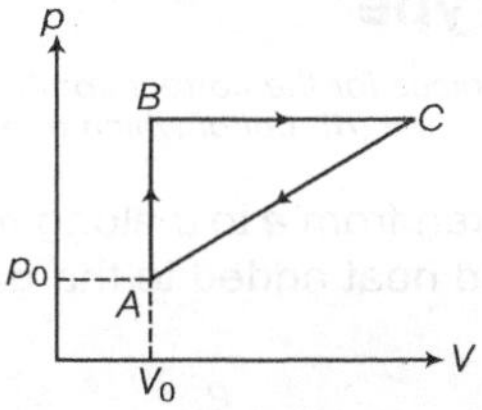

a. Work done during the cycle is $\frac{p_0V_0}{2}$

b. Heat rejected during the cycle is $9\,p_0V_0$

c. Heat injected during the cycle is $\frac{19}{2}\,p_0V_0$

d. Change in internal energy during the cycle is $-\,8\,p_0V_0$

Comprehension Type

Direction *(Q. Nos. 16-20) This section contains 2 paragraphs, each describing theory, experiments, data, etc. Five questions related to the paragraphs have been given. Each question has only one correct answer among the four given options (a), (b), (c) and (d).*

Passage I

A stone of mass 20 kg is dropped in a tank containing 200 kg water from a height of 15 m. Initially, stone and water are at same temperature.

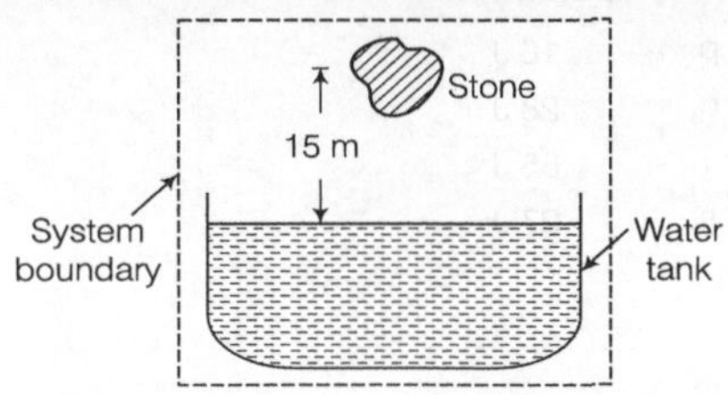

16 ΔQ, ΔW and ΔU, when the stone is about to enter the water, are respectively

a. 2943 J, – 2943 J, 0
b. – 2943 J, 2943 J, 0
c. 0, 2943 J, 0
d. 0, 0, 0

17 ΔQ, ΔW and ΔU, when stone dips into the tank and comes to rest at bottom, are respectively

a. – 2943 J, 2943 J, 0
b. 0, – 2943 J, 2943 J
c. 0, 2943 J, 0
d. 0, 0, 2943 J

18 ΔQ, ΔW and ΔU, when both water and stone comes to their original temperature are respectively

a. 0, 0, 2943 J
b. 0, 0, 0
c. – 2943 J, 0, 2943 J
d. 2943 J, 0, – 2943 J

Passage II

A fluid system undergoes a non-flow frictionless process following $p = \frac{5}{V} + 1.5$, where p is in bar and V is in m^3. During the process, volume changes from $0.15\ m^3$ to $0.05\ m^3$ and the system rejects 45 kJ of heat.

19 Work done during the process is (Take, $\log_e 3 = 1.098$)

a. – 564 kJ
b. 564 kJ
c. 483 kJ
d. – 483 kJ

20 Change in internal energy during the process is

a. – 45 kJ
b. 45 kJ
c. 519 kJ
d. – 519 kJ

Matching List Type

Direction (Q. No. 21) *Choices for the correct combination of elements from Column I and Column II are given as options (a), (b), (c) and (d), out of which one is correct.*

21 When a gas is taken from a to c along the curved path, the work done by the gas (W_{ac}) is -35 J and heat added to the gas is -63 J. Along path abc, work done (W_{abc}) is -54 J.

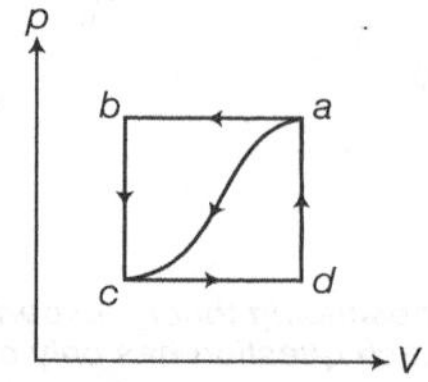

Also, $p_c = \frac{1}{2} p_b$ and $(\Delta E_{\text{int}})_d - (\Delta E_{\text{int}})_c = 12$ J. Now, match the entries of Column I and Column II and mark the correct option from the codes given below.

	Column I		Column II
i.	W_{cda}	p.	16 J
ii.	$(\Delta E_{\text{int}})_{cda}$	q.	28 J
iii.	$\Delta E_{\text{int}, a} - \Delta E_{\text{int}, c}$	r.	55 J
iv.	Q_{da}	s.	27 J

Codes

	i	ii	iii	iv
a.	r	q	s	p
b.	r	s	q	p
c.	s	r	q	p
d.	s	r	p	q

One Integer Value Correct Type

Direction (Q. Nos. 22 and 23) *This section contains 2 questions. When worked out will result in an integer from 0 to 9 (both inclusive).*

22 A packet of diatomic ideal gas is taken above h height from the earth's surface.

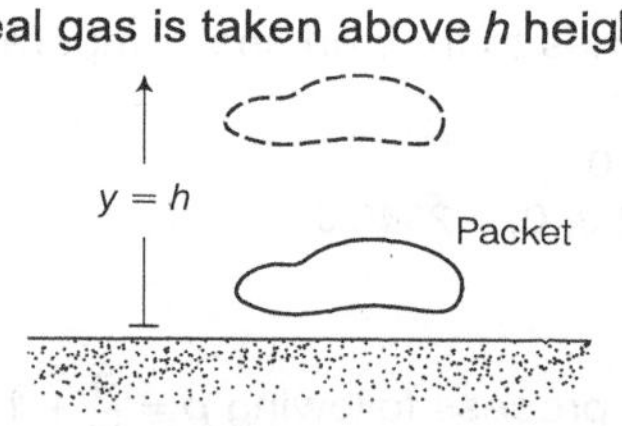

During its ascent, volume of packet change and as air is a poor conductor, heat from packet does not leave to atmosphere. Rate of change of tempeature of packet with respect to altitude h is given by

$$\frac{dT}{dh} = -N \cdot \left(\frac{mg}{7k}\right), \text{ find the value of } N.$$

where, k = Boltzmann's constant.

23 Air is expanded isentropically in a nozzle from 13.8 bar to 150°C to a pressure of 6.9 bar. The inlet velocity to the nozzle is very small and the process occurs in steady state, steady flow conditions. The exist velocity from nozzle, when inlet velocity is 10 ms^{-1} is found to be $-k \times 130$ ms^{-1}. Find the value of k, assuming nozzle is horizontal, also take $C_p = 1.005$.

Working Space

DPP-6 Thermodynamics-II
(Thermodynamic Processes)

Subjective Questions

Working Space

Direction (Q. Nos. 1-2) *These questions are subjective in nature, need to be solved completely on notebook.*

1 One mole of an ideal gas is taken round the cyclic process *ABCA* as shown in the figure.

Calculate:

i. Work done by the gas in complete process
ii. Heat rejected in the process *CA*
iii. Heat absorbed by the gas during process *AB*
iv. Net heat absorbed by the gas in process *BC*
v. Maximum temperature attained by the gas in entire cycle

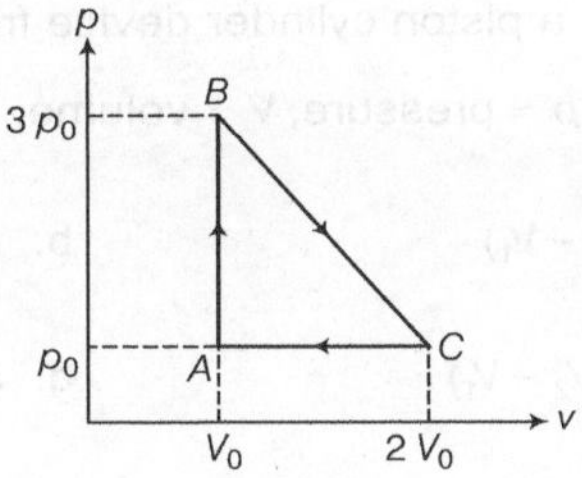

2 An ideal gas 0.1 mole, $\gamma = \frac{5}{3}$ is carried around the cycle shown in the figure.

Process *AB* is isochoric process, *BC* is adiabatic process and *CA* is isobaric process.

$p_A = 1$ atm, $T_C = 455$ K, $T_A = 300$ K, $T_B = 600$ K

Find the pressure and volume at points *A*, *B* and *C*. Also, find the work done by the gas in the cycle.

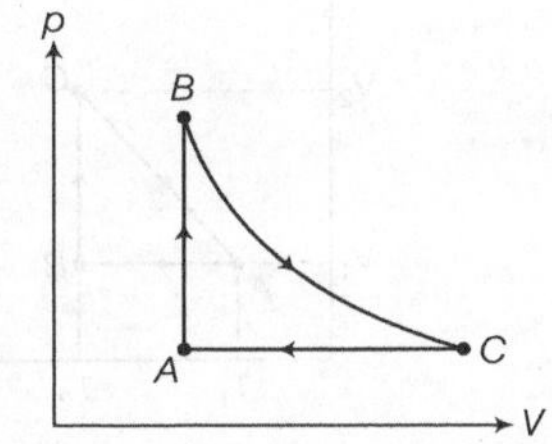

Only One Option Correct Type

Direction (Q. Nos. 3-14) *This section contains 12 multiple choice questions. Each question has four choices (a), (b), (c) and (d), out of which ONLY ONE is correct.*

3 A sample of 10 g of oxygen is heated through 100°C at a constant pressure. Then, choose the correct option is (Take, $C_V = 5.03$ cal mol^{-1}-K^{-1} and $R = 1.98$ cal mol^{-1}-K^{-1})

a. Heat supplied is 62 cal
b. Work done by gas is 219 cal
c. Change in internal energy of gas is 157 cal
d. No work is done by the gas

4 Certain quantity of an ideal gas at 17°C is suddenly compressed to one-tenth of its inital volume.

Ratio of final temperature of the gas, if gas is monatomic with the final temperature of gas, when the gas is diatomic, is

a. $10^{\frac{2}{5}}$ b. $10^{\frac{7}{5}}$
c. $10^{\frac{5}{3}}$ d. $10^{\frac{4}{15}}$

5 Two air samples with same composition are initially at same temperature and pressure. Both samples are compressed from volume V to $\frac{V}{2}$, first sample isothermally and second sample adiabatically. Ratio of final pressures will be

a. 1 b. 2

c. 2^{γ} d. $2^{1-\gamma}$

6 In the cycle $ABCA$, choose the correct option.

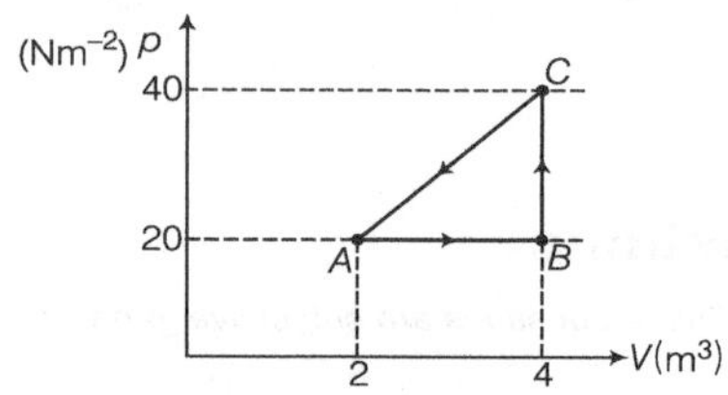

a. Work done by the gas is 20 J

b. Work done on the gas is 20 J

c. In process A to B, work done is negative

d. In process B to C, work done is positive

7 A gas expands in a piston cylinder device from volume V_1 to V_2, the process being $p = \frac{a}{V} + b$, where p = pressure, V = volume. The work done in the process is

a. $a \ln\left(\frac{V_1}{V_2}\right) + b(V_2 - V_1)$ b. $-a \ln\left(\frac{V_2}{V_1}\right) - b(V_2 - V_1)$

c. $-a \ln\left(\frac{V_1}{V_2}\right) - b(V_2 - V_1)$ d. $a \ln\left(\frac{V_2}{V_1}\right) + b(V_2 - V_1)$

8 V-T cycle for 1 mole of ideal gas is as shown in the figure.

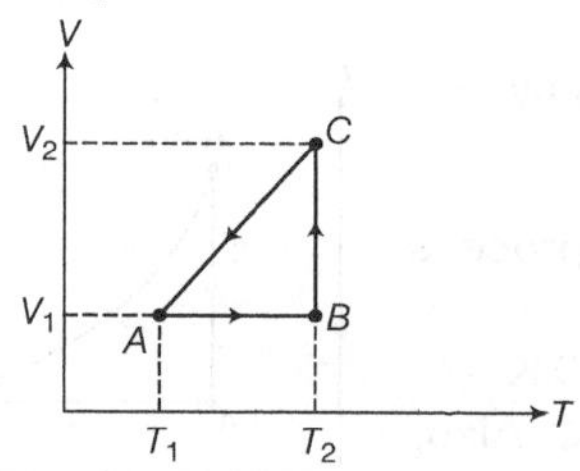

Work done in process AB, BC and CA respectively are

a. $0, RT_2 \ln\left(\frac{V_1}{V_2}\right), R(T_1 - T_2)$ b. $R(T_1 - T_2), 0, RT_2, \ln\left(\frac{V_1}{V_2}\right)$

c. $0, RT_2 \ln\left(\frac{V_2}{V_1}\right), R(T_1 - T_2)$ d. $0, RT_2 \ln\left(\frac{V_2}{V_1}\right), R(T_2 - T_1)$

9 p-V variation for a fixed mass of an ideal gas in cycle $ABCA$ as follows.

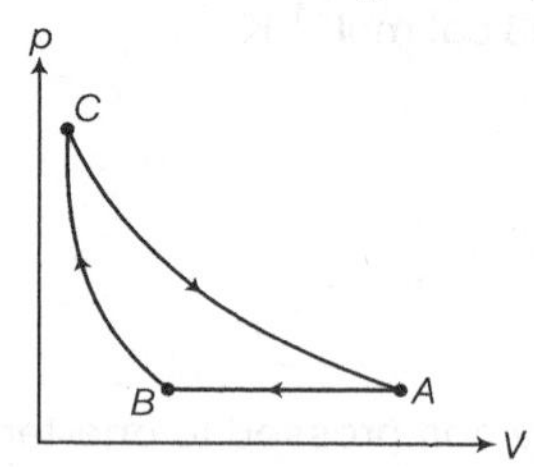

Correct set of processes is

	CB	CA	AB
a.	Isothermal expansion	Adiabatic compression	Compression at constant pressure
b.	Adiabatic expansion	Isothermal expansion	Pressure reduction
c.	Isothermal compression	Adiabatic expansion	Compression at constant pressure
d.	Adiabatic compression	Isothermal expansion	Compression at constant pressure

Working Space

10 Two moles of an ideal gas undergoes a process AB as shown in the figure.

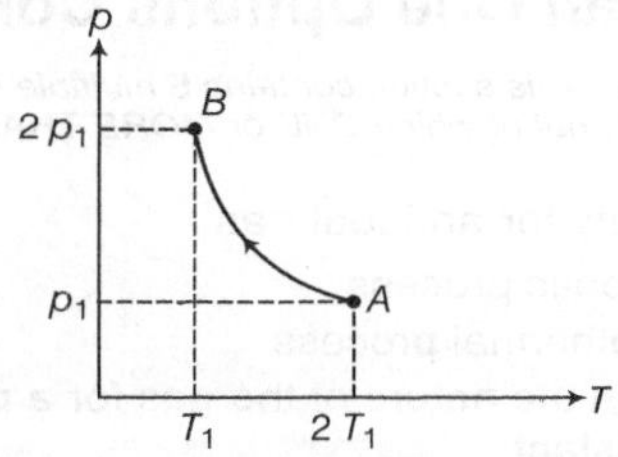

If $T_A = 300$ K and for path AB $P.T$ = a constant, then correct option is

a. Work done by the gas is 1200 R J
b. Work done on the gas is 1200 R J
c. Work done by the gas is 800 R J
d. Work done on the gas is 800 R J

11 In cycle $ABCA$ for an ideal gas,

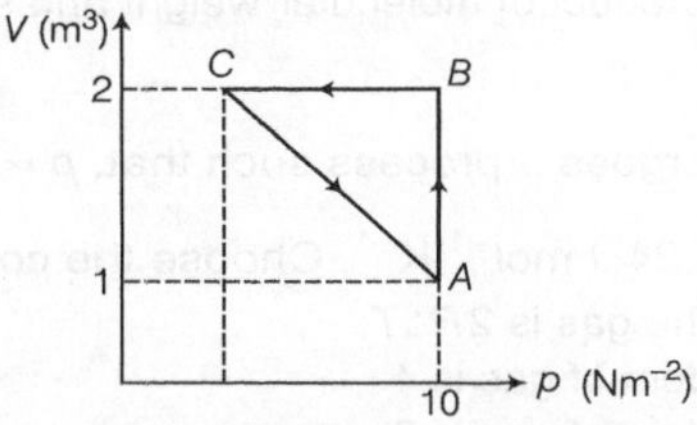

If the net heat supplied to gas in the cycle is 5 J, then the work done by the gas in process C to A is

a. – 5 J
b. – 10 J
c. – 15 J
d. – 20 J

12 Two cylinders A and B are fitted with pistons contains equal amounts of an ideal diatomic gas at 300 K. The piston A is free to move while that of B is held fixed. The same amount of heat is given to gas in each cylinder. If rise of temperature of the gas in A is 30 K, then the rise of temperature of gas in B is

a. 30 K
b. 18 K
c. 50 K
d. 42 K

13 An ideal gas expands isothermally from volume V_1 to V_2 and then compressed to original volume V_1 adiabatically. Initial pressure is p_1 and final pressure is p_2. The total work done is W. Then,

a. $p_2 > p_1, W > 0$
b. $p_2 < p_1, W < 0$
c. $p_2 > p_1, W < 0$
d. $p_2 = p_1, W = 0$

14 A cyclic process, $ABCA$ shown in V-T diagram is performed with constant mass of an ideal gas.

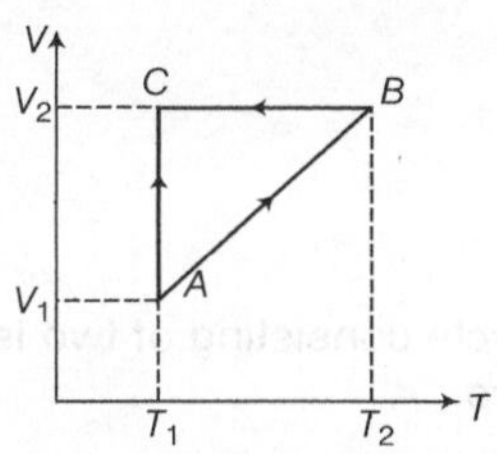

Corresponding p-V diagram is

a. 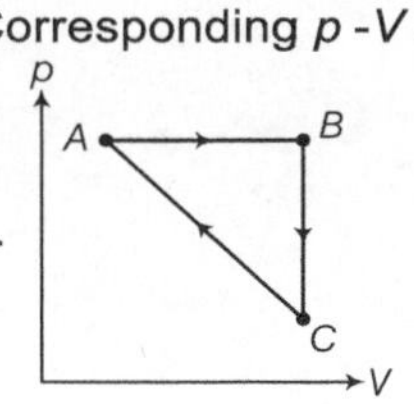

b. 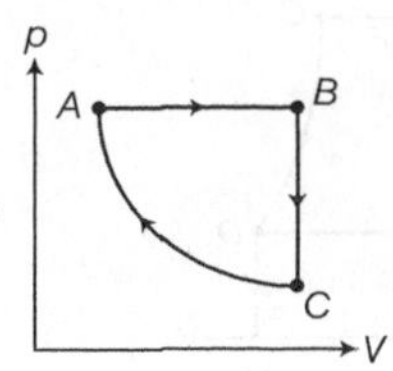

c. 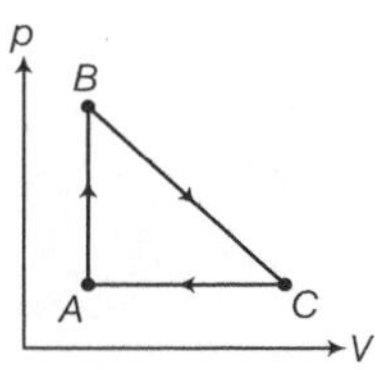

d. 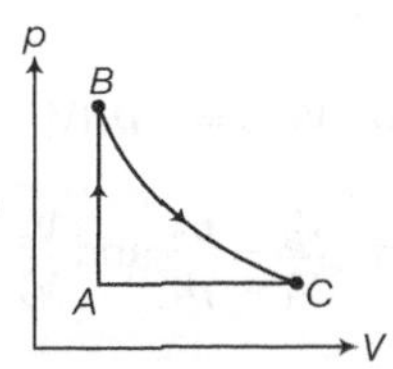

One or More than One Options Correct Type

Working Space

Direction (Q. Nos. 15-19) This section contains 5 multiple choice questions. Each question has four choices (a), (b), (c) and (d), out of which ONE or MORE THAN ONE are correct.

15 Molar heat capacity for an ideal gas

a. is zero for adiabatic process
b. is infinite for isothermal process
c. depends only on the nature of the gas for a process in which either volume or pressure is constant
d. is equal to the product of molecular weight and specific heat capacity for any process

16 An ideal gas undergoes a process such that, $p \propto \frac{1}{T}$. The molar heat capacity of this process is 33.24 J mol^{-1}-K^{-1}. Choose the correct options.

a. Work done by the gas is $2R\Delta T$
b. Degree of freedom of gas is 4
c. Degree of freedom of gas is 3
d. Ratio of specific heats for gas is 1.5

17 A sample of gas is taken in a piston cylinder arrangement. A 5.79 kJ of energy is transferred to the gas by heat and piston is adjusted, so that state of gas changes from A to B along a semi-circle as shown.

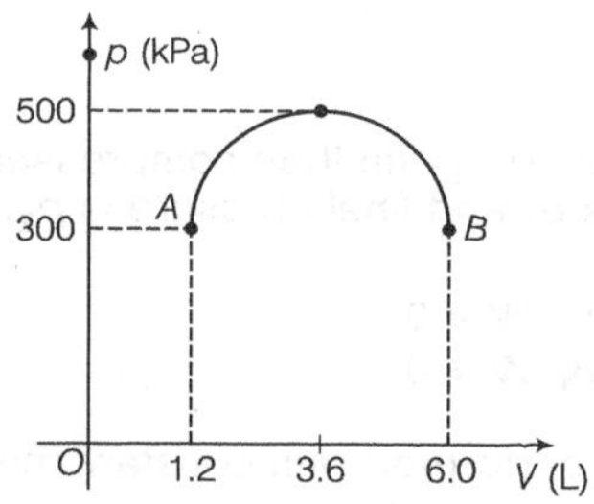

Then, choose the correct options.

a. Work done on the gas is 2190 J
b. Work done by the gas is 2190 J
c. Change in internal energy is 3.6 kJ
d. Change in internal energy is – 3.6 kJ

18 An ideal gas is carried through a thermodynamic cycle consisting of two isobaric and two isothermal processes as shown in the figure.

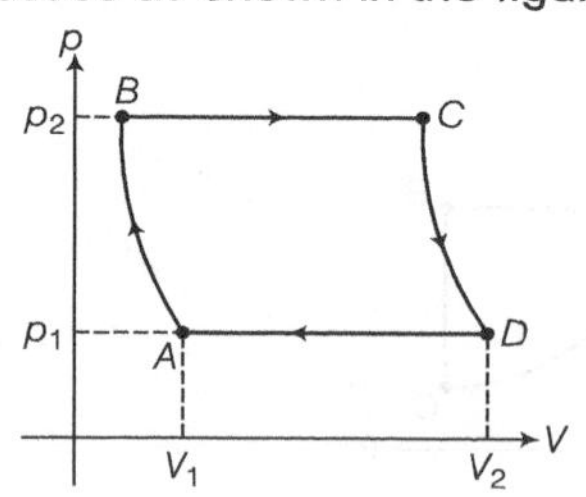

Then, choose the correct options.

a. $\frac{V_B}{V_1} = \frac{p_1}{p_2}$ and $\frac{V_2}{V_c} = \frac{p_2}{p_1}$

b. $W_{net} = -p_1(V_2 - V_1)\log_e\left(\frac{p_2}{p_1}\right)$

c. $W_{net} = p_1(V_2 - V_1)\log_e\left(\frac{p_2}{p_1}\right)$

d. $\frac{V_B}{V_1} = \frac{p_2}{p_1}$ and $\frac{V_c}{V_2} = \frac{p_2}{p_1}$

Working Space

19 One mole of an ideal gas at 0°C is initally taken through a cycle as shown in the figure. If p_i = Initial pressure and V_i = Initial volume, then, choose the correct options.

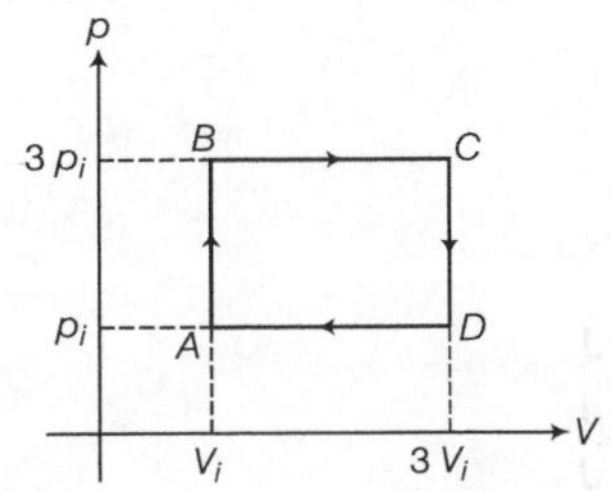

a. $\Delta E_{int} = 0$
b. $Q = -W$
c. $W = -4p_iV_i$
d. $W = -9.08$ kJ

Matching List Type

Direction *(Q. Nos. 20-22) Choices for the correct combination of elements from Column I and Column II are given as options (a), (b), (c) and (d), out of which one is correct.*

20 Match the process in Column I with the predictions of first law given in Column II.

	Column I		Column II
i.	Isothermal	p.	$Q = W = \Delta E_{int} = 0$
ii.	Adiabatic	q.	$\Delta E_{int} = -W$
iii.	Isobaric	r.	$\Delta E_{int} = 0, Q = W$
iv.	Free expansion	s.	$\Delta E_{int} + W = \Delta E_{int} + p\Delta V$

Codes

	i	ii	iii	iv
a.	r	q	s	p
b.	p	q	r	s
c.	r	q	p	s
d.	q	p	s	r

21 One mole of a diatomic gas is compressed to one-fourth of its original volume following a process VT^2 = constant. Initial pressure of gas is p_0. Match the entries of Column I with entries of Column II for this process.

	Column I		Column II
i.	$\lvert \Delta Q \rvert$	p.	$7\,p_0V_0$
ii.	$\lvert \Delta U \rvert$	q.	$2\,p_0V_0$
iii.	$\lvert \Delta W \rvert$	r.	$2.5\,p_0V_0$
iv.	$\lvert \Delta p \cdot V_0 \rvert$	s.	$0.5\,p_0V_0$

Codes

	i	ii	iii	iv
a.	p	q	r	s
b.	q	r	p	s
c.	s	r	q	p
d.	s	q	r	p

Working Space

22 Consider processes $A \to B \to C$, $A \to C$ and $C \to D \to A$.

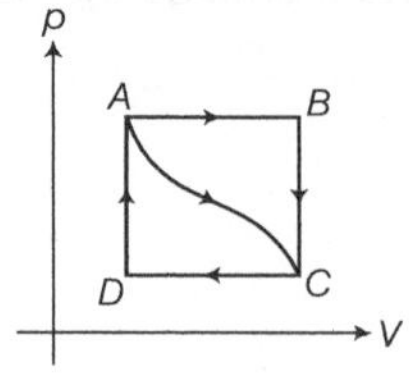

Given, ΔE_{int} (A to C) = + 800 J

$\Delta W_{(ABC)}$ = − 500 J

ΔE_{int}(D to A) = + 500 J

Now, match entries of Column I with entries of the Column II.

	Column I		Column II
i.	$\Delta Q(A \to B \to C)$	p.	− 1400 J
ii.	$\Delta W(C \to D \to A)$	q.	− 1300 J
iii.	$\Delta E_{int}(C \to D)$	r.	100 J
iv.	$\Delta Q(C \to D)$	s.	+ 1300 J

Codes

	i	ii	iii	iv
a.	s	r	p	q
b.	s	r	q	p
c.	p	r	q	s
d.	s	q	r	p

Comprehension Type

Direction (Q. Nos. 23-27) This section contains 2 paragraphs, each describing theory, experiments, data, etc. Five questions related to the paragraphs have been given. Each question has only one correct answer among the four given options (a), (b), (c) and (d).

Passage I

One mole of an ideal gas at pressure p_1 and volume V_1 expands as

$$pV^n = \text{constant}$$

23 When the gas expands to some volume V_2, the process is shown on p-V diagram.

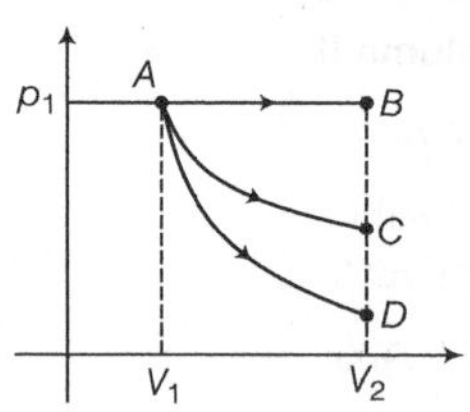

Values of n for AB, AC and AD process respectively are (γ = ratio of specific heat of gas)

a. 0, 1, γ
b. 0, γ, 1
c. 0, 1, 1
d. 0, γ, γ

24 Work done by the gas is least in the process for which

a. $n = 0$
b. $n = 1$
c. $n = \gamma$
d. work done is same in all processes

25 Change in internal energy is maximum in the process for which

a. $n = 0$
b. $n = 1$
c. $n = \gamma$
d. change of internal energy is same in all process

Passage II

Consider a thermodynamic cycle as shown in which a monatomic ideal gas is taken from state A to B to C to A.

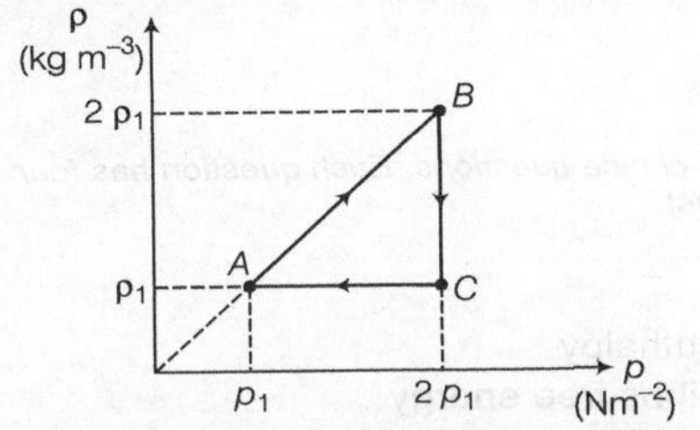

26 Ratio of work done in the process, AB, BC and CA ($W_{AB} : W_{BC} : W_{CA}$) is

a. log 2 : 1 : 1
b. − log 2 : 1 : 0
c. log 0.5 : 1 : 0
d. log 0.5 : 1 : 1

27 Ratio of heat rejected by gas to the total heat supplied is

a. $\frac{2}{5}(1 - \ln 2)$
b. $\frac{5}{2}(1 - \ln 2)$
c. $\frac{2}{5}\ln 2$
d. $\left(\frac{3 + \log 4}{5}\right)$

Statement Type

Direction *(Q. Nos. 28-30) This section is based on Statement I and Statement II. Select the correct answer from the codes given below.*

a. Both Statement I and Statement II are correct and Statement II is the correct explanation of Statement I
b. Both Statement I and Statement II are correct but Statement II is not the correct explanation of Statement I
c. Statement I is correct but Statement II is incorrect
d. Statement II is correct but Statement I is incorrect

28 **Statement I** Molar specific heat for and ideal monatomic gas for process $p = \frac{a}{T}$ is equal to $\frac{7}{2}R$.

Statement II $dQ = dU + dW$

29 **Statement I** Work done along an adiabatic between two given isothermals is independent of the particular adiabatic.

Statement II Work done in an adiabatic process is equal to $\frac{nR}{\gamma - 1} \cdot (T_i - T_f)$.

30 **Statement I** In free expansion of a gas, temperature of gas decreases.

Statement II When work is done on the expense of internal energy, temperature of gas decreases.

DPP-7 Thermodynamics-III (Engines and Fridges)

Only One Option Correct Type

Working Space

Direction (Q. Nos. 1-10) *This section contains 10 multiple choice questions. Each question has four choices (a), (b), (c) and (d), out of which ONLY ONE is correct.*

1 Second law of thermodynamics defines

a. internal energy
b. enthalpy
c. entropy
d. gibbs free energy

2 For a reversible adiabatic process, change in entropy is

a. 0
b. 1
c. ∞
d. None of these

3 For any reversible process, the change in entropy of the system and surroundings is

a. 0
b. 1
c. ∞
d. None of these

4 Which of these statements is correct according to Clausius statement of second law of thermodynamics.

a. It is possible to transfer heat from a body at lower temperature to a body at higher temperature
b. It is possible to transfer heat from a body at lower temperature to a body at higher temperature using a refrigeration cycle
c. It is impossible to transfer heat from a body at lower temperature to a body at higher temperature without the aid of an external source
d. None of the above

5 Entropy may be expressed as a function of

a. p and T
b. T and V
c. Q and W
d. All of these

6 When heat is absorbed by a gas, the change in entropy is

a. zero
b. positive
c. negative
d. unity

7 A heat engine receives heat at rate of 1500 $kJmin^{-1}$ and gives an output of 8.2 kW. Rate of heat rejection by the engine is

a. 1500 $kJmin^{-1}$
b. 8.2 kJs^{-1}
c. 16.8 kJs^{-1}
d. 8.2 $kJmin^{-1}$

8 During a process a system receives 30 kJ of heat from a reservoir and does 60 kJ of work. Then, choose the correct option(s).

a. It is possible to reach initial state by a quasistatic isothermal process
b. It is possible to reach initial state using a constant volume process
c. It is possible to reach initial state using a constant pressure process
d. It is possible to reach initial state using an adiabatic process

Working Space

9 Consider a heat engine working with a hot reservoir of the furnace gases at 2100°C, when the cooling water is available at –35.7°C. Then, choose the correct option(s).

a. Engine efficiency can be made more than 90% without altering operating temperatures
b. Engine efficiency always remains less than 90%
c. Engine efficiency is much lower practically due to irreversibility in the actual cycle
d. Engine efficiency is 90% but such temperatures are not available practically

10 Heat is transferred to a heat engine from a furnace at a rate of 80 mW. If the rate of heat rejection is 50 mW, then efficiency of the engine is

a. above 40% b. below 40%
c. equal to 40% d. equal to 35%

One or More than One Options Correct Type

Direction *(Q. Nos. 11-15) This section contains 5 multiple choice questions. Each question has four choices (a), (b), (c) and (d), out of which ONE or MORE THAN ONE are correct.*

11 A gas is contained in a small container, which is connected with a large container *via* a stop-cock.

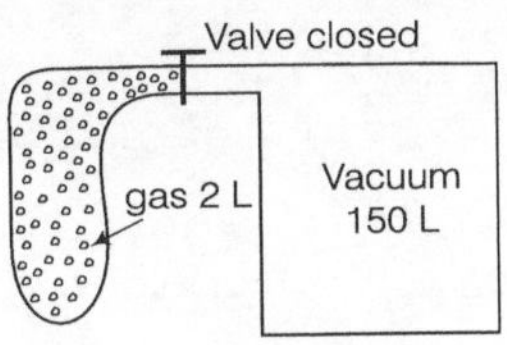

The 2 L cylinder contains 5.0 mol of gas at 300 K. Valve is now opened and gas is discharged into a 150 L vacuum chamber. Given, $\log_e 76 = 4.3$.

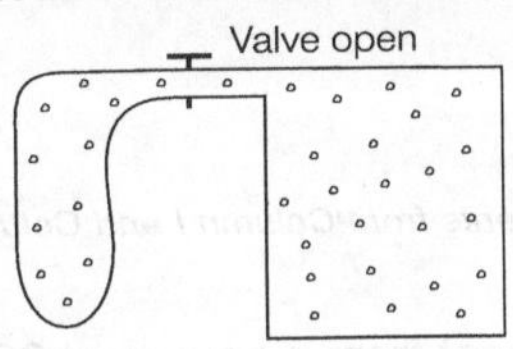

Choose the correct option (s).

a. Due to expansion of gas, a 54 kJ of energy becomes unavailable
b. 5 kJ of work can be extracted from a reversible isothermal expansion
c. Due to expansion, temperature as well as entropy of gas reduces
d. Due to expansion, temperature of gas reduces but entropy increases

12 In the given cycle, when we go from path *A* (from 1 to 2) and then back to initial state by path *B* (2 to 1).

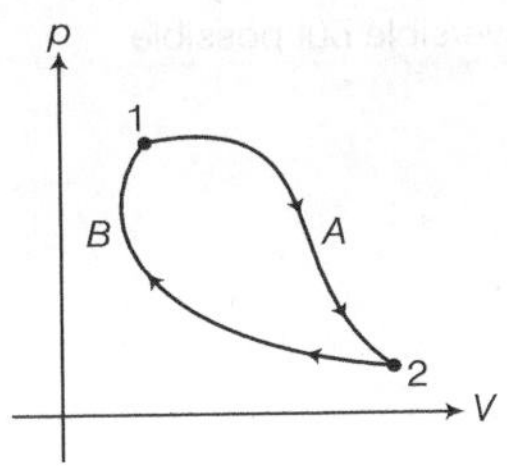

Choose the correct options.

a. Eentropy increases as we go from 1 to 2
b. Entropy increases as we go from 2 to 1
c. Entropy change from 1 to 2 is negative of entropy change from 2 to 1
d. Entropy change for complete cycle is zero

Working Space

13 A large object at 273 K and another large object at 373 K. For a small amount of energy of about 8 J be spontaneously transferred from cold object to hot object. Then, choose the correct options.

a. $\Delta S_{\text{hot object}} = 0.0214\ \text{JK}^{-1}$

b. $\Delta S_{\text{cold object}} = -0.0293\ \text{JK}^{-1}$

c. $\Delta S_{\text{hot}} + \Delta S_{\text{cold}} < 0$

d. Spontaneous transfer of heat from cold to hot object cannot be occur

14 Two objects with mass m and $2m$ with specific heats C and $1.5\ C$ are at temperatures 27°C and 127° C respectively. Both objects are placed in thermal contact with each other and are perfectly insulated, so that no heat is lost to surroundings. Now, choose the correct options.

a. Final temperature after equilibrium is reached is 375 K

b. Entropy change for system of masses is positive

c. The process is irreversible

d. The process is reversible

15 Four pair coins are tossed simultaneously. In terms of entropy, choose correct options:

a. "Most ordered" state has a probability of $\frac{1}{8}$

b. "Most ordered" state has a probability of $\frac{1}{2}$

c. "Most probable" state has a probability of $\frac{3}{8}$

d. "Most disordered" state has a probability of $\frac{3}{8}$

Matching List Type

Direction *(Q. No. 16) Choices for the correct combination of elements from Column I and Column II are given as options (a), (b), (c) and (d), out of which one is correct.*

16 A heat engine receives reversibly 420 kJ/cycle of heat from a source at 327°C and rejects heat reversibly to a sink at 27°C. There are no other heat transfers. Match the amounts of heat rejected in Column I with correct entries given in Column II and mark the correct option from codes given below.

	Column I		Column II
i.	210 kJ/cycle rejected	p.	Cycle is impossible
ii.	105 kJ/cycle rejected	q.	Cycle is reversible
iii.	315 kJ/cycle rejected	r.	Cycle is irreversible but possible

Codes

	i	ii	iii
a.	p	q	r
b.	r	p	q
c.	q	r	p
d.	q	p	r

Working Space

Comprehension Type

Direction *(Q. Nos. 17-19) This section contains a paragraph, describing theory, experiments, data, etc. Three questions related to the paragraphs have been given. Each question has only one correct answer among the four given options (a), (b), (c) and (d).*

Passage

Two identical gas samples are initially at same temperature and pressure. They are separately taken into two cylinders of volume V (each) and are compressed to volume $\frac{V}{2}$, one isothermally and other adiabatically.

17 If final pressures in the process are p_1 and p_2, then the ratio of p_1/p_2 is

a. 2^{γ}
b. 2
c. $\frac{1}{2}$
d. $2^{1-\gamma}$

18 Change in entropy of gas

a. is more for adiabatic process
b. is negative for adiabatic process
c. is zero for isothermal process
d. is "$-nR\log 2$" for isothermal process

19 Entropy change for surroundings is

a. positive for adiabatic process
b. negative for adiabatic process
c. positive for isothermal process
d. zero for isothermal process

One Integer Value Correct Type

Direction *(Q. No. 20) This section contains a question. When worked out will result in an integer from 0 to 9 (both inclusive).*

20 A Carnot heat engine draws heat from a reservoir at temperature T_1 and rejects heat to another reservoir at temperature T_3. The carnot engine derives a Carnot refrigerator which absorbs heat at temperature T_2 and rejects heat at temperature T_3. Given, $T_1 = 600\text{ K}$, $T_2 = 300\text{ K}$, heat supplied to engine Q_1 = heat absorbed by refrigerator Q_2. Find COP of refrigerator.

SET - 1

Revisal Problems for JEE Main

Only One Option Correct Type

Working Space

Direction *(Q. Nos. 1-20) This section contains 20 multiple choice questions. Each question has four choices (a), (b), (c) and (d), out of which ONLY ONE option is correct.*

1 A temperature scale of certain thermometer is given by the relation,

$$T = A\cdot 2^X + B$$

where, T is temperature in °C, A and B are constants and X is the property of fluid in the thermometer.

If at ice point and steam point, X has the value of 2.0 and 3.0 respectively. Value of temperature when X is -1.0 will be

a. 125.5°C
b. 75.5 °C
c. 112.5°C
d. 123.5°C

2 A thermocouple with test junction at t° C on gas thermometer scale and reference junction at ice point gives the emf as

$$e = 0.20t - 5\times10^{-4}t^2 \text{ (mV)}$$

When the gas thermometer reads 70°C, reading on the thermocouple will be

a. 70°C
b. 73°C
c. 67°C
d. 77°C

3 One gram mole of oxygen at 27°C and one atmospheric pressure is enclosed in a cubical insulated vessel. The vessel is moved with a constant speed v_0. When vessel is suddenly stopped, a rise of 1°C in temperature of gas is observed. Speed of vessel v_0 is

a. 483.4 ms^{-1}
b. 108 ms^{-1}
c. 47 ms^{-1}
d. 36 ms^{-1}

4 A rod of length 40 cm has coefficient of linear expansion $\alpha_1 = 6\times10^{-6}$ per °C. Another rod of length l has coefficient of linear expansion $\alpha_2 = 4\times10^{-6}\,°C^{-1}$. If difference in lengths of the two rods always remains same. Then, the value of l is

a. 26 cm
b. 60 cm
c. 80 cm
d. 32 cm

5 Three rods of same area of cross-section are made from same metal, are used to form sides of an isosceles triangle ABC with $\angle B = 90°$. Points A and B of triangle are maintained at temperatures T and $\sqrt{2}T$ respectively.

In steady state, temperature of point C is T_C. There is a point D on BA whose temperature is same as T_C. The ratio $\frac{BD}{DA}$ is

a. 1
b. $\sqrt{2}$
c. $\frac{1}{\sqrt{2}}$
d. 2

6 A gas mixture undergoes an adiabatic process. T and $\frac{1}{V}$ graph of gas is a parabola as shown in the figure.

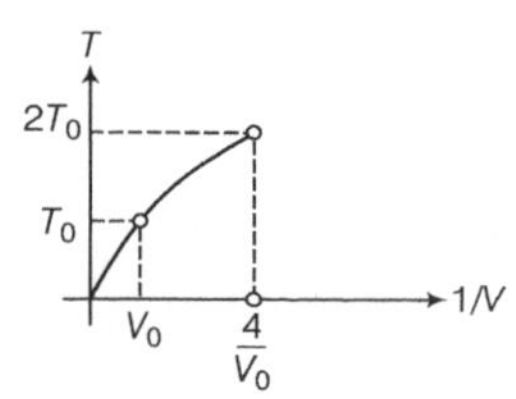

Ratio of v_{rms} and speed of sound in the mixture is

a. $\sqrt{\frac{3}{2}}$
b. $\sqrt{2}$
c. $\sqrt{\frac{2}{3}}$
d. $\sqrt{3}$

Working Space

7 Three rods of equal length and cross-section area are joined in series.

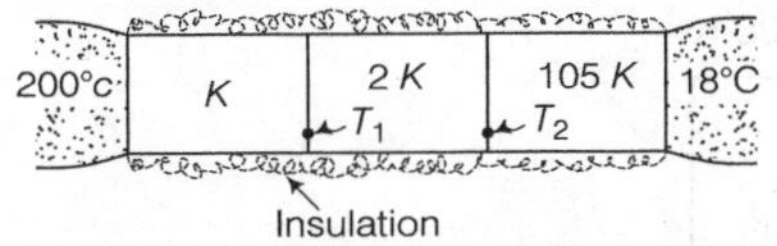

If ends of combination are kept at 200°C and 18°C, then temperature of inner layers T_1 and T_2 in steady state respectively are

a. 120°C, 80°C
b. 116°C, 80°C
c. 120°C, 74°C
d. 116°C, 74°C

8 The absolute temperature of a star increases from T to $2T$ while the radius of star shrinks from R to $R/2$. As a result power radiated by star

a. increases by a factor of 2
b. decreases by factor of 4
c. increases by a factor of 4
d. remains constant

9 A body cools from 85°C to 80°C in 10 minutes, when the temperature of the surrounding is 32.5°C. How much time will it take to cool down by the same amount (85°C to 80°C), if the temperature of the surroundings is 57.5°C ?

a. 20 min
b. 5 min
c. $\frac{20}{3}$ min
d. 40 min

10 A gas undergoes a change of state from A to B *via* two paths (path 1 and path 2).

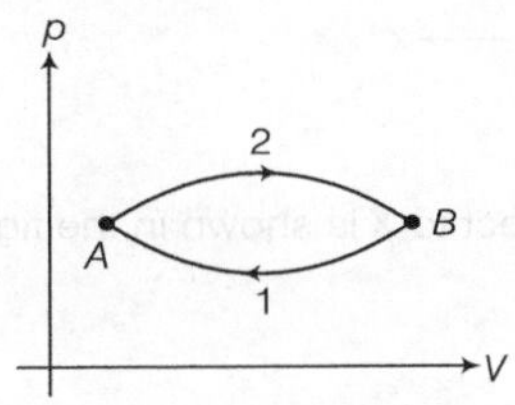

Ratio of molar heat capacities in path 1 and path 2 is

a. < 1
b. > 1
c. 1
d. Data insufficient

11 For an ideal gas undergoing a reversible adiabatic compression, (U = Internal energy T = Temperature, $\gamma = C_p/C_v$). Choose the correct options.

a. UV^{γ} = constant
b. Up^{γ} = constant
c. $VU^{\frac{1}{\gamma-1}}$ = constant
d. $TU^{\gamma-1}$ = constant

12 Two moles of an ideal gas undergoes a cyclic process $1 \rightarrow 2 \rightarrow 3 \rightarrow 1$. If net heat exchanged in the process is –300 J, then the work done by the gas in the process $2 \rightarrow 3$ is given by,

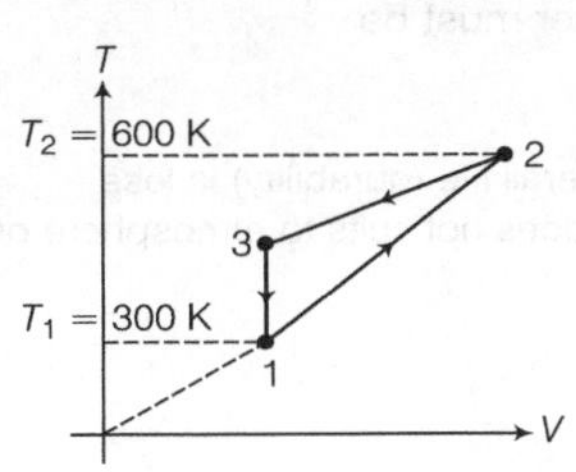

a. –5000J
b. 5000 J
c. –3000J
d. None of these

13 In a thermodynamic process, helium gas obeys the law $p^{-2/5}.T$ = constant. The heat given to gas, when temperature of 2 moles of gas is raised from T to $4T$ is

a. RT
b. $9RT$
c. 18 RT
d. zero

Working Space

14 In the isolated chamber shown, all walls are perfect insulators. Chambers A and B contains one mole of O_2 at 600 K and one mole of He at 300 K.

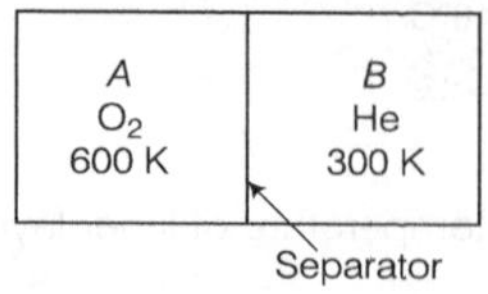

At some time t seperator is removed. After stabilisation,

a. change in internal energy of He is 1817 J
b. change in internal energy of O_2 is 207 J
c. temperature of gas mixture in steady state is 487 K
d. temperature of gas mixture in steady state is 450 K

15 One mole of an ideal monatomic gas is taken along a polytropic process with pV^X = constant.

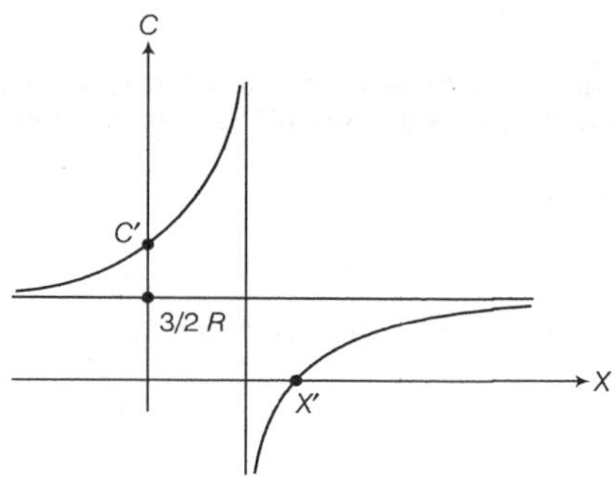

Variation of molar heat capacity C of such gas with respect to X is shown in the figure. Values of C' and X' are respectively :

a. $\frac{5}{2}R, \frac{5}{2}$ b. $\frac{5}{2}R, \frac{5}{3}$
c. $\frac{7}{2}R, \frac{7}{2}$ d. $\frac{5}{2}R, \frac{7}{2}$

16 For an ideal gas, a process follows the rule $U = A + B\,(p.V)$,

where, U = internal energy in A, B are constants, p = pressure and V = volume of the gas. Then, ratio of specific heat for the gas under going reversible adiabatic process is

a. $\frac{B+1}{B}$ b. $\frac{A+B}{B}$
c. $\frac{A+1}{A}$ d. $\frac{A+B}{A}$

17 An advertisor in local news paper, claims that their newly developed engine develops 15 kWh of mechanical work, when it works between two reservoirs, taking 105 mJ at 400K and rejecting 42 mJ at 200K. Your suggestion to a buyer must be

a. engine is suitable to buy
b. engine is not suitable to buy
c. engine is taking heat at very high temperature, so its overall life (durability) is less
d. engine is rejecting heat at much low temperature, so it does not suits to atmosphere of our country.

18 van der Waals' gas equation is $p = \frac{RT}{\left(\frac{V}{n} - b\right)} - \frac{a}{(V/n)^2}$

with symbols have their usual meanings. Ratio of constants a and b in terms of critical temperature (T_c) of gas is

a. $-\frac{8}{27} RT_c \frac{\text{N-m}^4}{\text{mol}^2}$ b. $\frac{27}{8} RT_c \frac{\text{N-m}}{\text{mol}}$
c. $\frac{27}{64} RT_c \frac{\text{m}^3}{\text{mol}}$ d. $\frac{1}{8} RT_c \frac{\text{N-m}^4}{\text{mol}^2}$

Working Space

19 An ideal gas undergoes process *BDA*, which is shown on *p-V* diagram as shown in the figure.

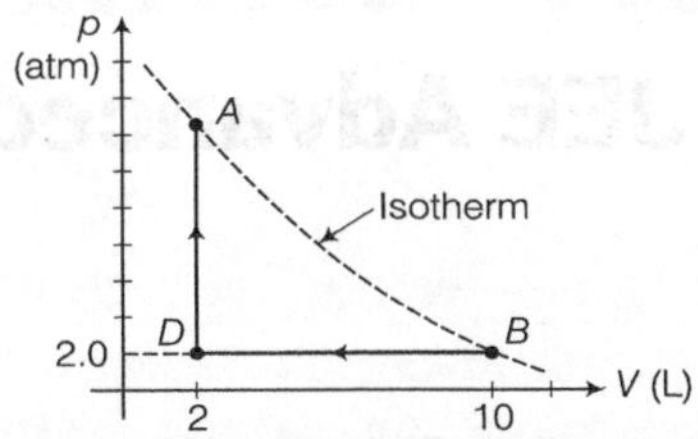

In the process *BDA*.

a. 1.6×10^3 J heat flows into the gas volume

b. 1.6×10^3 J heat flows out of the gas volume

c. 1.6×10^3 J of internal energy is converted into work

d. 1.6×10^3 J of work is stored in the form of internal energy of gas

20 The Gaint star "Betelgeuse" emits radiant energy at rate of 10^4 times more than our Sun whereas its surface temperature is only half that of our Sun. If Sun's radius is 7×10^8m, then radius of 'Betelgeuse' is

a. 7×10^8 m b. 3.5×10^8 m c. 3×10^{11}m d. 1.5×10^{11}m

Statement Type

Direction (Q. Nos. 21-25) ***This section is based on Statement I and Statement II. Select the correct answer from the codes given below.***

a. Both Statement I and Statement II are correct and Statement II is the correct explanation of Statement I

b. Both Statement I and Statement II are correct but Statement II is not the correct explanation of Statement I

c. Statement I is correct but Statement II is incorrect

d. Statement II is correct but Statement I is incorrect

21 **Statement I** Reversible ceiling fan is set to pull the air up in winter season and to drive the air down in summers to get a more pleasent feeling.

Statement II Warm air is near the ceiling in both seasons.

22 **Statement I** A certain mass of wood and same mass of iron left in the sun for some time. Iron piece is hotter to touch.

Statement II Metal absorbs more heat than wood.

23 **Statement I** An ideal gas, *n*-moles, undergoes the reversible process *ab* as shown.

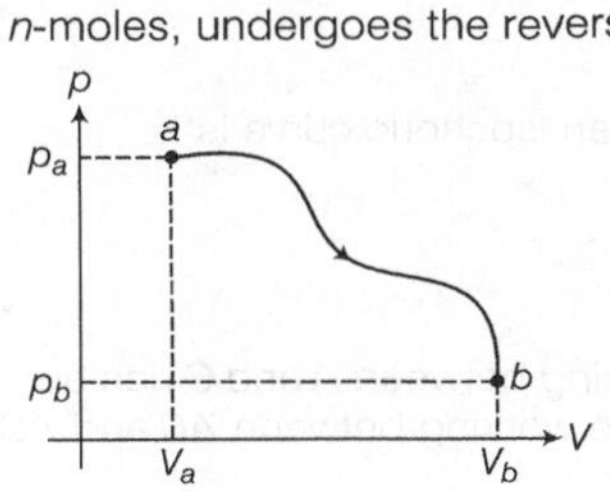

If $T_a = T_b$, then entropy change of the gas in this process is $\Delta S = nR \ln\left(\frac{V_b}{V_a}\right)$.

Statement II Entropy is a state variable.

24 **Statement I** Mon, di and poly atomic gasses are compressed separately to equal compression ratio p_2/p_1, then the monatomic gas has maximum final volume.

Statement II Monatomic gas has least degree of freedom.

25 **Statement I** Molar specific heat at constant volume for a gas cannot be negative.

Statement I At constant volume, heat is added in form of internal energy.

SET - 2

Revisal Problems
for JEE Advanced

Only One Option Correct Type

Working Space

Direction (Q. Nos. 1-7) *This section contains 7 multiple choice questions. Each question has four choices (a), (b), (c) and (d), out of which ONLY ONE is correct.*

1 A system is taken from 25° C to 125° C by a process in which specific heat of system is given by $C = (0.4 + 0.004\,T)$ kJ / kg° C. If mass of system is 6 kg, then heat required for the process is

a. 420 kJ
b. 330 kJ
c. 320 kJ
d. 384 kJ

2 A system undergoes three different processes as shown in table given, (few values are missing).

Process	ΔQ	ΔW	U_i	U_f
A	35	– 15	...	– 10
B	– 15	...	...	60
C	...	– 20	80	100

All values are in joules. Choose the correct options.

a. For cycle ABC, $\Delta U = 0$
b. Process A is isobaric
c. Process B is isochoric
d. Process C is adiabatic

3 An ideal gas is expanded to twice of its original volume, then

a. isothermal work is more than adiabatic work
b. adiabatic work is equal to isothermal work
c. adibatic work is more than isothermal work
d. Data is insufficient to reach any conclusion

4 A copper rod is in contact with a heat reservoir at 127 ° C and other with a reservoir at 27° C. During a certain time interval, 1200 cal of heat is conducted through the rod in steady state. Then, entropy change of rod is

a. – 3 cal K^{-1}
b. + 4 cal K^{-1}
c. + 1 cal K^{-1}
d. zero

5 In T-S diagram ratio of slope of an isobaric curve and an isochoric curve is

a. γ
b. $\frac{1}{\gamma}$
c. γ^2
d. zero

6 Consider four heat reservoirs A, B, C, D an engine working between A and C has an efficiency which is mean of efficiencies of same engine working between AB and AD. Then, absolute temperature of C will be

a. geometric mean of temperatures of A and D
b. arithmatic mean of temperatures of A and D
c. harmonic mean of temperatures of A and D
d. sum of temperatures of A and D

7 At atmospheric pressure, 1 g of water becomes 1671 cm^3 of steam, when boiled. Latent heat of vapourisation of water is 539 cal g^{-1}, 1 cal = 4.18 J and 1 atm pressure $= 1.013 \times 10^5$ Nm^{-2}.
Change in internal energy of water in the process is around

a. 170 J
b. 2235 J
c. 2083 J
d. 539 J

Working Space

One or More than One Options Correct Type

***Direction** (Q. Nos. 8-12) This section contains 5 multiple choice questions. Each question has four choices (a), (b), (c) and (d), out of which ONE or MORE THAN ONE are correct.*

8 A room air conditioner is modeled as shown below.

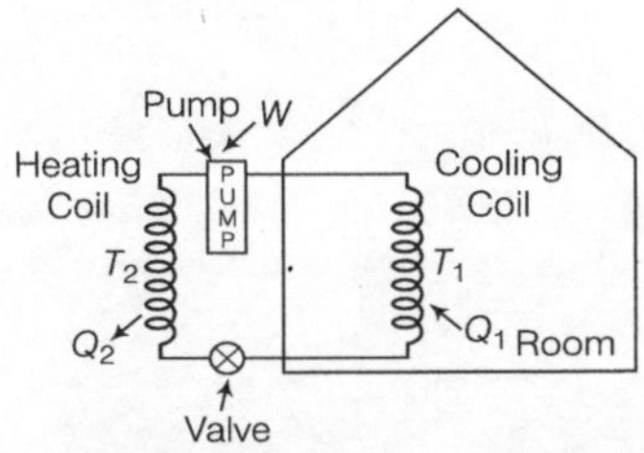

A cooling coil absorbs Q_1 amount of heat at T_1 temperature and rejects Q_2 amount of heat at T_2 temperature. A work W is input into the system every cycle by an electric pump. The air conditioner unit consumes electric power P. On the basis of above information, choose, the correct alternatives

a. Assuming air-conditioner runs on a Carnot cycle, maximum rate of heat removal from the room is $\frac{Q_1}{\Delta t} = P\left(\frac{T_1}{T_2 - T_1}\right)$

b. If $x = \frac{P}{K}$, where K = thermal conductivity of room (walls, floor and ceiling), the coldest possible temperature of room is $T_1 = T_2 - \frac{x}{2}\left(\sqrt{1 + \left(\frac{4T_2}{x}\right)} - 1\right)$

c. For $K = 173\ \text{W}\,°\text{C}^{-1}$, $T_2 = 40\,°\text{C}$, $T_1 = 25\,°\text{C}$, minimum power requirement for air-conditioner is 130 W

d. For $K = 173\ \text{W}\,°\text{C}^{-1}$, $T_2 = 40\,°\text{C}$, $T_1 = 25\,°\text{C}$, minimum power of air-conditioner is 200 W

9 Consider two identical homogenous balls A and B at same temperatures.

One of ball A is hung from an insulating thread while other B is kept on an horizontal surface. Some quantity of heat is given to both balls A and B, if all types of heat losses are negligible, then which of these are true?

a. Finally both A and B are at same temperatures
b. Finally A is at higher temperature than B
c. Finally temperature difference of A and B is 0
d. Finally temperature difference of A and B is proportional to coefficient of expansion of material of balls

10 Two kilogram of air undergoes following cycle; (Given, $\ln\frac{1}{5} = -1.609$)

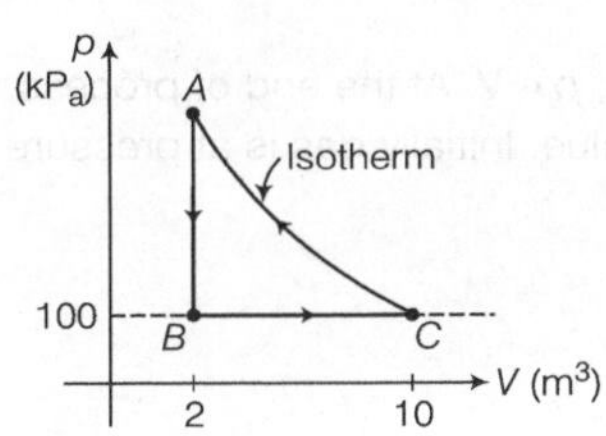

Then, choose the correct options.

a. Work done in process CA is – 1609 kJ
b. Work done in the complete cycle is 800 kJ
c. Work done in the complete cycle is – 809 kJ
d. Work done in the complete cycle is – 1609 kJ

Working Space

11 p-T graph for an ideal gas is as shown in the figure.

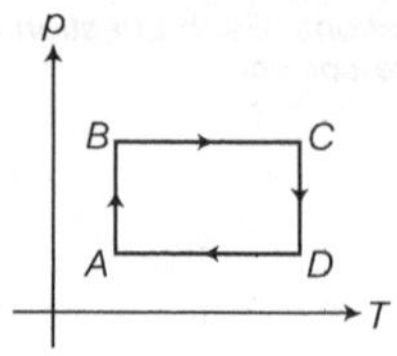

Then, choose the correct options.

a. V-T graph for cyclic $ABCD$ is

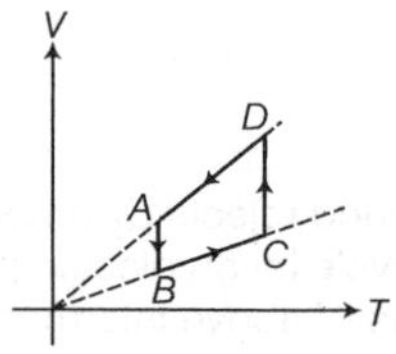

b. ρ-T graph for cycle $ABCD$ is

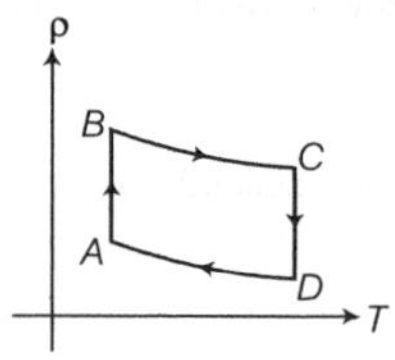

c. p-V graph for the cycle $ABCD$ is

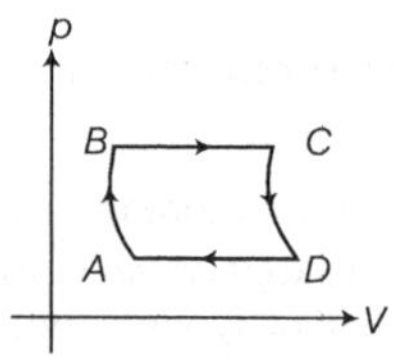

d. S-T graph for the cycle $ABCD$ is

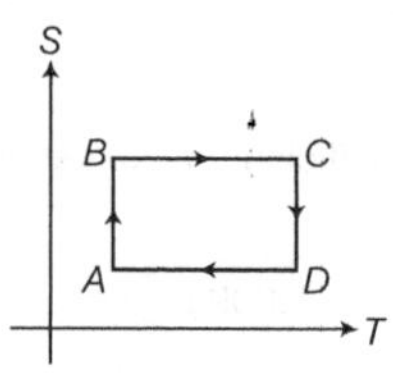

12 An ideal diatomic gas undergoes a process such that, $p \propto V$. At the end of process, rms speed of gas molecules is found double of its inital value. Initially gas is at pressure p_1 and have volume V_1. Choose the correct options.

a. Molar specific heat for the process is $\frac{5}{2}R$

b. Molar specific heat for the process is $3R$

c. Heat added to the gas during the process is $9\,p_1V_1$

d. Heat added to the gas during the process is $\frac{15}{2}p_1V_1$

Working Space

Comprehension Type

***Direction** (Q. Nos. 13-28) This section contains 8 paragraphs, each describing theory, experiments, data, etc. Sixteen questions related to the paragraphs have been given. Each question has only one correct answer among the four given options (a), (b), (c) and (d).*

Passage I

A cylinder is fitted with a frictionless piston held with constant pressure and it contains one mole of an ideal gas. Cylinder walls and piston are perfectly insulated.
Cylinder is also fitted with a resistor of resistance $R = 2\,k\Omega$ which is connected to a fully charged capacitor, $C = 75\,mF$. Initial potential difference across the capacitor is $\frac{640}{3}$ V with switch S opened.

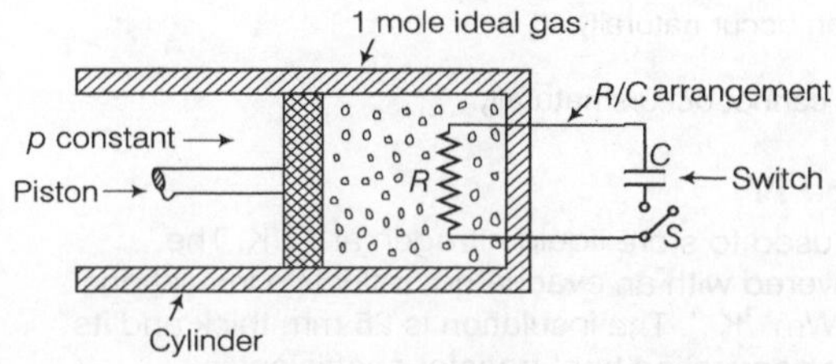

Switch S is closed for 2.5 In 4 minutes, temperature of gas is found increased by 72 K, $R = 8.3$ Jmole^{-1}K^{-1}. Now, answer the following questions.

13 Work done by the gas is
a. 0.2 kJ b. 0.4 kJ c. 1.6 kJ d. 0.8 kJ

14 Rise in internal energy of gas is
a. 1 kJ b. 2 kJ c. 3 kJ d. 4 kJ

Passage II

A reversible engine operates between temperatures T_1 and T $(T_1 > T)$. The energy rejected from this engine is received by a second reversible engine at the same temperature T. The second engine rejects energy at some temperature T_2 $(T_2 < T)$.

15 If the engines produce same amount of work output, then T is
a. GM of T_1 and T_2 b. AM of T_1 and T_2 c. HM of T_1 and T_2 d. None of these

16 If the engines have same efficiency, then T is
a. GM of T_1 and T_2 b. AM of T_1 and T_2
c. HM of T_1 and T_2 d. None of these

Passage III

Consider an isolated gas like system consisting of 10 distinguishable atoms, (each atom may have different colour), in a box. Each atom is moving with same speed v.

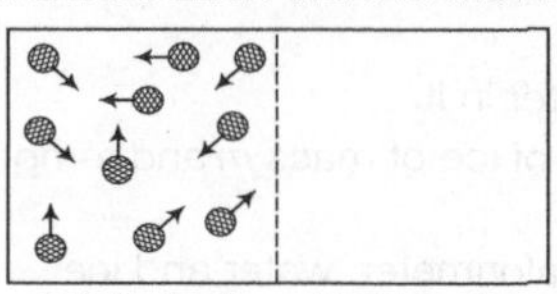

State A, $N_L = 10$, $N_R = 0$

Each unique arrangement of atoms with in the box is a microstate of this system.
Now, imagine two possible states;

State *A* All the atoms are in left hand side of the box.

State *B* Atoms are uniformly distributed throughout the complete volume of the box.

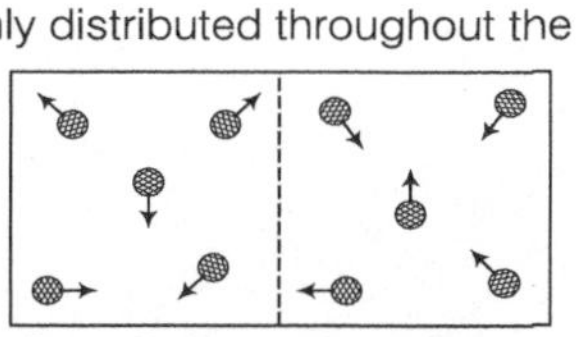

State B - Atoms are uniformly distributed, $N_L = N_R = 5$

Working Space

17 Assume that system is initially in state *A* and after some time it is found in state *B*. Then, choose the correct option.

a. Change in entropy is "$k_B \log_e 252$" and process can occur naturally
b. Change in entropy is "$-k_B \log_e 252$" and process cannot occur naturally
c. Change in entropy is "$\frac{1}{2} k_B \log_e 252$" and process can occur naturally
d. Change in entropy is "$-\frac{1}{2} k_B \log_e 252$" and process cannot occur naturally

18 Let the system is initially in state *B* and after some time, it is found in state *A*. Then, choose the correct option.

a. Change in entropy is '$k_B \log_e 252$' and process can occur naturally
b. Change in entropy is '$-k_B \log_e 252$' and process cannot occur naturally
c. Change in entropy is '$\frac{1}{2} k_B \log_e 252$' and it can occur naturally
d. Change in entropy is "$-\frac{1}{2} k_B \log_e 252$" and it cannot occurs naturally

Passage IV

A spherical thin walled metallic container is used to store liquid nitrogen at 77 K. The container has a diameter of 0.5 m and is covered with an evacuated, reflective insulation composed of silica powder with $k = 0.0017 \text{ Wm}^{-1}\text{K}^{-1}$. The insulation is 25 mm thick and its outer surface is exposed to air at 300 K. The convection heat transfer coefficient is $20 \text{ Wm}^{-2}\text{K}^{-1}$. The latent heat of vaporisation and the density fo liquid nitrogen are 2×10^5 Jkg^{-1} and 804 kgm^{-3}

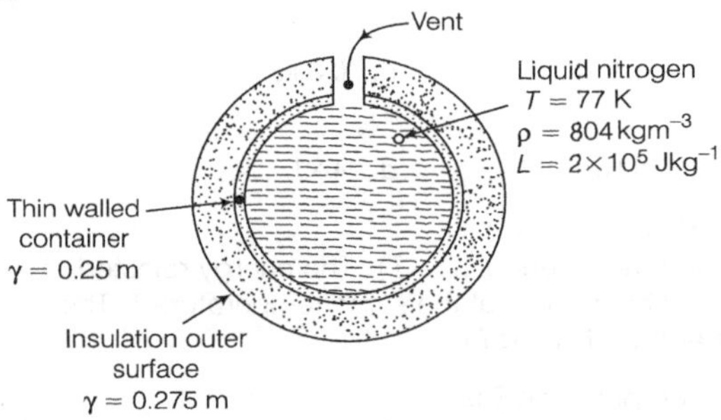

19 Heat flow rate into the container from outside will be

a. around 20 W
b. around 13 W
c. around 100 W
d. around 130 W

20 Rate at which liquid nitrogen boils off through the vent will be

a. 50.64 litres/day
b. 21.2 litres/day
c. 16.53 litres/day
d. 7.0 litres/day

Passage V

A calorimeter of mass *m* contains an equal mass of water in it.

The temperature of water and calorimeter is t_2. A block of ice of mass *m* and temperature $t_3 < 0° \text{C}$ is gentely dropped into the calorimeter.

Let C_1, C_2 and C_3 be the specific heats of materials of calorimeter, water and ice respectively and *L* be the latent heat of ice.

21 Whole of mixture in the calorimeter becomes ice, when

a. $C_1t_2 + C_2t_2 + L + C_3t_3 > 0$
b. $C_1t_2 + C_2t_2 + L + C_3t_3 < 0$
c. $C_1t_2 + C_2t_2 + L - C_3t_3 < 0$
d. None of these

22 Whole of mixture in calorimeter becomes water, when

a. $(C_1 + C_2)t_2 - C_3|t_3| + L > 0$
b. $(C_1 + C_2)t_2 + C_3|t_3| + L > 0$
c. $(C_1 + C_2)t_2 - C_3|t_3| - L > 0$
d. $(C_1 + C_2)t_2 + C_3|t_3| - L > 0$

Working Space

Passage VI

One mole of an ideal monatomic gas undergo a thermodynamic cyclic process *ABCDA* as shown on internal energy v/s density graph.

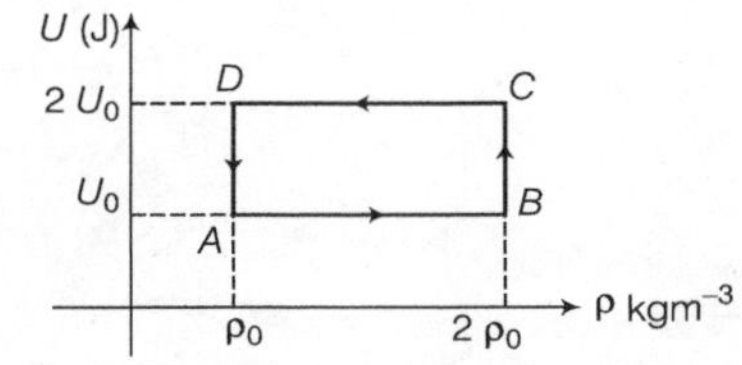

23 Corresponding *p-V* graph is

a.
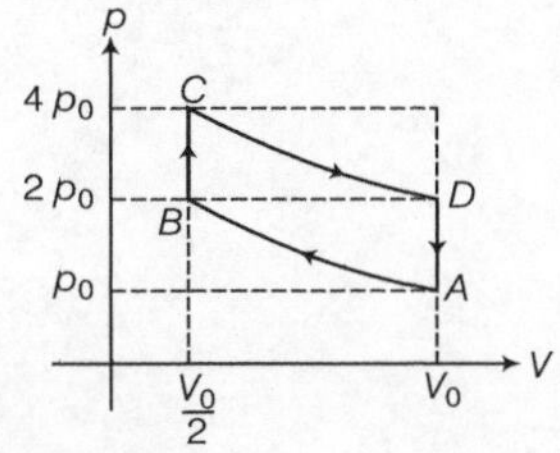

b.
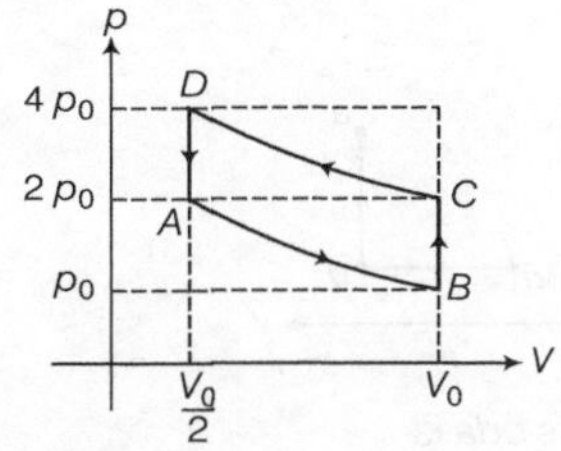

c.
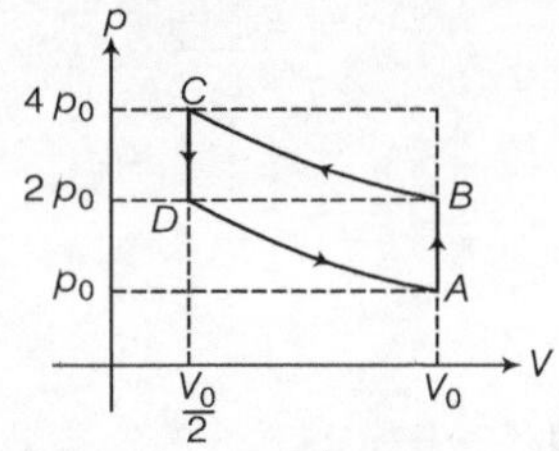

d.
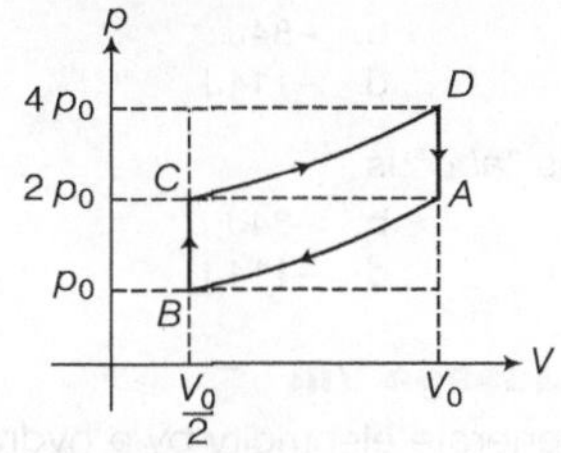

24 The corresponding temperature-pressure graph (*T-p* graph) is

a.
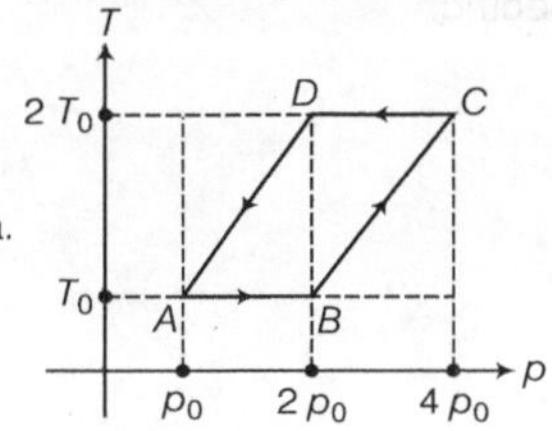

b.
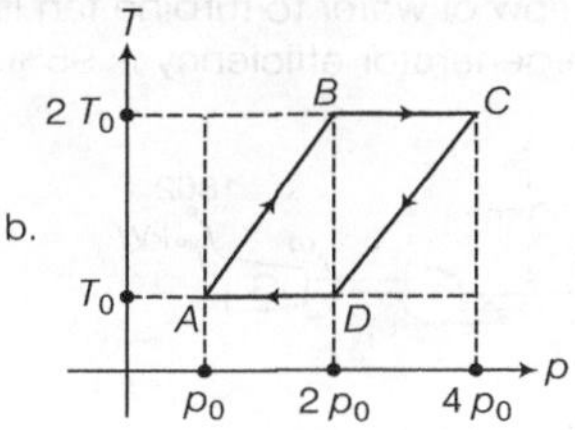

c.
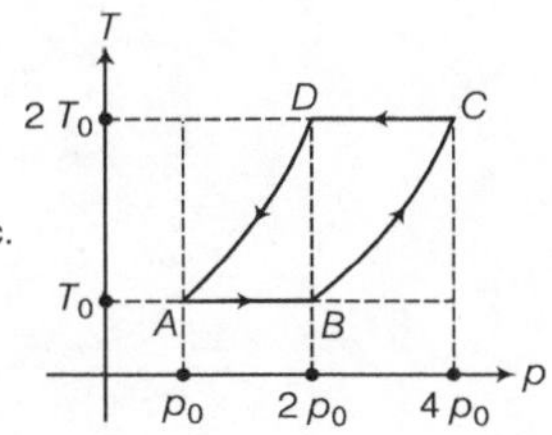

d.
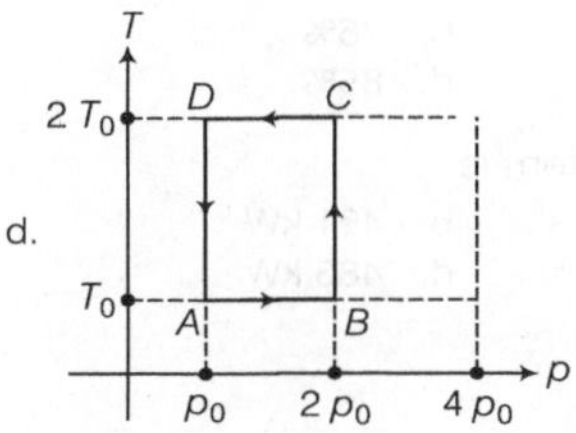

Passage VII

Consider three processes :

I Process *abc* $\Delta E_{\text{int } a} - \Delta E_{\text{int } b} = 15\,\text{J}$
(or ΔU_{ab})

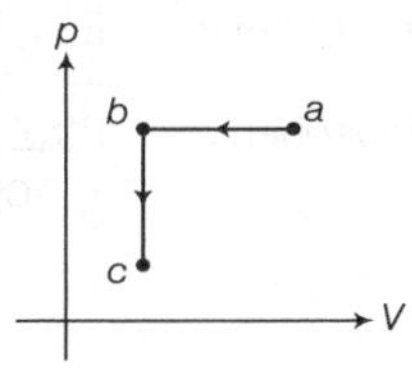

Working Space

II Process *ac* $Q_{ac} = -85\,J$

$W_{ac} = -55\,J$

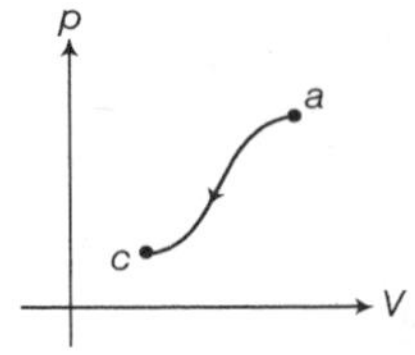

III Process *cda* $p_a = 2.2 p_d$

$W_{cda} = 38\,J$

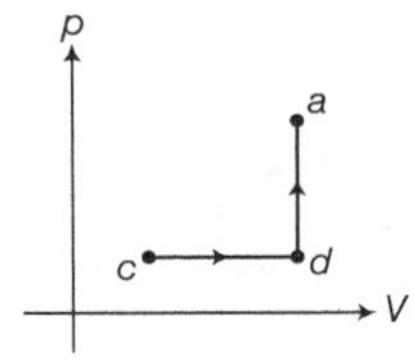

25 Heat added to the gas in the process *cda* is

a. 30 J
b. −84J
c. 68 J
d. −114J

26 Heat added to the gas in the process "*abc*" is

a. 30 J
b. −84J
c. 68 J
d. −114J

Passage VIII

Water from a large dam is used to generate electricity by a hydraulic turbine and generator.

Depth of water in lake is 50 m. Rate of flow of water to turbine fan is 5000 kgs^{-1}. Electric power is generated is 1862 kW and the generator efficiency is 95%.

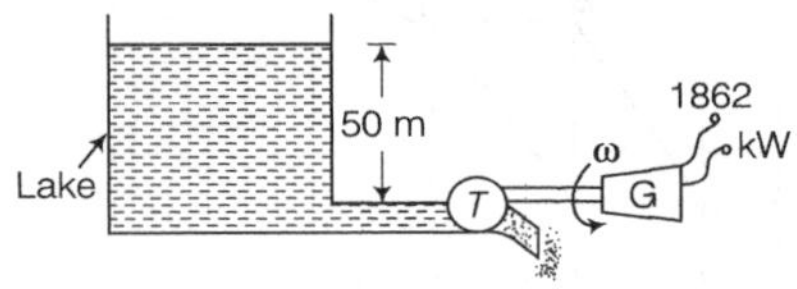

27 Overall efficiency of turbine generator system is

a. 80%
b. 76%
c. 95%
d. 85%

28 Loss of power in turbine generator system is

a. 102 kW
b. 491 kW
c. 593 kW
d. 485 kW

Matching List Type

Direction (Q. Nos. 29-32) ***Choices for the correct combination of elements from Column I and Column II are given as options (a), (b), (c) and (d), out of which one is correct.***

29 An insulated and frictionless piston cylinder arrangement contains one mole of ideal gas. A heating coil of resistance r provided inside the cylinder in which current I flows such that temperature of gas varies as $T = T_0 + at + bt^2$. (T = temperature, t = time).

Movement of piston keep the pressure of gas constant. Atmospheric pressure above cylinder is p_0.

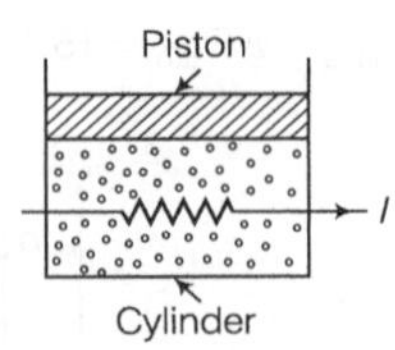

Working Space

Now, match entries of Column I and Column II and choose the correct options.

	Column I		Column II
i.	Current I	p.	$\frac{R}{pA}\cdot(2bt + a)$
ii.	Speed of piston	q.	$\left(\frac{5R}{2r}(2bt + a)\right)^{1/2}$
iii.	Acceleration of piston	r.	$\frac{R}{pA}\cdot(2b)$
iv.	C_p for the gas	s.	$\left(\frac{5R}{3r}(2bt + a)\right)^{1/2}$

Codes

	i	ii	iii	iv		i	ii	iii	iv
a.	q	r	p	s	b.	p	q	r	s
c.	r	p	q	s	d.	q	p	r	s

30 Few process are listed in Column I, column II is the result of the process. Match entries of Column I with Column II, there may be more than one correct match.

	Column I		Column II
i.	Isothermal expansion	p.	$\Delta U > 0$
ii.	Adiabatic expansion	q.	$\Delta T < 0$
iii.	Isobaric heating	r.	$\Delta V > 0$
iv.	Isochoric cooling	s.	$\Delta p < 0$

Codes

	i	ii	iii	iv		i	ii	iii	iv
a.	r,s	q	p	s	b.	r	q,r	p,r	q
c.	r,s	q,r,s	p,r	q,s	d.	r,s	q,r	p	q,s

31 Match processes (in Column I) with change in entropy of a unit mass of gas (in Column II) that occurs, when gas undergoes the process.

	Column I		Column II
i.	Constant volume process	p.	$R\cdot\ln(V_2/V_1)$
ii.	Constant pressure process	q.	$C_p\cdot\ln(V_2/v_1)$
iii.	Constant temperature process	r.	$C_V\cdot\ln(p_2/p_1)$
iv.	Reversible adiabatic process	s.	0
v.	Polytropic process	t.	$C_p\ln\frac{V_2}{V_1} + C_V\ln\frac{p_2}{p_1}$

Codes

	i	i	iii	iv	v
a.	s	q	p	t	r
b.	t	s	q	p	r
c.	p	q	t	s	r
d.	r	q	p	s	t

32 A plot of $\log_e p$ v/s $\log_e V$ is given, match the processes whose polytropic index is given in Column I with their graphs in in Column II.

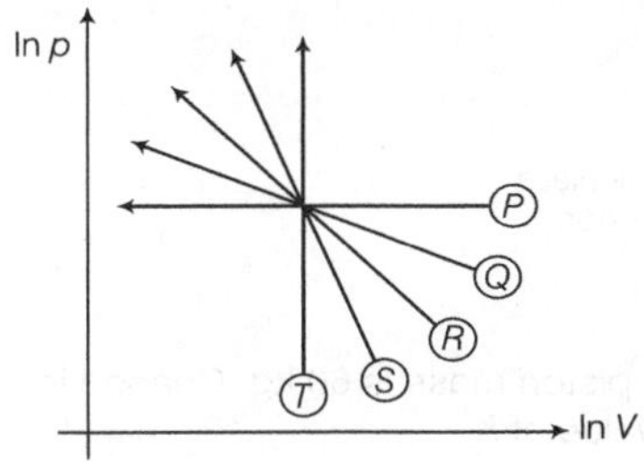

	Column I		Column II
i.	$n=0$	p.	P
ii.	$n \neq 0, 1, \gamma$	q.	Q
iii.	$n=1$	r.	R
iv.	$n=\gamma, \gamma=C_p/C_V$	s.	S
v.	$n=\infty$	t.	T

Codes

	i	ii	iii	iv	v
a.	p	q	r	s	t
b.	p	q	s	r	t
c.	q	r	s	t	p
d.	s	t	p	q	r

Working Space

One Integer Value Correct Type

Direction *(Q. Nos. 33-37) This section contains 5 questions. When worked out will result in an integer from 0 to 9 (both inclusive).*

33 A room is maintained at 20° C by a heater of resistance 20 Ω connected to 200 V mains. Outside temperature is 15.24°C. Heat is lost through a glass window of area $A\,\text{m}^2$. Find A, if glass is 0.2 cm thick. (Thermal conductivity of glass is 0.2 cal $\text{m}^{-1}\,\text{s}^{-1}\,\text{C}^{-1}$, mechanical equivalent of heat is 4.2 J cal^{-1}).

34 A body cools from 50°C to 40°C in 5 min. The surrounding temperature is 20°C. By how much ° C does the temperature decrease in the next 5 min? (Round-off to nearest integer).

35 An ideal gas is trapped inside a narrow test tube of area $A = 20\times10^{-6}\text{m}^2$ as shown.

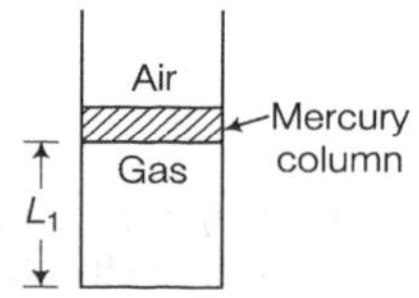

Length of gas column below a mercury column of mass 0.002 kg is L_1. If the tube is quickly inverted, then gas now occupies a length L_2 of the tube. Ratio $\frac{L_2}{L_1}$ is $\frac{101}{K\times11}$, find the value of K.

36 Two moles of a monatomic ideal gas is taken through a cyclic process as shown.

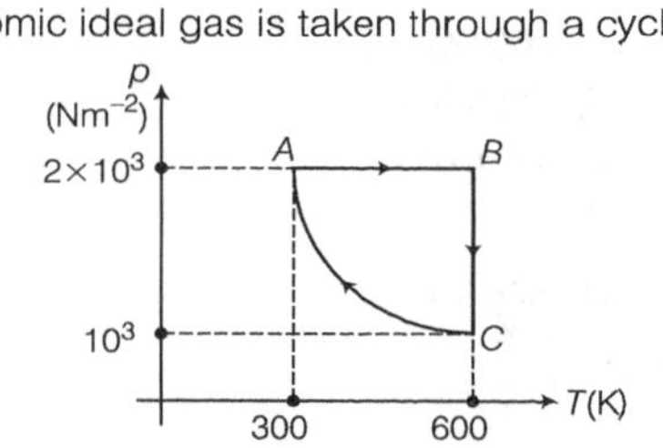

Process CA follows pT = constant. If cycle efficiency is $1-\frac{3x}{12\ln2+15}$, then find the value of x.

37 In the following arrangement, piston moves up by 5 cm, when 200 J of heat is added to the system.

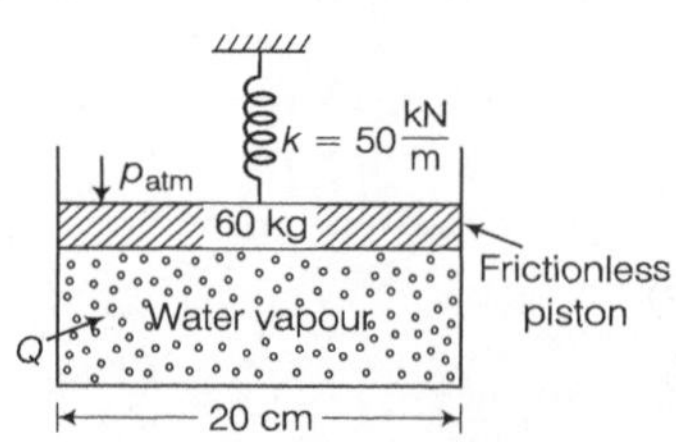

Spring is initially unstretched. Cylinder dia is 20 cm, piston mass is 60 kg. Change in internal energy of system is $-10\,K$ Joules. Find the value of K.

JEE Main & AIEEE Archive

(Compilation of Last 13 Years' Questions)

1 Consider an ideal gas confined in an isolated closed chamber. As the gas undergoes an adiabatic expansion, the average time of collision between molecules increases as V^q, where V is the volume of the gas. The value of q is $\left(\gamma = \frac{C_p}{C_v}\right)$ **(JEE Main 2015)**

a. $\frac{3\gamma + 5}{6}$

b. $\frac{3\gamma - 5}{6}$

c. $\frac{\gamma + 1}{2}$

d. $\frac{\gamma + 1}{2}$

2 An open glass tube is immersed in mercury in such a way that a length of 8 cm extends above the mercury level. The open end of the tube is then closed and sealed and the tube is raised vertically up by additional 46 cm. What will be length of the air column above mercury in the tube now? (Atmospheric pressure = 76 cm of Hg) **(JEE Main 2014)**

a. 16 cm b. 22 cm
c. 38 cm d. 6 cm

3 One mole of diatomic ideal gas undergoes a cyclic process ABC as shown in the figure. The process BC is adiabatic. The temperatures at A, B and C are 400 K, 800 K and 600 K, respectively.

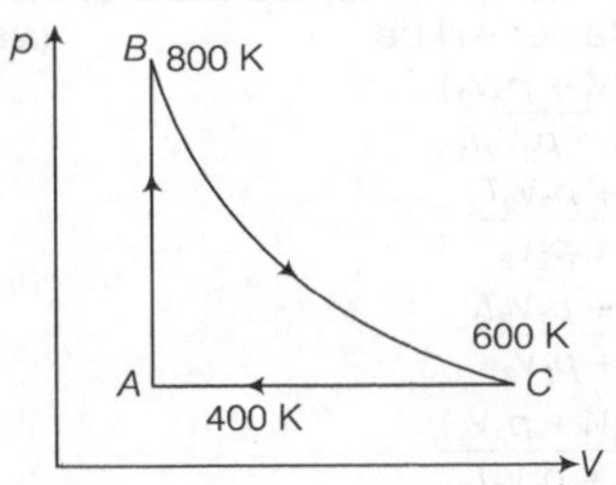

Choose the correct statement. **(JEE Main 2014)**

a. The change in internal energy in whole cyclic process is 250 R

b. The change in internal energy in the process CA is 700 R

c. The change in internal energy in the process AB is – 350 R

d. The change in internal energy in the process BC is – 500 R

4 The given p-V diagram represents the thermodynamic cycle of an engine operating with an ideal monatomic gas. The amount of heat, extracted from the source in a single cycle is **(JEE Main 2013)**

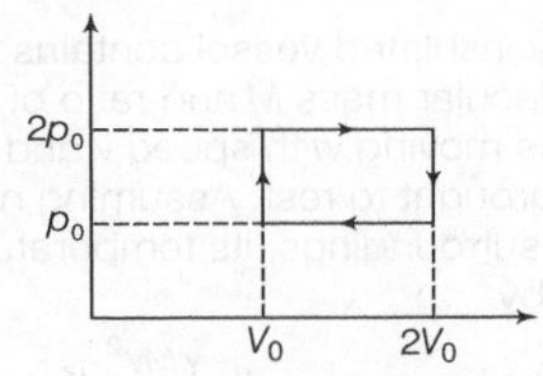

a. p_0V_0 b. $\left(\frac{13}{2}\right) p_0V_0$
c. $\left(\frac{11}{2}\right) p_0V_0$ d. $4p_0V_0$

5 Helium gas goes through a cycle $ABCDA$ (consisting of two isochoric and isobaric lines) as shown in the figure. Efficiency of this cycle is nearly (assume the gas to be close to ideal gas) **(AIEEE 2012)**

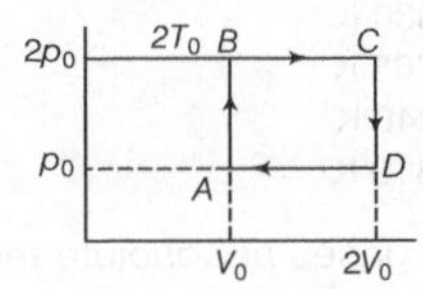

a. 15.4% b. 9.1%
c. 10.5% d. 12.5%

6 A Carnot engine, whose efficiency is 40%, takes in heat from a source maintained at a temperature of 500 K. It is desired to have an engine of efficiency 60%. Then, the intake temperature for the same exhaust (sink) temperature must be **(AIEEE 2012)**

a. efficiency of Carnot engine cannot be made larger than 50%

b. 1200 K

c. 750 K

d. 600 K

7 The specific heat capacity of a metal at low temperature (T) is given as $C_p(\text{kJ K}^{-1}\text{ kg}^{-1}) = 32\left(\frac{T}{400}\right)^3$. A 100 g vessel of this metal is to be cooled from 20 K to 4 K by a special refrigerator operating at room temperature (27°C). The amount of work required to cool the vessel is **(AIEEE 2011)**

a. equal to 0.002 kJ

b. greater than 0.148 kJ

c. between 0.148 kJ and 0.028 kJ

d. less than 0.028 kJ

8 A container with insulating walls is divided into two equal parts by a partition fitted with a valve. One part is filled with an ideal gas at a pressure p and temperature T, whereas the other part is completely evacuated. If the valve is suddenly opened, the pressure and temperature of the gas will be **(AIEEE 2011)**

a. $\frac{p}{2}, T$ b. $\frac{p}{2}, \frac{T}{2}$
c. p, T d. $p, \frac{T}{2}$

9 A thermally insulated vessel contains an ideal gas of molecular mass M and ratio of specific heats γ. It is moving with speed v and its suddenly brought to rest. Assuming no heat is lost to the surroundings, its temperature increases by **(AIEEE 2011)**

a. $\frac{(\gamma - 1)}{2\gamma R} M v^2$ K b. $\frac{\gamma M v^2}{2R}$ K
c. $\frac{(\gamma - 1)}{2R} M v^2$ K d. $\frac{(\gamma - 1)}{2(\gamma + 1)R} M v^2$ K

10 A Carnot engine operating between temperatures T_1 and T_2 has efficiency $\frac{1}{6}$. When T_2 is lowered by 62 K, its efficiency increases to 1/3. Then, T_1 and T_2 are respectively **(AIEEE 2011)**

a. 372 K and 330 K
b. 330 K and 268 K
c. 310 K and 248 K
d. 372 K and 310 K

11 Three perfect gases at absolute temperatures T_1, T_2 and T_3 are mixed. The masses of molecules are m_1, m_2 and m_3 and the number of molecules are n_1, n_2 and n_3 respectively. Assuming no loss of energy, the final temperature of the mixture is **(AIEEE 2011)**

a. $\frac{n_1T_1 + n_2T_2 + n_3T_3}{n_1 + n_2 + n_3}$
b. $\frac{n_1T_1^2 + n_2T_2^2 + n_3T_3^2}{n_1T_1 + n_2T_2 + n_3T_3}$
c. $\frac{n_1^2T_1^2 + n_2^2T_2^2 + n_3^2T_3^2}{n_1T_1 + n_2T_2 + n_3T_3}$
d. $\frac{(T_1 + T_2 + T_3)}{3}$

12 The potential energy function for the force between two atoms in a diatomic molecule is approximately given by $U(x) = \frac{a}{x^{12}} - \frac{b}{x^6}$, where a and b are constants and x is the distance between the atoms. If the dissociation energy of the molecule is $D = [U(x = \infty) - U_{\text{at equilibrium}}]$, D is **(AIEEE 2010)**

a. $\frac{b^2}{2a}$ b. $\frac{b^2}{12a}$
c. $\frac{b^2}{4a}$ d. $\frac{b^2}{6a}$

Direction *(Q.Nos. 13-15) are based on the following figure.*

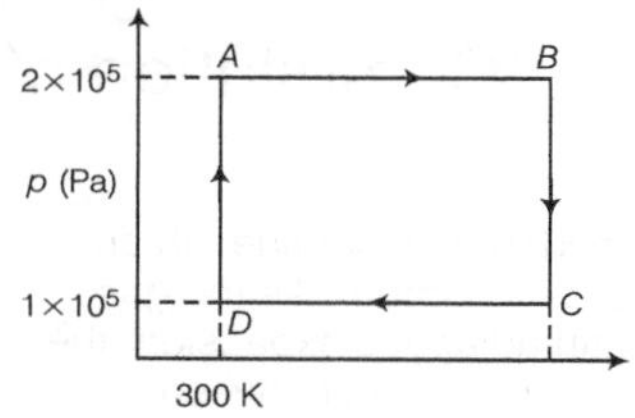

13 Assume the gas to be ideal, the work done on the gas in taking it from A to B is **(AIEEE 2009)**

a. 200 R b. 300 R
c. 400 R d. 500 R

14 The work done on the gas in taking it from D to A is (see above figure) **(AIEEE 2009)**

a. − 414 R b. + 414 R
c. − 690 R d. + 690 R

15 The net work on the gas in the cycle $ABCDA$ is (see above figure)

a. zero b. 276 R **(AIEEE 2009)**
c. 1076 R d. 1904 R

16 One kg of a diatomic gas is at a pressure of 8×10^4 Nm^{-2}. The density of the gas is 4 kgm^{-3}. What is the energy of the gas due to its thermal motion? **(AIEEE 2009)**

a. 3×10^4 J b. 5×10^4 J
c. 6×10^4 J d. 7×10^4 J

17 An insulated container of gas has two chambers separated by an insulating partition. One of the chambers has volume V_1 and contains ideal gas at pressure p_1 and temperature T_1. The other chamber has volume V_2 and contains ideal gas at pressure p_2 and temperature T_2. If the partition is removed without doing any work on the gas, the final equilibrium temperature of the gas in the container will be **(AIEEE 2008, 04)**

a. $\frac{T_1T_2(p_1V_1 + p_2V_2)}{p_1V_1T_2 + p_2V_2T_1}$
b. $\frac{p_1V_1T_1 + p_2V_2T_2}{p_1V_1 + p_2V_2}$
c. $\frac{p_1V_1T_2 + p_2V_2T_1}{p_1V_1 + p_2V_2}$
d. $\frac{T_1T_2(p_1V_1 + p_2V_2)}{p_1V_1T_1 + p_2V_2T_2}$

18 If C_p and C_V denote the specific heats of nitrogen per unit mass at constant pressure and constant volume respectively, then **(AIEEE 2007)**

a. $C_p - C_V = \frac{R}{28}$
b. $C_p - C_V = \frac{R}{14}$
c. $C_p - C_V = R$
d. $C_p - C_V = 28R$

19 When a system is taken from state i to state f along the path iaf, it is found that $Q = 50$ cal and $W = 20$ cal. Along the path ibf, $Q = 36$ cal. W along the path ibf is **(AIEEE 2007)**

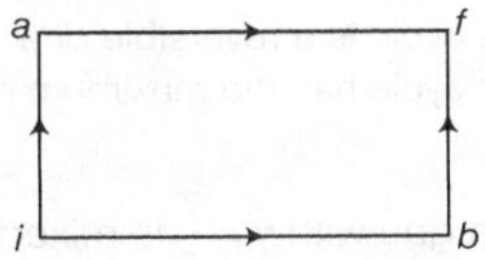

a. 6 cal
b. 16 cal
c. 66 cal
d. 14 cal

20 A Carnot engine, having an efficiency of $\eta = \frac{1}{10}$ as heat engine is used as a refrigerator. If the work done on the system is 10 J, the amount of energy absorbed from the reservoir at lower temperature is **(AIEEE 2007)**

a. 99 J b. 90 J
c. 1 J d. 100 J

21 Two rigid boxes $\frac{1}{2}$ containing different ideal gases are placed on a table. Box A contains one mole of nitrogen at temperature T_0, while box B contains one mole of helium at temperature (7/3) T_0. The boxes are then put into thermal contact with each other and heat flows between them until the gases reach a common final temperature (ignore the heat capacity of boxes). Then, the final temperature of the gases, T_f, in terms of T_0 is **(AIEEE 2006)**

a. $T_f = \frac{3}{7} T_0$
b. $T_f = \frac{7}{3} T_0$
c. $T_f = \frac{3}{2} T_0$
d. $T_f = \frac{5}{2} T_0$

22 The work of 146 kJ is performed in order to compress one kilo mole of a gas adiabatically and in this process, the temperature of the gas increases by 7°C. The gas is ($R = 8.3$ J mol^{-1} K^{-1}) **(AIEEE 2006)**

a. diatomic
b. triatomic
c. a mixture of monatomic and diatomic
d. monatomic

23 A system goes from A to B *via* two processes I and II. If ΔU_1 and ΔU_2 are the change in internal energies in the processes I and II respectively, then **(AIEEE 2005)**

a. $\Delta U_1 = \Delta U_2$
b. relation between ΔU_1 and ΔU_2 cannot be determined
c. $\Delta U_2 > \Delta U_1$
d. $\Delta U_2 < \Delta U_1$

24 The temperature-entropy diagram of a reversible engine cycle is given in the figure. Its efficiency is **(AIEEE 2005)**

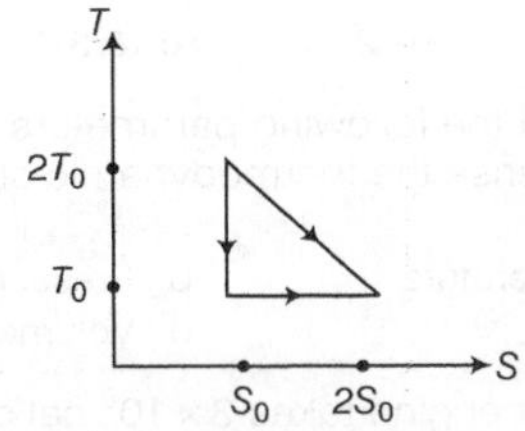

a. $\frac{1}{2}$ b. $\frac{1}{4}$
c. $\frac{1}{3}$ d. $\frac{2}{3}$

25 Which of the following is incorrect regarding the first law of thermodynamics? **(AIEEE 2005)**

a. It is not applicable to any cyclic process
b. It is a restatement of the principle of conservation of energy
c. It introduces the concept of the internal energy
d. Both (a) and (c)

26 A gaseous mixture consists of 16 g of helium and 16 g of oxygen. The ratio $\frac{C_p}{C_V}$ of the mixture is **(AIEEE 2005)**

a. 1.59 b. 1.62
c. 1.4 d. 1.54

27 One mole of ideal monoatomic gas $\left(\gamma = \frac{5}{3}\right)$ is mixed with one mole of diatomic gas $\left(\gamma = \frac{7}{5}\right)$. What is γ for the mixture ? γ denotes the ratio of specific heat at constant pressure to that at constant volume. **(AIEEE 2004)**

a. $\frac{3}{2}$
b. $\frac{23}{15}$
c. $\frac{35}{23}$
d. $\frac{4}{3}$

28 Which of the following statements is correct for any thermodynamic system? **(AIEEE 2004)**

a. The internal energy changes in all processes
b. Internal energy and entropy are state functions
c. The change in entropy can never be zero
d. The work done in an adiabatic process is always zero

29 "Heat cannot be itself flow from a body at lower temperature to a body at higher temperature" is a statement or consequence of **(AIEEE 2003)**

a. second law of thermodynamics
b. conservation of momentum
c. conservation of mass
d. first law of thermodynamics

30 During an adiabatic process, the pressure of a gas is found to be proportional to the cube of its absolute temperature. The ratio C_p/C_V for the gas is **(AIEEE 2003)**

a. 4/3 b. 2 c. 5/3 d. 3/2

31 Which of the following parameters does not characterise the thermodynamic state of matter? **(AIEEE 2003)**

a. Temperature b. Pressure
c. Work d. Volume

32 A Carnot engine takes 3×10^6 cal of heat from a reservoir at 627°C and gives it to a sink at 27°C. The work done by the engine is **(AIEEE 2003)**

a. 4.2×10^6 J b. 8.4×10^6 J
c. 16.8×10^6 J d. zero

33 Even Carnot engine cannot give 100% efficiency, because we cannot **(AIEEE 2002)**

a. prevent radiation
b. find ideal sources
c. reach absolute zero temperature
d. eliminate friction

34 Which of the following statements is incorrect? **(AIEEE 2002)**

a. All reversible cycles have same efficiency
b. Reversible cycle has more efficiency than an irreversible one
c. Carnot cycle is a reversible one
d. Carnot cycle has the maximum efficiency in all cycles

35 1 mol of a gas with $\gamma = \frac{7}{5}$ is mixed with 1 mol of a gas with $\gamma = \frac{5}{3}$, then the value of γ for the resulting mixture is **(AIEEE 2002)**

a. $\frac{7}{5}$ b. $\frac{2}{5}$
c. $\frac{24}{16}$ d. $\frac{12}{7}$

36 At what temperature is the rms velocity of a hydrogen molecule equal to that of an oxygen molecule at 47°C ? **(AIEEE 2002)**

a. 80 K b. –73 K
c. 3 K d. 20 K

JEE Advanced & IIT JEE Archive

(Compilation of Last 10 Years' Questions)

1 An ideal monatomic gas is confined in a horizontal cylinder by a spring loaded piston (as shown in the figure). Initially the gas is at temperature T_1, pressure p_1 and volume V_1 and the spring is in its relaxed state. The gas is then heated very slowly to temperature T_2, pressure p_2 and volume V_2. During this process, the piston moves out by a distance x. Ignoring the friction between the piston and the cylinder.
Which of the following statement(s) is/are correct? **(2015 Adv., Only One Option Correct Type)**

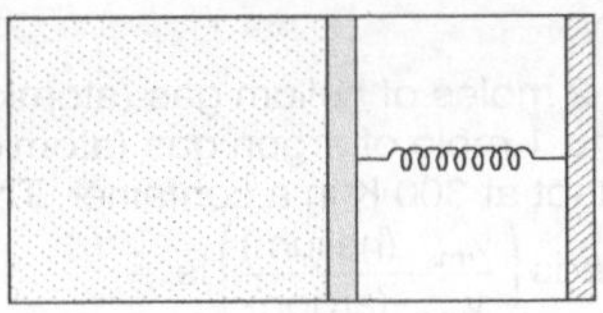

a. If $V_2 = 2V_1$ and $T_2 = 3T_1$, then the energy stored in the spring is $\frac{1}{4}p_1V_1$

b. If $V_2 = 2V_1$ and $T_2 = 3T_1$, then the change in internal energy is $3p_1V_1$

c. If $V_2 = 3V_1$ and $T_2 = 4T_1$, then the work done by the gas is $\frac{7}{3}p_1V_1$

d. If $V_2 = 3V_1$ and $T_2 = 4T_1$, then the heat supplied to the gas is $\frac{17}{6}p_1V_1$

2 A container of fixed volume has a mixture of one mole of hydrogen and one mole of helium in equilibrium at temperature T. Assuming the gases ae ideal, choose the correct statements.
(2015 Adv., One or More than One Options Correct Type)

a. The average energy per mole of the gas mixture is $2RT$

b. The ratio of speed of sound in the gas mixture to that in helium gas is $\sqrt{6/5}$

c. The ratio of the rms speed of helium atoms to that of hydrogen molecules is 1/2

d. The ratio of the rms speed of helium atoms to that of hydrogen molecules is $1/\sqrt{2}$

Passage for Q. Nos. (3-4)

In the figure a container is shown to have a movable (without friction) piston on top. The container and the piston are all made of perfectly insulating material allowing no heat transfer between outside and inside the container. The container is divided into two compartments by a rigid partition made of a thermally conducting material that allows slow transfer of heat.

The lower compartment of the container is filled with 2 moles of an ideal monatomic gas at 700 K and the upper compartment is filled with 2 moles of an ideal diatomic gas at 400 K. The heat capacities per mole of an ideal monatomic gas are $C_V = \frac{3}{2}R, C_p = \frac{5}{2}R$, and those for an ideal diatomic gas are $C_V = \frac{5}{2}R, C_p = \frac{7}{2}R$.

(2014 Adv., Comprehension Type)

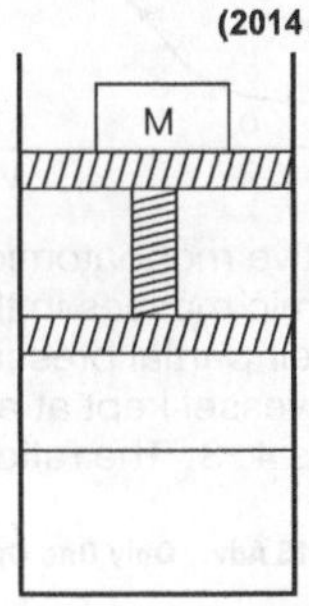

3 Consider the partition to be rigidly fixed so that it does not move. When equilibrium is achieved, the final temperature of the gases will be

a. 550 K b. 525 K
c. 513 K d. 490 K

4 Now consider the partition to be free to move without friction so that the pressure of gases in both compartments is the same. Then, total work done by the gases till the time they achieve equilibrium will be

a. 250 R b. 200 R
c. 100 R d. −100 R

5 A student is performing an experiment using a resonance column and a tuning fork of frequency 244 s^{-1}. He told that the air in the tube has been replaced by another gas (assume that the column remains filled with the gas). If the minimum height at which resonance occurs is (0.350 ± 0.005) m, the gas in the tube is
(2014 Adv., One or More than One Options Correct Type)

Useful Information

$\sqrt{167RT} = 640\ J^{1/2}\ mole^{-1/2}$;
$\sqrt{140RT} = 590\ J^{1/2}\ mole^{-1/2}$.

The molar masses M in grams are given in the options. Take the value of $\sqrt{10/M}$ for each gas as given there.

a. Neon ($M = 20$, $\sqrt{10/20} = 7/10$)
b. Nitrogen ($M = 28$, $\sqrt{10/28} = 3/5$)
c. Oxygen ($M = 32$, $\sqrt{10/32} = 9/16$)
d. Argon ($M = 36$, $\sqrt{10/36} = 17/32$)

JEE Archive

6 A thermodynamic system is taken from an initial state i with internal energy $U_i = 100$ J to the final state f along two different paths iaf and ibf, as schematically shown in the figure. The work done by the system along the paths af, ib and bf are $W_{af} = 200$ J, $W_{ib} = 50$ J and $W_{bf} = 100$ J respectively. The heat supplied to the system along the path iaf, ib and bf are Q_{iaf}, Q_{ib} and Q_{bf} respectively. If the internal energy of the system in the state b is $U_b = 200$ J and $Q_{iaf} = 500$ J. Find the ratio Q_{bf} / Q_{ib}.

(2014 Adv., Integer Type).

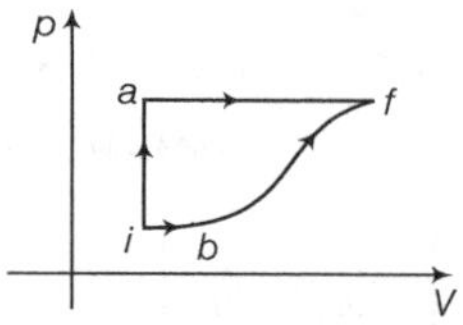

7 Two non-reactive monoatomic ideal gases have their atomic masses in the ratio 2 : 3. The ratio of their partial pressures, when enclosed in a vessel kept at a constant temperature, is 4 : 3. The ratio of their densities is

(2013 Adv., Only One Option Correct Type)

a. 1 : 4 b. 1 : 2
c. 6 : 9 d. 8 : 9

8 One mole of a monatomic ideal gas is taken along two cyclic processes $E \to F \to G \to E$ and $E \to F \to H \to E$ as shown in the p-V diagram. The processes involved are purely isochoric, isobaric, isothermal or adiabatic. Match the paths given in Column I with the magnitudes of the work done in Column II and select the correct answer using the codes given below.

(2013 Adv., Matching Type)

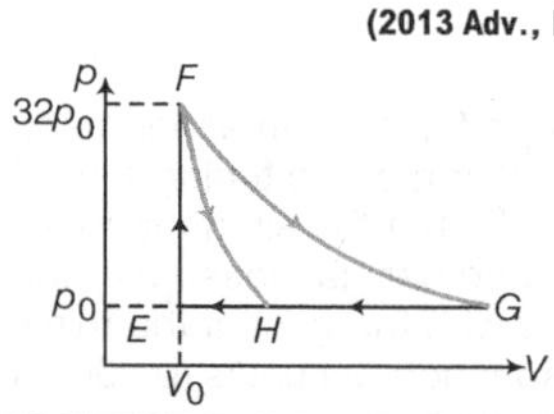

	Column I		Column II
i.	$G \to E$	p.	$160\, p_0 V_0 \ln 2$
ii.	$G \to H$	q.	$36\, p_0 V_0$
iii.	$F \to H$	r.	$24\, p_0 V_0$
iv.	$F \to G$	s.	$31\, p_0 V_0$

Codes

	i	ii	iii	iv
a.	s	r	q	p
b.	s	r	p	q
c.	r	p	q	s
d.	p	r	q	s

9 The figure below shows the variation of specific heat capacity (C) of a solid as a function of temperature (T). The temperature is increased continuously from 0 to 500 K at a constant rate. Ignoring any volume change, the following statement(s) is (are) correct to reasonable approximation.

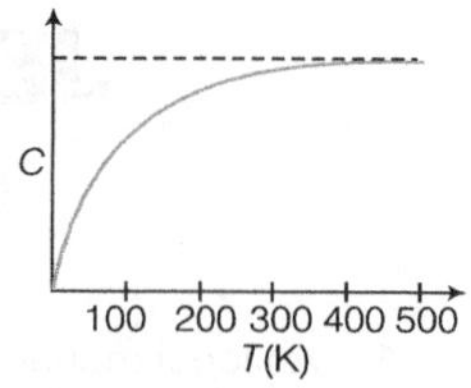

(2013 Adv., One or More than One Options Correct Type)

a. The rate at which heat is absorbed in the range 0-100 K varies linearly with temperature T
b. Heat absorbed in increasing the temperature from 0-100 K is less than the heat required for increasing the temperature from 400-500 K
c. There is no change in the rate of heat absorbtion in the range 400-500 K
d. The rate of heat absorption increases in the range 200-300 K

10 A mixture of 2 moles of helium gas (atomic mass = 4 amu) and 1 mole of argon gas (atomic mass = 40 amu) is kept at 300 K in a container. The ratio of the rms speeds $\left(\frac{v_{rms}\text{ (helium)}}{v_{rms}\text{ (argon)}}\right)$ is

(2012, Only One Option Correct Type)

a. 0.32 b. 0.45
c. 2.24 d. 3.16

11 5.6 L of helium gas at STP is adiabatically compressed to 0.7 L. Taking the initial temperature to be T_1, the work done in the process is

(2011, Only One Option Correct Type)

a. $\frac{9}{8}RT_1$ b. $\frac{3}{2}RT_1$
c. $\frac{15}{8}RT_1$ d. $\frac{9}{2}RT_1$

12 One mole of a monatomic ideal gas is taken through a cycle $ABCDA$ as shown in the p-V diagram. Column II gives the characteristics involved in the cycle. Match them with each of the processes given in Column I. **(2011 Matching Type)**

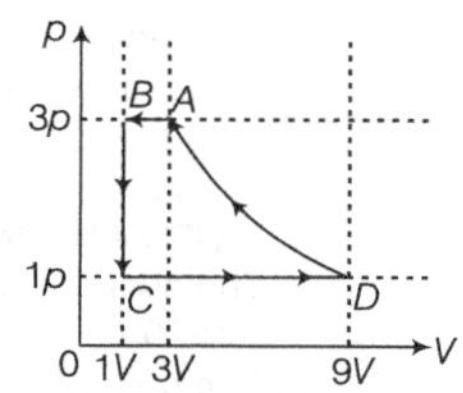

	Column I		Column II
i.	Process $A \to B$	p.	Internal energy decreases
ii.	Process $B \to C$	q.	Internal energy increases
iii.	Process $C \to D$	r.	Heat is lost
iv.	Process $D \to A$	s.	Heat is gained
		t.	Work is done on the gas

Codes

	i	ii	iii	iv
a.	p	p,q	r	s,t
b.	p,q	q,s,t	r,s,t	p,r,s,t
c.	p,q,r	q,t	q,s	s,t
d.	p,r,t	p,r	q,s	r,t

13 One mole of an ideal gas in initial state A undergoes a cyclic process $ABCA$ as shown in the figure. Its pressure at A is p_0. Choose the correct option(s) from the following.

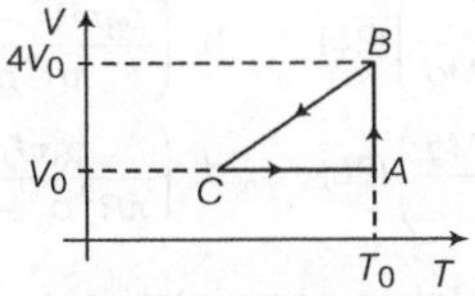

(2010, One or More than One Options Correct Type)

a. Internal energies at A and B are the same

b. Work done by the gas in process AB is $p_0V_0 \ln 4$

c. Pressure at C is $\frac{p_0}{4}$

d. Temperature at C is $\frac{T_0}{4}$

14 A diatomic ideal gas is compressed adiabatically to $\frac{1}{32}$ of its initial volume. If the initial temperature of the gas is T_i (in kelvin) and the final temperature is aT_i, the value of a is

(2010, Integer Type)

15 A real gas behaves like an ideal gas, if its

(2010, Only One Option Correct Type)

a. pressure and temperature are both high

b. pressure and temperature are both low

c. pressure is high and temperature is low

d. pressure is low and temperature is high

16 C_V and C_p denote the molar specific heat capacities of a gas at constant volume and constant pressure, respectively. Then,

(2009, One or More than One Options Correct Type)

a. $C_p - C_V$ is larger for a diatomic ideal gas than for a monatomic ideal gas

b. $C_p + C_V$ is larger for a diatomic ideal gas than for a monatomic ideal gas

c. $\frac{C_p}{C_V}$ is larger for a diatomic ideal gas than for a monatomic ideal gas

d. $C_p \cdot C_V$ is larger for a diatomic ideal gas than for a monatomic ideal gas

17 The figure shows the p-V plot an ideal gas taken through a cycle $ABCDA$. The part ABC is a semi-circle and CDA is half of an ellipse. Then,

(2009, One or More than One Options Correct Type)

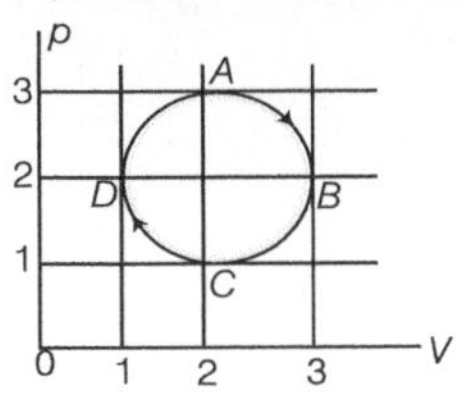

a. the process during the path $A \rightarrow B$ is isothermal

b. heat flows out of the gas during the path $B \rightarrow C \rightarrow D$

c. work done during the path $A \rightarrow B \rightarrow C$ is zero

d. positive work is done by the gas in the cycle $ABCDA$

18 An ideal gas is expanding such that $pT^2 =$ constant. The coefficient of volume expansion of the gas is

(2008, Only One Option Correct Type)

a. $\frac{1}{T}$ b. $\frac{2}{T}$

c. $\frac{3}{T}$ d. $\frac{4}{T}$

19 Column I contains a list of processes involving expansion of an ideal gas. Match this with Column II describing the thermodynamic change during this process. Indicate your answer by darkening the appropriate bubbles of the 4×4 matrix given in the ORS.

(2008, Matching Type)

	Column I		Column II
i.	An inuslated container has two chambers separated by a valve. Chamber I contains an ideal gas and the Chamber II has vacuum. The valve is opened. 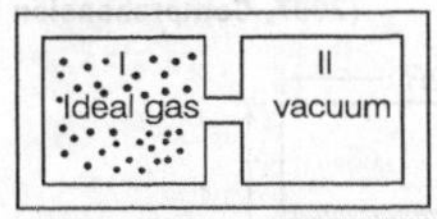	p.	The temperature of the gas decreases
ii.	An ideal monatomic gas expands to twice its original volume such that its pressure $p \propto \frac{1}{V^2}$, where V is the volume of the gas.	q.	The temperature of the gas increases or remains constant
iii.	An ideal monatomic gas expands to twice its original volume such that its pressure $p \propto \frac{1}{V^{4/3}}$, where V is its volume.	r.	The gas loses heat
iv.	An ideal monatomic gas expands such that its pressure p and volume V follows the behaviour shown in the graph. 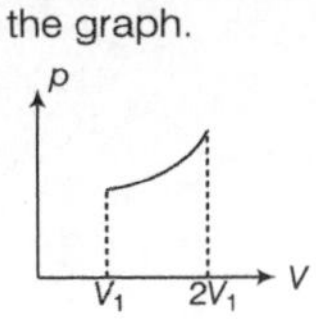	s.	The gas gains heat

Codes

	i	ii	iii	iv
a.	q	p,r	p,s	q,s
b.	p,q	q,s	q,r	p,q,s
c.	p	q	q,s	p,r
d.	p,q	q,s	r,s	s

20 **Statement I** The total translational kinetic energy of all the molecules of a given mass of an ideal gas is 1.5 times the product of its pressure and its volume.

Statement II The molecules of a gas collide with each other and the velocities of the molecules change due to the collision.

(2007, Statement Type)

a. Both Statement I and Statement II are correct and Statement II is the correct explanation of Statement I
b. Both Statement I and Statement II are correct but Statement II is not the correct explanation of Statement I
c. Statement I is correct but Statement II is incorrect
d. Statement II is correct but Statement I is incorrect

Passage for Q. Nos. (21-23)

A fixed thermally conducting cylinder has a radius R and height L_0. The cylinder is open at its bottom and has a small hole at its top. A piston of mass M is held at a distance L from the top surface, as shown in the figure. The atmospheric pressure is p_0. **(2007, Comprehension Type)**

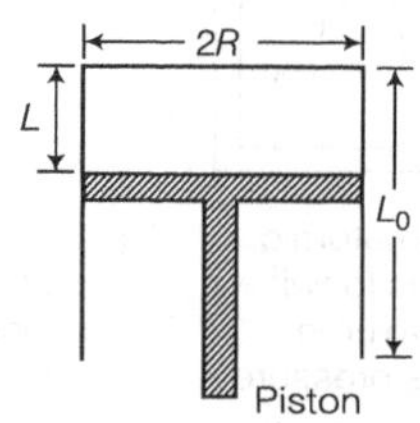

21 The piston is now pulled out slowly and held at a distance $2L$ from the top. The pressure in the cylinder between its top and the piston will then be

a. p_0
b. $\frac{p_0}{2}$
c. $\frac{p_0}{2} + \frac{Mg}{\pi R^2}$
d. $\frac{p_0}{2} - \frac{Mg}{\pi R^2}$

22 While the piston is at a distance $2L$ from the top, the hole at the top is sealed. The piston is then released, to a position, where it can stay in equilibrium. In this condition, the distance of the piston from the top is

a. $\left(\frac{2p_0\pi R^2}{\pi R^2 p_0 + Mg}\right)(2L)$ b. $\left(\frac{p_0\pi R^2 - Mg}{\pi R^2 p_0}\right)(2L)$

c. $\left(\frac{p_0\pi R^2 + Mg}{\pi R^2 p_0}\right)(2L)$ d. $\left(\frac{p_0\pi R^2}{\pi R^2 p_0 - Mg}\right)(2L)$

23 The piston is taken completely out of the cylinder. The hole at the top is sealed. A water tank is brought below the cylinder and put in a position so that, the water surface in the tank is at the same level as the top of the cylinder as shown in the figure. The density of the water is ρ. In equilibrium, the height H of the water column in the cylinder satisfies,

a. $\rho g(L_0 - H)^2 + p_0(L_0 - H) + L_0 p_0 = 0$
b. $\rho g(L_0 - H)^2 - p_0(L_0 - H) - L_0 p_0 = 0$
c. $\rho g(L_0 - H)^2 + p_0(L_0 - H) - L_0 p_0 = 0$
d. $\rho g(L_0 - H)^2 - p_0(L_0 - H) + L_0 p_0 = 0$

24 Match the following columns for the given process.

(2006, Matching Type)

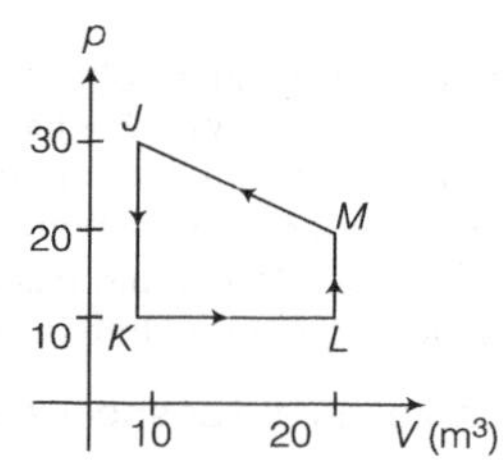

	Column I		Column II
i.	Process $J \to K$	p.	$Q > 0$
ii.	Process $K \to L$	q.	$W < 0$
iii.	Process $L \to M$	r.	$W > 0$
iv.	Process $M \to J$	s.	$Q < 0$

Codes

	i	ii	iii	iv
a.	p,q,r	p	p,q	r,s
b.	s	p,r	p	q,s
c.	p	p,q	q	r,s
d.	p,s	p,q	r,s	r

Answers *with* Explanations

Answers

1. Heat, Temperature and Heat Transfer

DPP-1 Heat, Temperature and Expansion of Solids

1. 32Ω 2. 445°C
3. (i) 55 min (ii) $-(5\times10^{-3})\%$
4. $3\times10^{7}\text{Nm}^{-2}$
5. *d* 6. *c* 7. *b* 8. *a* 9. *d* 10. *a* 11. *a* 12. *c* 13. *a* 14. *d*
15. *b* 16. *a,b,c,d* 17. *a* 18. *a,b,c* 19. *a,d* 20. *b,d* 21. *d* 22. *a* 23. *a*
24. *a* 25. *c* 26. *a* 27. 4 28. 6

DPP-2 Expansion of Liquids and Gases

1. (i) 320 kPa (ii) 2 kPa
2. (i) 99.4 cm^3 (ii) 0.943 cm
3. (i) 1.06×10^{21} kg (ii) 59.8 K
4. 472 K 5. *b* 6. *c* 7. *b* 8. *a* 9. *d* 10. *a,b* 11. *a* 12. *a,d* 13. *a,c*
14. *a,b,c,d* 15. *d* 16. *b* 17. *d* 18. *a* 19. *a* 20. *c* 21. *c* 22. *d*

DPP-3 Calorimetry

1. 1.44 kg 2. 5.87×10^{4} °C 3. 6.4 min 4. 46.3 min 5. *d* 6. *d* 7. *b*
8. *c* 9. *b* 10. *a* 11. *c* 12. *c* 13. *c* 14. *a* 15. *b* 16. *a,b,c* 17. *a,d*
18. *a,b,c* 19. *a,b,c* 20. *a,b,c* 21. *a* 22. *d* 23. *a* 24. *a* 25. *c* 26. *c* 27. *d*

DPP-4 Heat Transfer-I (Conduction of Heat)

1. 166 s 2. 26.48 °C, 0.52 °C 41.6 W 3. ≤67 °C 4. 10.2 h 5. *b* 6. *b* 7. *b* 8. *c*
9. *c* 10. *c* 11. *d* 12. *b* 13. *d* 14. *d* 15. *c* 16. *a,b,c,d* 17. *a,c*
18. *b,d* 19. *d* 20. *b,c* 21. *a* 22. *c* 23. *b* 24. *c* 25. *c* 26. *b* 27. 4
28. 2

DPP-5 Heat Transfer-II (Convective and Radiative Heat Transfer)

1. (i) $\frac{dQ}{dt}=hA\Delta T$ (ii) $2.8\times10^{-4}\text{ Jh}^{-1}$ 2. 300 °C 3. 1.4 kg 4. 4 °C 5. *c* 6. *d* 7. *a*
8. *c* 9. *b* 10. *c* 11. *a* 12. *b* 13. *c* 14. *a* 15. *a,b* 16. *a,d*
17. *a,b,c* 18. *a,d* 19. *a,b,c,d* 20. *d* 21. *b* 22. *c* 23. *b* 24. *d* 25. *b*
26. *d* 27. *a* 28. *c* 29. 4 30. 3

Revisal Problems (JEE Main)

1. *c* 2. *d* 3. *d* 4. *c* 5. *c* 6. *a* 7. *b* 8. *c* 9. *d* 10. *b*
11. *d* 12. *b* 13. *d* 14. *b* 15. *c* 16. *b* 17. *a* 18. *d* 19. *c* 20. *a*
21. *a* 22. *b* 23. *d* 24. *c* 25. *c*

Revisal Problems (JEE Advanced)

1. *a* 2. *c* 3. *d* 4. *b* 5. *a* 6. *c* 7. *a* 8. *a* 9. *a* 10. *d*
11. *a* 12. *a,d* 13. *b,c,d* 14. *a,b* 15. *a,d* 16. *a,b* 17. *d* 18. *d* 19. *b* 20. *c*
21. *a* 22. *c* 23. 3 24. 3

JEE Main & AIEEE Archive

1. *c* 2. *a* 3. *a* 4. *b*

JEE Advanced & IIT JEE Archive

1. 2 2. *a* 3. *d* 4. *c* 5. *a,c,d* 6. 3 7. 9 8. 8 9. 9 10. *c*
11. *d* 12. 273 K 13. *a* 14. *c* 15. *c* 16. *a*
17. (i) 50 °C (ii) 0.05 J (iii) 19999.95 J

2. Kinetic Theory of Gases and Thermodynamics

DPP-1 Kinetic Theory of Gases-I (Real and Ideal Gases)

1. (i) 5 cms^{-1}
(ii) 5.45 cms^{-1}
(iii) 1.59 ≈ one and half trip nearly in 1 second (iv) 6 cms^{-1}
2. (i) $6\times10^3\ \text{ms}^{-1}$ (ii) 24000 K (iii) $\sqrt{3}$
3. (i) 2011 J (ii) $\frac{1}{3}$ rd (iii) $\approx(2.70\times10^{25})e^{-433}$

4. 160 N 5. *d* 6. *c* 7. *c* 8. *d* 9. *d* 10. *c* 11. *d* 12. *a* 13. *b*
14. *b* 15. *a* 16. *a,c,d* 17. *a,c,d* 18. *a,b,c,d* 19. *b,c,d* 20. *c,d* 21. *d* 22. *a* 23. *d*
24. *d* 25. *d* 26. *a* 27. *a*

DPP-2 Kinetic Theory of Gases-II (Gas Laws)

1. 36900 Å^3 2. 0.776 atm, 0.024 atm 3. 2.7 cm
4. (ii) $\frac{3}{2}p$ 5. *c* 6. *c* 7. *d* 8. *d* 9. *a* 10. *a* 11. *b* 12. *d*
13. *c* 14. *d* 15. *a,d* 16. *b,d* 17. *a,b* 18. *a,b* 19. *b,d* 20. *a* 21. *d* 22. *c*
23. *c*

DPP-3 Kinetic Theory of Gases-III (Degree of Freedom and Energy of Gas)

1. (i) 5.65×10^{-21} J/molecule (ii) 3400 J/molecule (iii) 851 J
2. (i) 6.21×10^{-2} J/molecule (ii) 10.35×10^{-21} J/molecule (iii) $\frac{3}{5}$th part, $\frac{2}{5}$ part
3. (i) 8×10^4K 4. (i) $\beta=\frac{2}{3}\left(\frac{C_v}{C_p-C_v}\right)$ (ii) $f=3A-B$ (iii) $\gamma=1+\frac{2}{f}$

5. *d* 6. *d* 7. *c* 8. *c* 9. *d* 10. *c*
11. *d* 12. *a* 13. *c* 14. *c* 15. *a,b,d* 16. *b,c,d* 17. *b,c,d* 18. *c,d* 19. *a,c,d* 20. *b*
21. *d* 22. *d* 23. *a* 24. *b* 25. *c* 26. 4 27. 4 28. *c*

DPP-4 Specific Heat of Solids, Liquids and Gases

1. *c* 2. *d* 3. *d* 4. *d* 5. *d* 6. *b* 7. *a* 8. *a* 9. *a* 10. *b*
11. *b* 12. *a* 13. *b* 14. *c* 15. *c* 16. *a,c* 17. *b,c,d* 18. *a,c* 19. *a,b* 20. *b,d*
21. *a*

DPP-5 Thermodynamics-I (Work, Heat and Internal Energy)

1. *d* 2. *b* 3. *d* 4. *b* 5. *c* 6. *c* 7. *a* 8. *c* 9. *d* 10. *c*
11. *a,c* 12. *b,c* 13. *b,c* 14. *a,b* 15. *a,b,c* 16. *d* 17. *d* 18. *c* 19. *a* 20. *c*
21. *c* 22. 2 23. 3

DPP-6 Thermodynamics -II (Thermodynamic Processes)

1. (i) $p_0 V_0$ J (ii) $\frac{5}{2} p_0 V_0$ J (iii) $3 p_0 V_0$ J (iv) $\frac{p_0 V_0}{2}$ (v) $\frac{25}{8}\left(\frac{p_0 V_0}{R}\right)$

2. 2 atm, 3.73 L, 52 J 3. *c* 4. *d* 5. *d* 6. *b* 7. *d* 8. *c* 9. *d* 10. *b*

11. *a* 12. *d* 13. *c* 14. *b* 15. *a,b,c,d* 16. *a,b,d* 17. *b,c* 18. *a,b* 19. *a,b,c,d* 20. *a*

21. *c* 22. *b* 23. *a* 24. *c* 25. *c* 26. *b* 27. *d* 28. *a* 29. *a* 30. *d*

DPP-7 Thermodynamics-III (Engines and Fridges)

1. *c* 2. *a* 3. *a* 4. *c* 5. *a* 6. *c* 7. *c* 8. *d* 9. *c* 10. *b*

11. *a,b* 12. *c,d* 13. *a,b,c,d* 14. *a,b,c* 15. *a,c,d* 16. *d* 17. *d* 18. *d* 19. *c* 20. 3

Revisal Problems (JEE Main)

1. *c* 2. *d* 3. *d* 4. *b* 5. *c* 6. *b* 7. *d* 8. *c* 9. *a* 10. *a*

11. *c* 12. *d* 13. *d* 14. *c* 15. *b* 16. *a* 17. *b* 18. *b* 19. *b* 20. *c*

21. *a* 22. *c* 23. *a* 24. *a* 25. *a*

Revisal Problems(JEE Advanced)

1. *a* 2. *d* 3. *a* 4. *d* 5. *b* 6. *b* 7. *c* 8. *a,b,c* 9. *b,d* 10. *a,c*

11. *a,b,c* 12. *b,c* 13. *c* 14. *a* 15. *b* 16. *a* 17. *a* 18. *b* 19. *b* 20. *d*

21. *c* 22. *b* 23. *b* 24. *a* 25. *c* 26. *d* 27. *b* 28. *a* 29. *d* 30. *c*

31. *d* 32. *a* 33. 1 34. 7 35. 9 36. 7 37. 5

JEE Main & AIEEE Archive

1. *c* 2. *a* 3. *d* 4. *b* 5. *a* 6. *c* 7. *c* 8. *a* 9. *c* 10. *d*

11. *a* 12. *c* 13. *c* 14. *b* 15. *c* 16. *b* 17. *a* 18. *a* 19. *a* 20. *b*

21. *c* 22. *a* 23. *a* 24. *c* 25. *d* 26. *b* 27. *a* 28. *b* 29. *a* 30. *d*

31. *c* 32. *b* 33. *c* 34. *a* 35. *c* 36. *d*

JEE Advanced & IIT JEE Archive

1. *b* 2. *d* 3. *d* 4. *d* 5. *d* 6. 2 7. *d* 8. *a* 9. *b,c,d* 10. *d*

11. *a* 12. *d* 13. *a,b* 14. 4 15. *d* 16. *b,d* 17. *b,d* 18. *c* 19. *a* 20. *b*

21. *a* 22. *d* 23. *c* 24. *b*

1. Heat, Temperature and Heat Transfer

DPP-1 Heat, Temperature and Expansion of Solids

1. $R = R_0(1 + \alpha t + \beta t^2)$

Put $R = 10\,\Omega$ at 0°C

$\Rightarrow \quad 10 = R_0(1 + \alpha \times 0 + \beta \times 0^2)$ ...(i)

Put $R = 16\,\Omega$ at 100°C

$\Rightarrow \quad 16 = R_0(1 + \alpha \times 100 + \beta \times 100^2)$ ...(ii)

Put $R = 30\,\Omega$ at 445°C

$\Rightarrow \quad 30 = R_0(1 + \alpha \times 445 + \beta \times 445^2)$ ...(iii)

From Eq. (i), we have

$$R_0 = 10\,\Omega$$

Substituting R_0 in Eqs. (ii) and (iii), we have

$$0.6 = 100\,\alpha + (100)^2 \cdot \beta \quad \text{...(iv)}$$

and $\quad 2 = 445\alpha + (445)^2 \cdot \beta$ ...(v)

Solving Eqs. (iv) and (v), we get

$$\alpha = 0.0064 = 0.64 \times 10^{-2}\ (^\circ\text{C}^{-1})$$

and $\quad \beta = -0.04 \times 10^{-4}\ (^\circ\text{C}^{-2})$

And when temperature is 500°C,

$$R = R_0(1 + \alpha t + \beta t^2)$$
$$= 10\,[1 + 0.64 \times 10^{-2} \times 500 - 0.04 \times 10^{-4} \times (500)^2]$$
$$= 10\,(1 + 3.2 - 1) = 32\,\Omega$$

2. Ratio of BP of sulphur and water is found as :

H_2O (BP) p_W	S_8 (BP) p_S	Ratio $\frac{p_S}{p_W}$
50	96.4	1.928
100	193	1.93
200	387	1.935
300	582	1.940

So, average value of

$$\frac{p_S}{p_W} = \frac{p_2}{p_1} = 1.928$$

As, pressure is thermodynamic property used to measure temperature.

So, $\quad p \propto T$

or $\quad \frac{p_1}{p_2} = \frac{T_1}{T_2}$ or $T_2 = \frac{p_2}{p_1} \times T_1$

Taking, T_1 = BP of water = 100°C = 373 K

and T_2 = BP of sulphur at 1 atm, we have

$\Rightarrow \quad T_2 = 373 \times 1.928 = 718\text{ K} = 445°\text{C}$

So, BP of sulphur is 445°C.

3. (i) $T_1 = 2\pi\sqrt{\frac{l_1}{g}}$ at 17°C and $T_2 = 2\pi\sqrt{\frac{l_2}{g}}$ at 28°C

As, $l_2 > l_1 \quad \therefore T_2 > T_1$

So, $$\frac{\Delta T}{T_1} = \frac{2\pi\sqrt{\frac{l_2}{g}} - 2\pi\sqrt{\frac{l_1}{g}}}{2\pi\sqrt{\frac{l_1}{g}}} = \frac{\sqrt{l_2} - \sqrt{l_1}}{\sqrt{l_1}}$$

$$= \frac{\sqrt{l_1 + \alpha l_1\,\Delta T} - \sqrt{l_1}}{\sqrt{l_1}} = (\sqrt{1 + \alpha \Delta T} - 1) = 1.04 \times 10^{-4}$$

$\therefore \quad \Delta T = (1.04 \times 10^{-4})T_1$

For one year,

$$\Delta T = 1.04 \times 10^{-4} \times 3.16 \times 10^7\text{ s} \approx 3286\text{ s} \approx 55\text{ min}$$

(ii) Change in radius due to heating does not cause any torque. So, angular momentum remains constant. Also, moment of inertia of cylinder about its own axis is

$$I = \frac{1}{2}mr^2$$

So, $\quad L_i = L_f$

$\Rightarrow \quad I_1\omega_1 = I_2\omega_2$

$$\Rightarrow \quad \omega_2 = \frac{I_1\omega_1}{I_2} = \frac{\frac{1}{2}mr_1^2\omega_1}{\frac{1}{2}mr_2^2} \Rightarrow \omega_2 = \frac{r_1^2\omega_1^2}{r_2^2}$$

So, $$\frac{\Delta\omega}{\omega_1} = \frac{\omega_2 - \omega_1}{\omega_1} = \frac{\left(\frac{r_1^2\omega_1}{r_2^2} - \omega_1\right)}{\omega_1}$$

$$= \frac{r_1^2}{r_2^2} - 1 = \frac{r_1^2}{(r_1 + \alpha r_1\Delta T)^2} - 1$$

$$= -\alpha\Delta T \cdot \left(\frac{2 + \alpha\Delta T}{(1 + \alpha\Delta T)^2}\right)$$

As, $\quad \alpha\Delta T << 1$

So, $\quad \frac{\Delta\omega}{\omega_1} = -2\alpha\Delta T$

Hence, percentage change in the angular speed per degree rise of temperature is

$$-\left(\frac{\Delta\omega}{\omega_1} \times 100\right) \times \frac{1}{\Delta T} = -2\alpha \times 100$$
$$= -2 \times 25 \times 10^{-6} \times 100$$
$$= -50 \times 10^{-4}$$
$$= -(5 \times 10^{-3})\%$$

Negative sign shows a reduction in angular speed.

4. To keep the aluminium rod of same length, the stress must compensate for the thermal expansion.

$\therefore \quad \text{Stress} = \frac{F}{A} = \alpha Y\Delta T$

$$= 25 \times 10^{-6} \times 70 \times 10^9 \times (35 - 18)$$
$$= 3 \times 10^7\text{ Nm}^{-2}$$

5. (d) For all the scales,

$$\frac{X - \text{LFP}}{\text{UFP} - \text{LFP}} = a\ (\text{constant})$$

6. (c) As, $\frac{C - 0}{100 - 0} = \frac{N - 100}{400 - 100}$

$$\Rightarrow \quad \frac{x}{100} = \frac{x - 100}{300}$$

$$\Rightarrow \quad x = \frac{x - 100}{3} \Rightarrow x = -50°\text{C}$$

7. (b) From graph,

For scale A, LFP = 30° A, UFP = 180° A

For scale B, LFP = 0° B, UFP = 100° B

So, at any temperature, $t_A = t_B$

$$\frac{t_A - \text{LFP}}{\text{UFP} - \text{LFP}} = \frac{t_B - \text{LFP}}{\text{UFP} - \text{LFP}}$$

$$\frac{t_A - 30}{150} = \frac{t_B}{100}$$

8. (a) When a substance is heated, KE of molecules increases causing an increase in vibrations randomly.

9. (d) As aluminium expands more

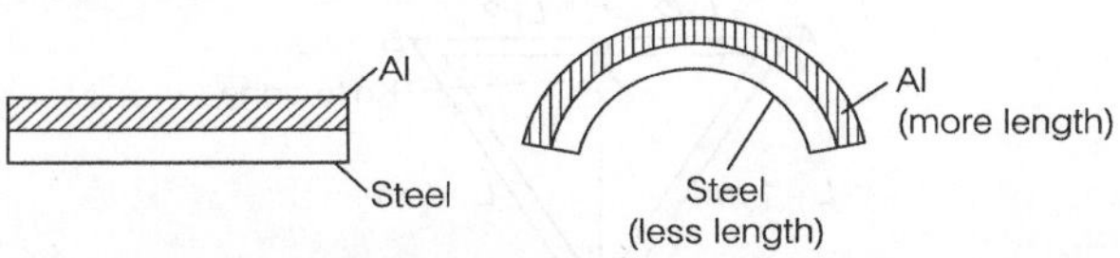

$\therefore$ Strip will bend with steel on concave side.

10. (a) When bob is heated, its volume increases.

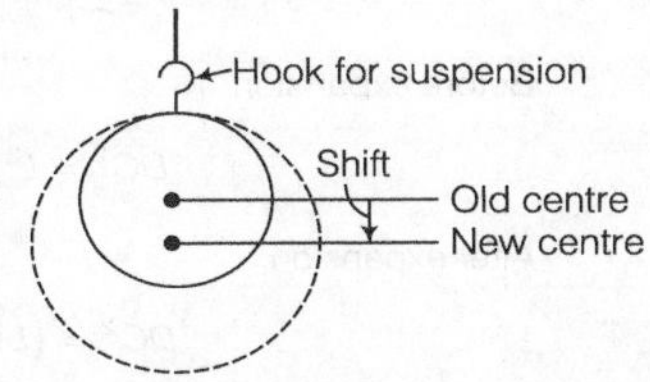

So, its centre of mass shifts downwards (but it is still at centre of bob) and effective length increases.

As, $T \propto \sqrt{l}$

$\therefore$ T also increases.

11. (a) As, surface to volume ratio for small spheres is more than that of large sphere. So, option (a) is correct. For dics (pizzas), there is no much variation in the ratio with volume.

12. (c) $I = \dfrac{1}{12} Ml^2$ so, $I' = \dfrac{1}{12} M(l + \Delta l)^2$

$\Rightarrow \quad I' = \dfrac{1}{12}(Ml^2 + 2Ml\Delta l + M(\Delta l)^2 \alpha)$

$I' \approx I + \dfrac{1}{12} Ml^2 2\alpha\Delta T$

$\therefore \quad I' - I = 2\alpha I\Delta T$

13. (a) It is required that $l_{iron} - l_{brass} = 10$ cm at all temperatures.

$\Rightarrow \quad l_{iron} - l_{brass} = 10$ cm $\quad \ldots$(i)

$\Rightarrow \quad l_{iron}(1 + \alpha_{iron}\Delta t) - l_{brass}(1 + \alpha_{brass}\Delta t) = 10$

Also, $\quad l_{iron}\,\alpha_{iron} = l_{brass}\,\alpha_{brass}$

$\Rightarrow \quad \dfrac{l_{iron}}{l_{brass}} = \dfrac{1.8}{1.2} = \dfrac{3}{2} \quad \ldots$(ii)

Solving Eqs. (i) and (ii), we have

$l_{brass} = 20$ cm, $l_{iron} = 30$ cm

14. (d) For a faulty thermometer, if reading is y whereas current reading is x then,

$$\frac{x - (\text{LFP})_{\text{correct scale}}}{(\text{UFP} - \text{LFP})_{\text{correct scale}}} = \frac{y - (\text{LFP})_{\text{incorrect scale}}}{(\text{UFP} - \text{LFP})_{\text{incorrect scale}}}$$

That means we can view them as two different scales.

So, $\quad \dfrac{60 - (-10)}{\text{UFP} - \text{LFP}} = \dfrac{50 - 0}{100 - 0}$

$\Rightarrow \quad (\text{UFP} - \text{LFP})_{\text{incorrect scale}} = 2 \times 70 = 140$

$\Rightarrow \quad y - (-10) = 140$

$\Rightarrow \quad y = 140 - 10 = 130°$ C

15. (b) $\gamma = \dfrac{\Delta V}{V\Delta T} = \dfrac{0.12}{100} \times \dfrac{1}{20} = \dfrac{0.0012}{20} = 6 \times 10^{-5}\,°\text{C}^{-1}$

So, $\quad \alpha = \dfrac{\gamma}{3} = \dfrac{6 \times 10^{-5}}{3} = 2 \times 10^{-5}\,°\text{C}^{-1}$

16. (a,b,c,d) We can think of atoms/molecules in a lattice structure of metal as balls connected with springs.

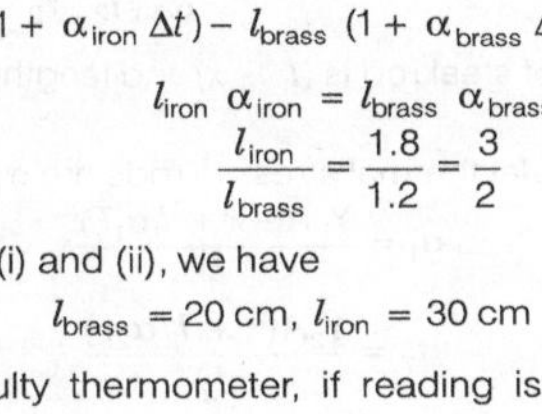

Heat is absorbed in form of PE and KE both. Result is increased in interatomic separation and vibrational amplitudes.

17. (a) $L_2 = L_1(1 + \alpha\Delta T)$

$5.050 = 5(1 + 24 \times 10^{-6} \times \Delta T) \Rightarrow T = 437°\text{C}$

For options (b) and (d), temperature below $-273°$C is required which is unattainable.

For option (c), temperature of 3000°C is required which is above melting point of aluminium.

18. (a,b,c) When α is a constant, $dL = \alpha L\Delta T$

$\Rightarrow \quad \dfrac{dL}{L} = \alpha\Delta T \Rightarrow [\log L]_{L_1}^{L_2} = [\alpha T]_{T_1}^{T_2}$

$\Rightarrow \quad \log\dfrac{L_2}{L_1} = \alpha(T_2 - T_1) \Rightarrow L_2 = L_1 e^{\alpha(T_2 - T_1)}$

When α is a function of temperature,

$\int_{L_1}^{L_2} dL = \int_{T_1}^{T_2} \alpha dT \Rightarrow \log\dfrac{L_2}{L_1} = \int_1^2 \alpha dT$

$\Rightarrow \quad L_2 = L_1 e^{\int_{T_1}^{T_2} \alpha dT}$

Also, option (a) is correct and option (d) is incorrect.

19. (a,d) Force on steel = Force on copper

$\sigma_{steel}A_{steel} + \sigma_{Cu}A_{Cu} = 0$

$\sigma_{steel} \cdot \dfrac{\pi}{4}d_{steel}^2 = -\sigma_{Cu}\dfrac{\pi}{4}\cdot(d_2^2 - d_1^2)$

$\sigma_{Cu} = \dfrac{\left(\dfrac{d}{2}\right)^2}{d^2 - \left(\dfrac{d}{2}\right)^2} \cdot \sigma_{steel} \Rightarrow \sigma_{cu} = \dfrac{-\left(\dfrac{1}{4}\right)}{1 - \left(\dfrac{1}{4}\right)} \cdot \sigma_{steel}$

$\sigma_{Cu} = -\dfrac{1}{3}\sigma_{steel}$

Also, strain in copper tube $= \dfrac{\sigma_{Cu}}{Y_{Cu}} + \alpha_{Cu}\Delta T$

20. (b,d) Let L_0 is length of each strip before heating.

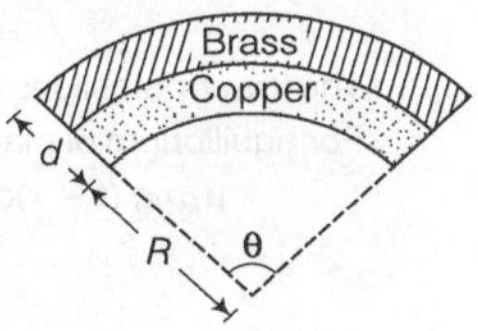

Length after heating will be

$L_{brass} = L_0(1 + \alpha_{brass}\Delta T)$

$= (R + d)\cdot\theta$

and

$L_{copper} = L_0(1 + \alpha_{copper}\Delta T) = R\theta$

$\Rightarrow \quad \dfrac{R + d}{R} = \dfrac{1 + \alpha_{brass}\Delta T}{1 + \alpha_{copper}\Delta T}$

$\Rightarrow \quad R = \dfrac{d}{(\alpha_{brass} - \alpha_{copper})\Delta T}$

$\therefore \quad R \propto \dfrac{1}{\Delta T}$ and $R \propto \dfrac{1}{|\alpha_{brass} - \alpha_{copper}|}$

21. (d) Statement I is incorrect but Statement II is correct.

Percentage increase in volume of sphere is largest.

22. (a) Both Statement I and Statement II are correct and Statement II is correct explanation of Statement I.

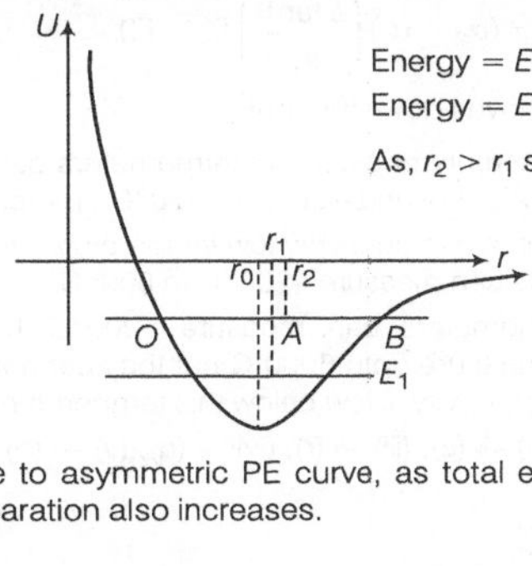

Energy $= E_1$, mean separation $= r_1$

Energy $= E_2$ mean separation $= r_2$

As, $r_2 > r_1$ so expansion occurs.

Due to asymmetric PE curve, as total energy increases, mean separation also increases.

23. (a) Let stationary line is at a distance of xL from top edge, when temperature is raising.

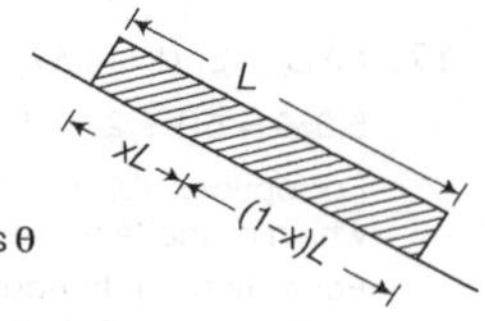

Normal force on plate on lower part

$$= mg(1-x)\cos\theta$$

and force of friction acting upon the roof is

$$f_{k_2} = \mu_k mg(1-x)\cos\theta$$

So, force of friction acting on upper part of plate down the roof is

$$f_{k_1} = \mu_k mgx\cos\theta$$

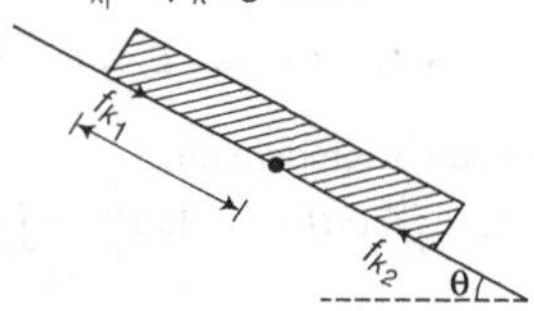

For equilibrium of plate,

$$-\mu_k mgx\cos\theta + \mu_k mg(1-x)\cos\theta = mg\sin\theta$$

$\Rightarrow$ $$x = \frac{1}{2} - \frac{\tan\theta}{2\mu_k}$$

So, stationary line is at a distance of

$$xL = \frac{L}{2}\left(1 - \frac{\tan\theta}{\mu_k}\right)$$ from top edge of plate.

24. (a) With temperature falling plate contracts more than roof.

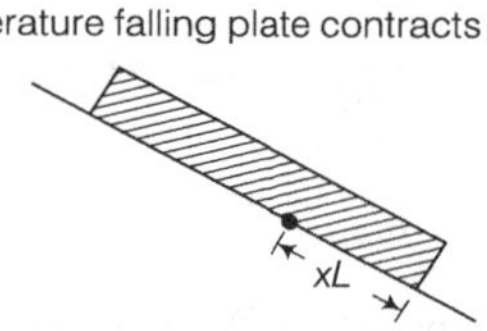

Upper part slides down and feels an upward frictional force

$$= \mu_k mg(1-x)\cos\theta$$

The lower part slides up and feels a force $= \mu_k mgx\cos\theta$

For equilibrium of plate,

$$\mu_k mg(1-x)\cos\theta - \mu_k mgx\cos\theta = mg\sin\theta$$

$\Rightarrow$ $$xL = \frac{L}{2}\left(1 - \frac{\tan\theta}{\mu_k}\right)$$

25. (c) As temperature starts to rise, stationary line moves down the roof due to expansion of plate.

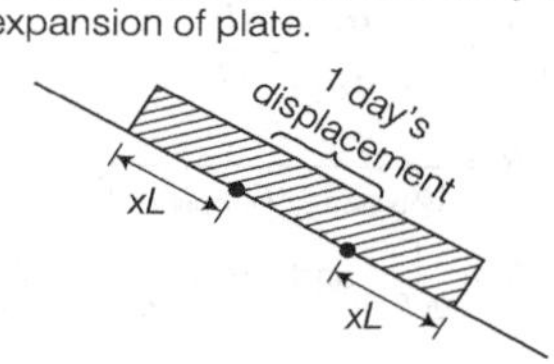

Its displacement is given by

$$\Delta L = (\alpha_2 - \alpha_1)(L - xL - xL)\Delta T$$

$$= (\Delta T)(\alpha_2 - \alpha_1)\left[L - 2\frac{L}{2}\left(1 - \frac{\tan\theta}{\mu_k}\right)\right]$$

$$= (\alpha_2 - \alpha_1)\left(\frac{L\tan\theta}{\mu_k}\right)(T_1 - T_2)$$

In this way plate moves down the roof.

26. (a) Liquid filled (mercury or alcohol) thermometers can measure –80°C to 350°C : Alcohol and – 50°C to 350°C : mercury.

As, gas can remain a gas upto very low temperature : the gas filled thermometers can measure –200° C to 800° C.

Resistance thermometers can measure –200°C to 1200°C. Pyrometers can measure upto 6000°C but they cannot measure below 800°C as emissivity is low below this temperature.

Hence, (i) → (t), (ii) → (s), (iii) → (r), (iv) → (q), (v) → (p)

27. (4) Given arrangement is

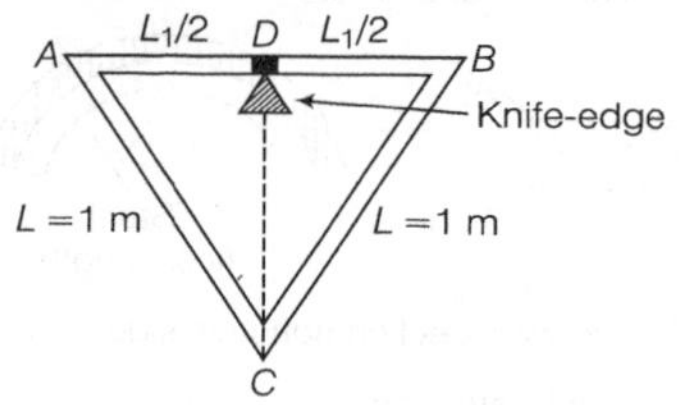

Before expansion,

$$DC^2 = L^2 - \left(\frac{L_1}{2}\right)^2$$

After expansion,

$$DC^2 = \{L(1+\alpha\Delta T)\}^2 - \left\{\frac{L_1}{2}(1+\alpha_1\Delta T)\right\}^2$$

As vertex C remains at same distance from D, equating both values of DC, we have

$$L^2 + \left(\frac{L_1}{2}\right)^2 = \{L(1+\alpha\Delta T)\}^2 - \left\{\frac{L_1}{2}(1+\alpha_1\Delta T)\right\}^2$$

$\Rightarrow$ $$\frac{L_1}{L} = 2\sqrt{\frac{\alpha}{\alpha_1}} = 2\sqrt{\frac{2.4\times10^{-5}}{6\times10^{-6}}} = 2\sqrt{4} = 4$$

28. (6) If allowed to expand the rods increase in length by $l_1\alpha_1 T$ and $l_2\alpha_2 T$. Because of walls, there is no free expansion. If one expands other, the other would compress. Let the junction shift by x towards steel side (towards left).

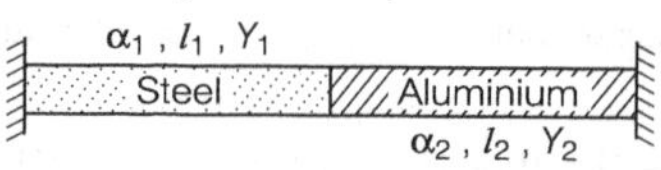

Then, length of steel rod is $(l_1 - x)$ and length of aluminium rod is $(l_2 + x)$.

In equilibrium, let thermal stress in rods are σ_1 and σ_2.

Then, $$\sigma_1 = \frac{Y_1A(-x + l_1\alpha_1 T)}{l_1}$$

and $$\sigma_2 = \frac{Y_2A(x + l_2\alpha_2 T)}{l_2}$$

At the joint in equilibrium, forces due to thermal stress balances each other.

$\Rightarrow$ $$F_1 = F_2 \Rightarrow \frac{F_1}{A} = \frac{F_2}{A}$$

$\Rightarrow$ $$\sigma_1 = \sigma_2$$

$\Rightarrow$ $$\frac{Y_1A(-x + l_1\alpha_1 T)}{l_1} = \frac{Y_2A(x + l_2\alpha_2 T)}{l_2}$$

$\Rightarrow$ $$x(Y_2 l_1 + Y_1 l_2) = l_1 l_2(Y_1\alpha_1 - Y_2\alpha_2)T$$

$\Rightarrow$ $$x = \frac{l_1 l_2(Y_1\alpha_1 - Y_2\alpha_2)\cdot T}{Y_2 l_1 + Y_1 l_2}$$

Substituting the values, we get

$$x = \frac{\left[\begin{array}{r}100\times7\times20\,(1.2\times10^{-5}\times20\times10^{10} - 2.4\\ \times10^{-5}\times7\times10^{10})\end{array}\right]}{(20\times7\times10^{10} + 7\times20\times10^{10})}$$

$$= \frac{100\times7\times20\times10^4\,(240-168)}{2\times20\times7\times10^{10}}$$

$\Rightarrow$ $$x = 36\times10^{-4}\text{ m}$$

$$= 36\times10^{-4}\times10^3\text{ mm}$$

$$= 3.6\text{ mm}$$

$\therefore$ $$\frac{N^2}{10} = 3.6$$

$\Rightarrow$ $$N^2 = 36$$

$\Rightarrow$ $$N = 6$$

DPP-2 Expansion of Liquids and Gases

1. (i) For a gas,

$$\frac{p_iV_i}{T_i} = \frac{p_fV_f}{T_f}$$

As, volume remains same

so, $\frac{p_i}{T_i} = \frac{p_f}{T_f} \Rightarrow p_f = \frac{T_f}{T_i} \cdot p_i$

$\Rightarrow$ $p_f = \frac{468}{295} \times 202 \text{ kPa} = 320 \text{ kPa}$

(ii) Change in volume of container,

$$\Delta V = \gamma V_i \Delta T$$
$$= 3\alpha \cdot V_i \cdot \Delta T$$
$$= 3 \times 11 \times 10^{-6} \times 125 \times 173$$
$$= 0.71 \text{ cm}^3$$

$\therefore$ p_f = final pressure inside the container $= \frac{T_f}{T_i} \cdot \frac{V_i}{V_f} \cdot p_i$

As, $\frac{V_i}{V_f} = \frac{125}{125.71} = 0.994$

So, final pressure reading differs only about 6%.

As, $p_f = \frac{T_f}{T_i} \cdot \frac{V_i}{V_f} \cdot p_i$

$$= \frac{468}{295} \times \frac{125}{125.71} \times 202$$
$$= 318 \text{ kPa}$$

$\therefore$ Change in pressure value calculated $= 320 - 318$
$= 2$ kPa

2. (i) $\Delta V = V_f r_f \Delta T - V_{Al} 3\alpha_{Al} \Delta T = (r_f - 3\alpha_{Al}) V_i \Delta T$

$$= (9 \times 10^{-4} - 0.720 \times 10^{-4})(2000)(60)$$
$$= 99.4 \text{ cm}^3$$

(ii) New volume,

$= 20000 + 9 \times 10^{-4} \times 2000 \times 60 = 2108 \text{ cm}^3$

So, fraction lost is

$$\frac{99.4 \text{ cm}^3}{2108 \text{ cm}^3} = 4.71 \times 10^{-2}$$

This fraction of cylinder will be empty on cooling.

Empty part of cylinder $= 4.71 \times 10^{-2} \times 20$

$= 0.943$ cm

3. (i) $pV = nRT$

$\Rightarrow$ $n = \frac{pV}{RT} \Rightarrow \frac{m}{M} = \frac{pV}{RT}$

$\Rightarrow$ $m = \frac{pVM}{RT}$

$\therefore$ Mass of helium initially inside the balloon is

$$m = \frac{1.013 \times 10^5 \times \frac{4}{3}\pi(6.37 \times 10^6)^3 \times 4 \times 10^{-3}}{8.314 \times 50}$$

$= 1.06 \times 10^{21}$ kg

(ii) $\frac{p_fV_f}{T_f} = n_fR$ and $\frac{p_iV_i}{T_i} = n_iR$

$\therefore$ $\frac{p_fV_f}{p_iV_i} = \frac{n_fRT_f}{n_iRT_i}$

$\Rightarrow$ $2.1 = \left(\frac{1.06 \times 10^{21} + 8 \times 10^{20}}{1.06 \times 10^{21}}\right) \frac{T_f}{50}$

$\Rightarrow$ $T_f = 59.8$ K

4. For the balloon, $\Sigma F_y = 0$

$\Rightarrow$ $\rho_{out} gV - \rho_{in} gV - 200\, g = 0$

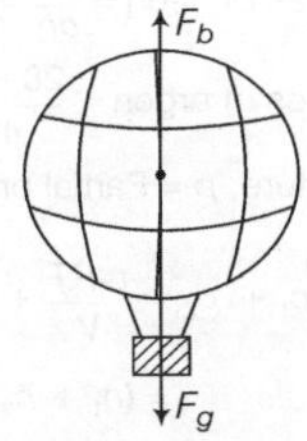

$\Rightarrow$ $(\rho_{out} - \rho_{in})\, 400 = 200$ ...(i)

Air density outside $= 1.25 \text{ kg m}^{-3}$

From $pV = nRT$,

$$\frac{n}{V} = \frac{p}{RT}$$

And $\rho_{\text{hot air}} = 1.25 \times \frac{283}{T_{in}}$

From Eq. (i), we get

$$1.25 \times \left(1 - \frac{283}{T_{in}}\right)(400) = 200$$

$\Rightarrow$ $1 - \frac{283}{T_{in}} = 0.4$

$\Rightarrow$ $T_{in} = 472$ K

5. (b) $\Delta V = V(\gamma_L - \gamma_S)\Delta T$

$\Rightarrow$ $\Delta V = 1000\,(1.82 - 0.3) \times 10^{-4} \times 100$

$= 15.2$ cc

6. (c) The flask expands first and then fluid gets heated. Thus, level of fluid initially falls, then increases.

7. (b) As, expansion of ball is less than that of liquid, we may neglect expansion of the ball.

So, when ball is immersed in liquid at 0°C, it displaces some volume V of fluid and

$$w_1 = w - V\rho_0 g = \text{True weight} - \text{Upthrust}$$

and $w_2 = w - V\rho_{60} g$

As, $\rho_{60} < \rho_0$

$\Rightarrow$ $w_2 > w_1$

8. (a) $k_1 = \left(\frac{\rho_1}{\rho_2}\right)_{0° C}$

$$k_2 = \left(\frac{\rho_1}{\rho_2}\right)_{60° C}$$

$\therefore$ $\frac{k_1}{k_2} = \left(\frac{\rho_1}{\rho_2}\right)_{0° C} \times \left(\frac{\rho_2}{\rho_1}\right)_{67° C}$

$$= \frac{1 + 60\gamma_1}{1 + 60\gamma_2}$$

9. (d) At constant pressure $\frac{V}{T}$ = constant (As, no change in piston cylinder arrangement is made, pressure over piston remains same).

$\therefore$ $\frac{V_2}{V_1} = \frac{T_2}{T_1}$

$\Rightarrow$ $\frac{Ah_2}{Ah_1} = \frac{T_2}{T_1}$

$\Rightarrow$ $h_2 = h_1\left(\frac{T_2}{T_1}\right) = 1.0 \times \frac{400}{300} = \frac{4}{3}$ m

10. (a,b) $T = 27°C = 300$ K,

Let m = mass of neon gas in the mixture, then mass of argon $= 28 - m$

Number of gram moles of neon $= \frac{m}{20} = n_1$

Number of gram moles of argon $= \frac{28-m}{40} = n_2$

Total pressure of mixture, p = Partial pressure of neon + partial pressure of argon

$$\therefore \quad p = p_1 + p_2 = \frac{n_1RT}{V} + \frac{n_2RT}{V} = (n_1+n_2)\frac{RT}{V}$$

Substituting the given values, we have

$$1\times10^5 = \left(\frac{m}{20} + \frac{28-m}{40}\right)\left(\frac{8.314\times300}{0.02}\right)$$

$$\Rightarrow \quad m = 4.074 \text{ g}$$

So, mass of neon = 4.074 g and mass of argon $= 28 - m = 23.926$ g

11. (a) pT^3 = constant

$$\frac{nRT}{V}\cdot T^3 = \text{constant}$$

$$\Rightarrow \quad T^4V^{-1} = \text{constant}$$

$$\Rightarrow \quad 4T^3V^{-1} - T^4\cdot V^{-2}\frac{dV}{dT} = 0$$

$$\Rightarrow \quad \frac{4T^3}{V} = \frac{T^4}{V^2}\cdot\frac{dV}{dT} \Rightarrow \frac{4}{T} = \frac{dV}{V\cdot dT}$$

So, $\gamma = \frac{dV}{V\cdot dT}$ = volumetric expansion coefficient of gas $= \frac{4}{T}$.

12. (a,d) As, $\gamma_{Hg} > \gamma_{Fe}$, density of mercury decreases more on heating so, Iron cube will sink more on heating.

Also, fraction of volume submerged $= \frac{V_{out}}{V_{cube}}$

Fractional change on heating is given by

$$\frac{(V_0/V_c)_{25°C} - (V_0/V_c)_{0°C}}{(V_0/V_c)_{0°C}} = \frac{(1+180\times10^{-6})\times25}{(1+35\times10^{-6})\times25} - 1 = 3.6\times10^{-3}$$

$$\therefore \quad \% \text{ Change} = 0.36\% \approx 0.40\%$$

13. (a,c) Density at 0°C $= \rho_0 = \frac{M}{V_0}$

Density at 35°C $= \rho = \frac{M}{V} = \frac{M}{V_0(1+r\Delta T)}$

$$= \frac{\rho_0}{(1+r\Delta T)} = \frac{0.68\times10^3}{1+950\times10^{-6}\times35} = 0.6581\times10^3 \approx 660 \text{ kg m}^{-3}$$

% change in density $= \frac{0.6581-0.68}{0.68}\times100 = -3\%$

14. (a,b,c,d) When water is heated to 277 K, hydrogen bonds are formed which causes it anamolous behaviour.

15. (d) Let A = cross-sectional area of tube.

Let pressure initially on each side of tube is atmospheric.

When tube is held in vertical position, let x = displacement of mercury column.

Then, pV = constant is applied to top part and bottom part.

$$p\times45\times A = p_1(45-x)A$$

and $$p\times45\times A = p_2(45+x)A$$

$$p_1 = \frac{45}{45-x}\cdot p, \quad p_2 = \frac{45}{45+x}\cdot p$$

Also, $p_1 - p_2 = 10$ cm of mercury column

$$\Rightarrow \quad \left(\frac{45}{45-x}\right)p - \left(\frac{45}{45+x}\right)p = 10$$

where, p = 76 cm of mercury column

$$\therefore \quad x = 2.95 \text{ cm} \approx 3 \text{ cm}$$

16. (b)

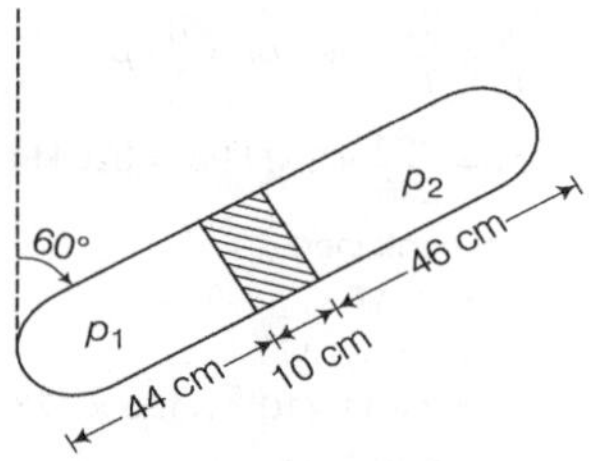

$p_1 - p_2$ = pressure due to 10 cm slanted mercury column

$$= 10\cos60° = \frac{10}{2} = 5 \text{ cm of mercury column}$$

pV = constant for each side of tube

$$\therefore \quad p\times45\times A = p_1\times44\times A$$

and $$p\times45\times A = p_2\times46\times A$$

$$p_1 = \frac{45p}{44}, \quad p_2 = \frac{45p}{46}$$

$$p_1 - p_2 = \frac{45p}{44} - \frac{45p}{46} = 5 \text{ cm of Hg column}$$

$$\Rightarrow \quad p = 112.4 \text{ cm of Hg column}$$

17. (d) Using pV^γ = constant and $\frac{pV}{T}$ = constant

We have, $TV^{\gamma-1}$ = constant

$$\therefore \quad \frac{T_1}{T_2} = \left(\frac{V_2}{V_1}\right)^{\gamma-1} = \left(\frac{AL_2}{AL_1}\right)^{\gamma-1}$$

Given, $\gamma = \frac{5}{3}$

$$\therefore \quad \frac{T_1}{T_2} = \left(\frac{L_2}{L_1}\right)^{2/3}$$

18. (a) Combining pV^γ = constant and $\frac{pV}{T}$ = constant, we have

$$TV^{\gamma-1} = \text{constant}$$

$$\therefore \quad T_2V_2^{\gamma-1} = T_1V_1^{\gamma-1} \Rightarrow T_2 = T_1\left(\frac{V_1}{V_2}\right)^{\gamma-1}$$

$$T_2 = T_1\left(\frac{5.6}{0.7}\right)^{\frac{5}{3}-1} = T_1(8)^{\frac{2}{3}}$$

$$\Rightarrow \quad T_2 = 4T_1$$

19. (a) At any instant, forces on rod are

(i) Gravitational pull $= mg$

(ii) Buoyant force = weight of fluid displaced $= \rho Axg$

where, A = area of cross-section of rod and x = depth inside fluid

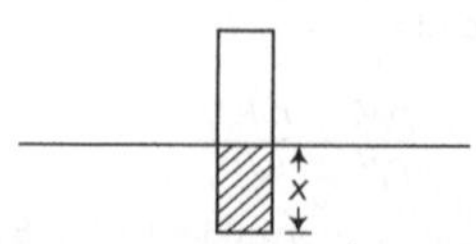

$\therefore$ Acceleration of rod as a function of x

$$= \frac{\text{Net force}}{\text{Mass of rod}} = \frac{mg - \rho Axg}{m} = g - \frac{\rho Axg}{2\rho AL} = g - \frac{gx}{2L}$$

20. (c) As, acceleration of rod, $a = g\left(1 - \frac{x}{2L}\right)$

We have, $v\frac{dv}{dx} = g\left(1 - \frac{x}{2L}\right) \quad \Rightarrow \quad v\,dv = g\left(1 - \frac{x}{2L}\right)dx$

Integrating, we have, $\int_0^v v\,dv = \int_0^{L/2} g\left(1 - \frac{x}{2L}\right)dx$

$$\Rightarrow \quad v_1 = \sqrt{\frac{7}{8}gL} \quad \ldots(i)$$

When fluid is heated, its mass remains same.

So, $M_i = M_f$

$\rho_1 V_1 = \rho_2 V_2$

As, $V_2 = \left(1 + \frac{1}{4}\right)V_1 = \frac{5}{4}V_1$

we have, $\rho_2 = \frac{4}{5}\rho_1$

So, acceleration of rod becomes,

$$a = g - \frac{\rho' Axg}{2\rho AL} \quad \text{where } \rho' = \rho_2 = \frac{4}{5}\rho$$

So, $$a = g - \frac{\frac{4}{5}\rho Axg}{2\rho AL} \Rightarrow a = g - \frac{2gx}{5L}$$

$$\Rightarrow \int_0^v v\,dv = \int_0^{L/2}\left(g - \frac{2gx}{5L}\right)dx = \left(gx - \frac{2g}{5L}\cdot\frac{x^2}{2}\right)\Bigg|_0^{L/2}$$

$$= g\frac{L}{2} - \frac{g}{5L}\cdot\frac{L^2}{4} = gL\left(\frac{1}{2} - \frac{1}{20}\right)$$

$$\frac{v^2}{2} = gL \times \frac{9}{20} \Rightarrow v^2 = \frac{9gL}{10}$$

$$\Rightarrow \quad v_2 = \sqrt{\frac{9}{10}gL}$$

So, $$\frac{v_2}{v_1} = \sqrt{\frac{9gL}{10}} \times \sqrt{\frac{8}{7gL}} = \sqrt{\frac{9 \times 8}{10 \times 7}} = \sqrt{\frac{36}{35}} = \frac{6}{\sqrt{35}}$$

21. (c) When temperature increases, mean separation between atoms of *A* increases while of *B* decreases (see graph) so, a expands while *B* contracts. At temperature $> T_2$, mean separation for *A* is very large and for *B*, mean separation is excessive above temperature T_1.

So, above T_1, both *A* and *B* liquifes but above T_2, only *A* liquifies.

Hence, (i) → (p), (ii) → (q), (iii) → (p,q), (iv) → (p)

22. (d) Statement I is incorrect but Statement II is correct.

Given, V_1 = volume at T_1,

V_2 = volume at T_2,

After mixing , temperature of mixtrue is

$$T = \frac{m_1T_1 + m_2T_2}{m_1 + m_2}$$

Volumes at this temperature are

$V_{0_1}(1 + \gamma T)$ and $V_{0_2}(1 + \gamma T)$

where, γ = coefficient of volumetric expansion, V_{0_1} and V_{0_2} are volumes at 0°C.

Sum of volumes after mixing $= V_{0_1}(1 + \gamma T) + V_{0_2}(1 + \gamma T)$

$= V_{0_1} + V_{0_2} + \gamma(V_{0_1} + V_{0_2})T$

$= V_{0_1} + V_{0_2} + \gamma(V_{0_1} + V_{0_2})\left(\frac{m_1T_1 + m_2T_2}{m_1 + m_2}\right)$

$= V_{0_1} + V_{0_2} + \gamma\left(\frac{m_1 + m_2}{d}\right)\cdot\left(\frac{m_1T_1 + m_2T_2}{m_1 + m_2}\right)$

$\left[\text{where } V_{0_1} = \frac{m_1}{d}, V_{0_2} = \frac{m_2}{d}, d = \text{density}\right]$

$= V_{0_1} + V_{0_2} + \gamma\left(\frac{m_1T_1}{d}\right) + \gamma\left(\frac{m_2T_2}{d}\right)$

$= V_{0_1} + \gamma(V_{0_1})T + V_{0_2} + \gamma(V_{0_2})T_2$

$= V_{0_1}(1 + \gamma T_1) + V_{0_2}(1 + \gamma T_2)$

$= V_1 + V_2$

= a constant

DPP-3 Calorimetry

1. Energy is absorbed at a constant rate *P*. For the period from 50 min to 60 min, $Q = mC\,\Delta T$.

$\Rightarrow \quad P(10\text{min}) = (10\text{ kg} + m_{ice})(4186)(20 - 0)$

$\Rightarrow \quad P(10\text{min}) = 83.7 + (8.37)\,m_{ice} \quad \ldots(i)$

From 0 to 50 min, $Q = m_{ice}\cdot L_f$

$\Rightarrow \quad P(50\text{min}) = m_{ice}(3.33 \times 10^5) \quad \ldots(ii)$

From Eqs. (i) and (ii), we have

$$\frac{m_{ice}(3.33 \times 10^5)}{50} = \frac{83.7 + (8.37)\,m_{ice}}{10}$$

$\Rightarrow \quad m_{ice} = 1.44$ kg

2. Loss of mechanical energy $= \frac{1}{2}mv^2 + \frac{GMm}{R}$

$= \frac{1}{2} \times 670 \times (1.4 \times 10^3)^2 + \frac{6.67 \times 10^{-11} \times 5.98 \times 10^{24} \times 670}{6.37 \times 10^6}$

$= 1.08 \times 10^{11}$ J

Half of this energy is received by aluminium.

Then, $(1.08 \times 10^{10}) \times \frac{1}{2} = mC\,\Delta T + mL + mC(T_f - T_{bp})$

$\Rightarrow \quad (1.08 \times 10^{10}) \times \frac{1}{2} = 9.71 \times 10^9 + 670\,(1170)(T_f - 2450)$

$\Rightarrow \quad T_f = 5.87 \times 10^4\,°\text{C}$

3. Resistance of kettle is $R_1 = \frac{V_1^2}{P_1}$ and new power on 200 V mains is $P = \frac{V^2}{R_1}$, as resistance remains same.

$\Rightarrow \quad P = \left(\frac{V^2}{V_1^2}\right)\cdot P_1$

$\Rightarrow \quad P = \left(\frac{200}{240}\right)^2 \times 750 = 520$ W

So, heat supplied to water = 520 J in one second.

Heat gained by water and kettle $= (C\,\Delta T)_{kettle} + (mC_w\Delta T)_{water}$

$= 400\,(100 - 20) + 500 \times 4.2 \times (100 - 20)$

$= 2500 \times 80$ J

So, time required $= \frac{2500 \times 80}{520} = 385\text{ s} = 6.4$ min

4. Heat gained by metal target

= 0.98 × KE of *n*-electrons per second × time

$\Rightarrow \quad mC\,\Delta T = 0.98 \times \frac{n}{2}mv^2 \times t$

$\Rightarrow \quad t = \frac{mC\,\Delta T}{0.98 \times \frac{n}{2}mv^2} = \frac{200 \times 0.5 \times (50)}{\frac{1}{2}(10^{18} \times 9.8 \times 10^{-31}) \times (2 \times 10^6)^2}$

$= \frac{50000}{1.8} = 2780\text{ s} = 46.3$ min

5. (d) Specific heat of water is 4186 $\text{Jkg}^{-1}\text{K}^{-1}$, which is quite high compared to other solids and liquids. So, water requires more heat to get hot and it rejects more heat in getting cold.

6. (d) Hydrogen has the largest specific heat capacity (not molar specific heat capacity). Its value is 14400 $\text{Jkg}^{-1}\text{K}^{-1}$.

7. (b) Heat required to melt ice

$= mL = 5g \times 80\ \text{calg}^{-1} = 400$ cal

Heat required to raise the temperature of 5 g water from 0°C to 100°C

$= ms\,\Delta t = 5 \times 1 \times 100 = 500$ cal

Heat rejected by condensation of steam (1 g) $= mL$

$= 1 \times 540 = 540$ cal

So, heat obtained from steam is able to melt ice completely but not able to raise the temperature of water formed at 100°C.

Now, out of 540 cal, 400 cal heat is absorbed in formation of water.

Now, mixture contains water (5g) at 0°C and water (1g) at 100°C.

Let final temperature of mixture is t°C. Then, heat lost (by 1 g of water) + excess heat = heat gained (by 5 g water)

$\Rightarrow \quad m_1 s(100 - t) + (540 - 400) = m_2 s(t - 0)$

$\Rightarrow \quad 1 \times 1(100 - t) + 140 = 5 \times 1 \times (t - 0)$

$\Rightarrow \quad 240 = 6t$

$\Rightarrow \quad t = \dfrac{240}{6} = 40°$ C

Note *Specific heat of water* = 1 cal g^{-1}°C^{-1}

8. (c) Heat rejected by steam on complete condensation

$= mL = 10\text{ g} \times 540\text{ cal g}^{-1} = 5400$ cal

Heat rejected by water at 100°C to cool down upto 0°C

$= ms\,\Delta t = 10\text{ g} \times \dfrac{1\text{ cal}}{\text{g°C}} \times (100 - 0)\text{°C} = 1000$ cal

So, maximum heat supplied to ice = 6400 cal

If m = mass of ice melted, then mL = heat from steam

$m = \dfrac{6400\text{ cal}}{80\text{ calg}^{-1}} = 80$ g

Ice remains in mixture = 100 − 80 = 20 g

Water in mixture = 80 g + 10 g = 90 g

9. (b) Heat required to melt ice at 0°C $= mL = 30\text{ g} \times 80\ \dfrac{\text{cal}}{\text{g}}$

= 2400 cal

Heat required by 30 g water to raise its temperature from 0°C to 100°C

$= ms\,\Delta t = 30 \times 1 \times 100 = 3000$ cal

Heat of condensation of steam $= mL$

$= 25\text{g} \times 540\text{ calg}^{-1} = 13500$ cal

As, 13500 cal > 5400 cal

So, resulting temperature of mixture = 100°C

10. (a) Heat lost by A (30°C to 26°C)

= Heat gained by B (20°C to 26°C)

$\Rightarrow \quad ms_A(30 - 26) = ms_B(26 - 20)$

$\Rightarrow \quad \dfrac{s_A}{s_B} = \dfrac{3}{2}$

11. (c) $W = n \times mgh = 2000 \times 10^3$ cal

$\Rightarrow n = \dfrac{2 \times 1000 \times 1000 \times 4.2\text{ Jcal}^{-1}}{50 \times 10 \times 2} = 4.2 \times 2000 = 8400$ times

12. (c) Thermal capacity = Mass × Specific heat

$\therefore$ Ratio of thermal capacities $= \dfrac{m_1 s_1}{m_2 s_2}$

As, both spheres have same material.

$\therefore \quad s_1 = s_2$ and $\rho_1 = \rho_2$

$$\frac{m_1}{m_2} = \frac{\frac{4}{3}\pi r_1^3 \cdot \rho}{\frac{4}{3}\pi r_2^3 \cdot \rho} = \left(\frac{r_1}{r_2}\right)^3 = \left(\frac{d_1}{d_2}\right)^3 = \left(\frac{1}{2}\right)^3$$

= 1 : 8

13. (c) $\dfrac{dQ}{dt} = mC \Rightarrow dQ = mCdt$

$\Rightarrow \quad Q = \int mCdt = m\int Cdt$

$= m\int_5^{15} (0.2 + 0.14t + 0.023t^2)\,dt$

$= 2\left(0.2t + \dfrac{0.14t^2}{2} + \dfrac{0.023t^3}{3}\right)_5^{15} = 82$ cal

14. (a) Work done = Heat supplied

$= mC_{\text{ice}}\Delta T + mL_{\text{ice}} + mC_{\text{steam}}\Delta T + mL_{\text{vapour}}$

$= (1 \times 0.5 \times 10) + (1 \times 80) + (1 \times 1 \times 100) + (1 \times 540)$

$= 725$ cal $= 725 \times 4.2 = 3045$ J

15. (b) $mC_{\text{ice}}(20) + mL = mC_{\text{water}}(20)$

$\Rightarrow \quad 2 \times 0.5 \times 20 + m' \times 80 = 5 \times 1 \times 20$

$\Rightarrow \quad m' = 1$ kg

Final mass of water = 1 + 5 = 6 kg

16. (a,b,c) Heat is absorbed, temperature of body raises and expansion occurs.

Rise of temperature is done to increase in KE and expansion is done to increase in PE of the body.

17. (a,d) OB shows fusion and CD shows vaporisation.

18. (a,b,c) Options (a), (b) and (c) are correct. Option (d) is incorrect as melting point variation in few cases is very insignificant, at low pressure but that does not mean there is no variation.

19. (a,b,c) Heat received by ice $= mL + mC\,\Delta T = 10700$ cal

Heat given by container, then

$$= -\int_{500}^{300} m_c(A + BT)\,dT = -m_C\left[AT + \frac{BT^2}{2}\right]_{500}^{300} = 21600m_C$$

Heat lost = Heat gained

$\therefore \quad 10700 = 21600m_C$

$\Rightarrow \quad m_C = \dfrac{10700}{21600} \approx 0.495 \approx 0.5$ kg

Also, heat lost by container ≈ 0.5 × 21600 ≈ 10800 cal

20. (a,b,c) When A and B are mixed,

$ms_A(13 - 10) = ms_B(15 - 13)$

$3s_A = 2s_B \Rightarrow \dfrac{s_A}{s_B} = \dfrac{2}{3}$

When B and C are mixed, $s_B \times 1 = s_C \times 4$

$\Rightarrow \quad \dfrac{s_B}{s_C} = 4 \Rightarrow \dfrac{3}{2}\dfrac{s_A}{s_C} = 4 \Rightarrow \dfrac{s_A}{s_C} = \dfrac{8}{3}$

When A and C are mixed,

$s_A(\theta - 10) = s_B(20 - \theta) \Rightarrow \theta = \dfrac{140}{11}$°C

21. (a) Both Statement I and Statement II are correct and Statement II is correct explanation of Statement I.

After the liquid and vapour densities are equal, fluid becomes supercritical and there is no distinction between liquid and vapour.

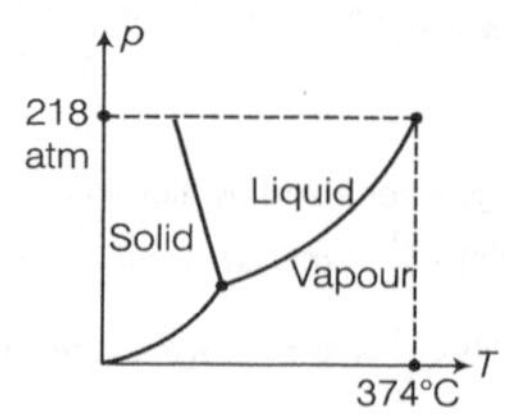

22. (d) Statement I is incorrect as, $C = \lim\limits_{\Delta t \to 0} \dfrac{\Delta Q}{m\Delta t}$ is not called average specific heat of the substance.

Statement II is correct.

23. (a) (i) → (q), (ii) → (s), (iii) → (r), (iv) → (p)
(Use basic definitions to get proper units).

24. (a) (i) → (q), (ii) → (s), (iii) → (r), (iv) → (p)
(Use basic definitions to get proper dimensions).

25. (c) In portion *A*, *C* or *E* heat absorbed $Q = mC\,\Delta T$

where, C = specific heat

$$\Rightarrow \quad \Delta T = \frac{Q}{mC}$$

which is comparable to $y = mx$

Hence, slope of line $= \dfrac{1}{mC} = \dfrac{1}{\text{Heat capacity}}$

Now, ratio of slopes is

$$m_A : m_C : m_E :: \frac{1}{\text{Heat capacity of ice}} : \frac{1}{\text{Heat capacity of water}} : \frac{1}{\text{Heat capacity of steam}}$$

$$\Rightarrow \quad m_A : m_C : m_E :: \frac{1}{2090} : \frac{1}{4190} : \frac{1}{2010} \quad \text{(from graph)}$$

26. (c) Latent heat is amount of energy which is used to change phase of substance.

∴ Heat absorbed is used to increase potential energy and separation of molecules.

27. (d) Ratio of L_V and $L_f = \dfrac{mL_v}{mL_f} = \dfrac{2260}{333}$ (from graph)

DPP-4 **Heat Transfer-I** (Conduction of Heat)

1. Let at any time, temperature of the disc be θ.

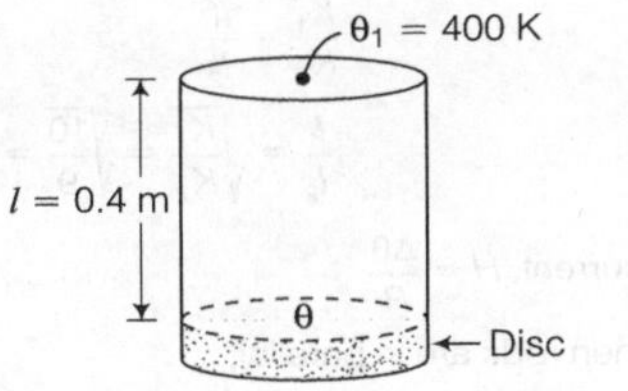

At this moment rate of heat flow $= \dfrac{dQ}{dt} = \dfrac{KA(\Delta\theta)}{l} = \dfrac{KA(\theta_1 - \theta)}{l}$

This heat is utilised in heating disc.

$$\Rightarrow \quad \frac{dQ}{dt} = ms\frac{d\theta}{dt}$$

From above these equations,

$$ms\frac{d\theta}{dt} = \frac{KA}{l}(\theta_1 - \theta) \Rightarrow \frac{d\theta}{\theta_1 - \theta} = \frac{KA}{msl}\cdot dt$$

Let in time *t* second, θ increases from 300 K to 350 K.

$$\Rightarrow \quad \int_{300}^{350} \frac{d\theta}{\theta_1 - \theta} = \frac{KA}{msl}\int_0^t dt$$

$$\Rightarrow \quad [-\log(\theta_1 - \theta)]_{300}^{350} = \frac{KA}{msl}\cdot t$$

$$\Rightarrow \quad t = \frac{msl}{KA}\cdot\log\left(\frac{\theta_1 - 300}{\theta_1 - 350}\right)$$

Substituting the values, we have

$$t = \frac{0.4 \times 600 \times 0.4}{10 \times 0.04}\cdot\log\left(\frac{400 - 300}{400 - 350}\right)$$

$$= 240\log(2) = 240 \times 0.6931$$

$$= 166.355 \approx 166 \text{ s}$$

2. Let θ_1 and θ_2 are temperatures of two interfaces as shown :

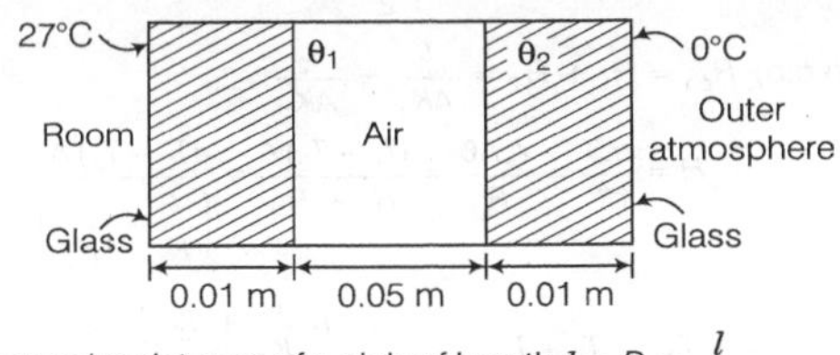

Thermal resistance of a slab of length $l = R = \dfrac{l}{KA}$

Here, $\quad R_1 = R_3 = \dfrac{0.01}{0.8 \times 1} = 0.0125 \text{ kW}^{-1}$

and $\quad R_2 = \dfrac{0.05}{0.08 \times 1} = 0.625 \text{ kW}^{-1}$

In steady state, rate of heat flow through all three layers will be same.

$$\Rightarrow \quad \frac{dQ_1}{dt} = \frac{dQ_2}{dt} = \frac{dQ_3}{dt}$$

$$\Rightarrow \quad \frac{27 - \theta_1}{0.0125} = \frac{\theta_1 - \theta_2}{0.625} = \frac{\theta_2 - 0}{0.0125}$$

Solving these, we get

$$\theta_1 = 26.48°\text{C} \quad \text{and} \quad \theta_2 = 0.52\ °\text{C}$$

and $\quad \dfrac{dQ}{dt} = \dfrac{27 - \theta_1}{0.0125} = \dfrac{27 - 26.48}{0.0125} = 41.6 \text{ W}$

3. We assume that, area of integrated circuit is small enough for energy to flow by conduction.

So, $\quad P = \dfrac{dQ}{dt} = \dfrac{KA(T_1 - T_2)}{l} \Rightarrow T_1 = T_2 + \dfrac{Pl}{KA}$

$$= 35 + \frac{1.5 \times 0.085 \times 10^{-3}}{0.075 \times 8.5 \times 6.25 \times 10^{-6}}$$

$$= 35 + 32 = 67°\text{C}$$

So, operating temperature for integrated circuit = 67°C or less.

4. If *m* is mass of ice formed due to extraction of *Q* amount of heat from the pond, then

$$mL = Q \quad \text{or} \quad \rho \cdot V \cdot L = Q$$

$$\rho ALx = Q$$

where, V = Volume of ice, $\quad L$ = Latent heat

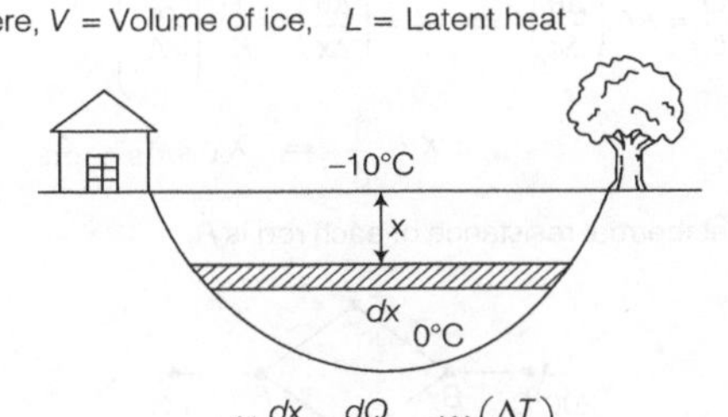

$$\Rightarrow \quad \rho AL\frac{dx}{dt} = \frac{dQ}{dt} = KA\left(\frac{\Delta T}{x}\right)$$

$$\Rightarrow \quad L \cdot \rho \cdot x dx = K\,\Delta T \cdot dt$$

$$\Rightarrow \quad L\rho\int_{0.04}^{0.08} x dx = K\,\Delta T\int_0^t dt$$

$$\Rightarrow \quad L\rho\left(\frac{x^2}{2}\right)_{0.04}^{0.08} = K\,\Delta T \cdot t \Rightarrow t\ \frac{L\rho\left(\dfrac{x^2}{2}\right)_{0.04}^{0.08}}{K \cdot \Delta T}$$

Substituting the values, we get

$$t = \frac{3.3 \times 10^5 \times 917 \times [(0.08)^2 - (0.04)^2]}{2 \times 2 \times 10} = 10.2 \text{ h}$$

5. (b) Let R_1 and R_2 be thermal resistances of inner and outer cylinder. As temperature difference across each cylinder is same, the resistances are in parallel. So, $R_{net} = \frac{R_1 R_2}{R_1 + R_2}$

$$\Rightarrow \quad \frac{l}{K_{eq}\pi(2R)^2} = \frac{\frac{l}{K_1 \pi R^2} \times \frac{l}{K_2(3\pi R^2)}}{\frac{l}{K_1 \pi R^2} + \frac{l}{K_2(3\pi R^2)}}$$

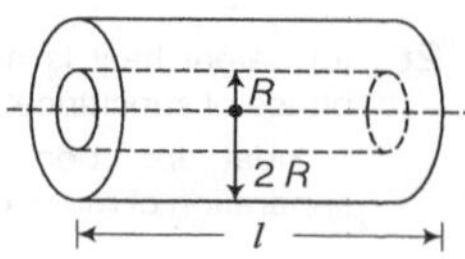

$$\Rightarrow \quad \frac{1}{4K_{eq}} = \frac{\frac{1}{3K_1K_2}}{\frac{1}{K_1} + \frac{1}{3K_2}}$$

$$\Rightarrow \quad K_{eq} = \frac{3K_2 + K_1}{4}$$

6. (b) Rods are joined as shown in the figure.

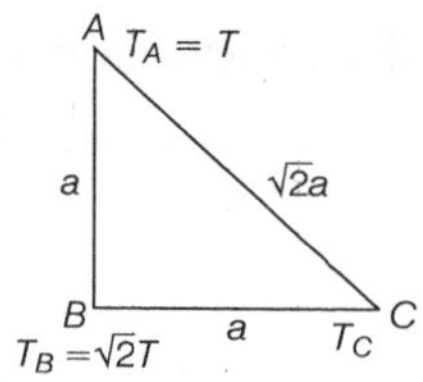

Rods are in steady state and only conduction is taking place.

$\therefore \quad \frac{dQ}{dt} = \frac{KA\,\Delta T}{l}$, for each rod.

Now, from AC to CB, rate of heat flow from A to C = rate of heat flow from C to B

$$\Rightarrow \quad \left(\frac{\Delta T}{\sqrt{2}a}\right)_{AC} = \left(\frac{\Delta T}{a}\right)_{CB}$$

$$\Rightarrow \quad \frac{T - T_C}{\sqrt{2}a} = \frac{T_C - \sqrt{2}T}{a} \quad \Rightarrow \quad \frac{T_C}{T} = \frac{3}{\sqrt{2} + 1}$$

7. (b) Let θ = junction temperature.

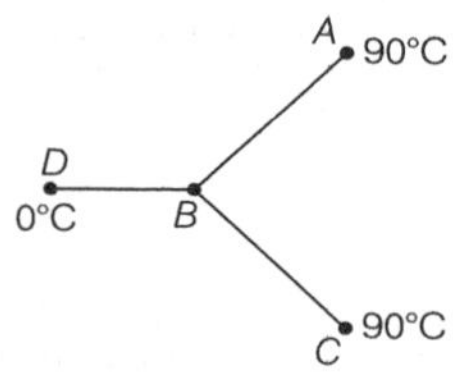

Then, rate of heat flow $A \to B$ + rate of heat flow $C \to B$ = rate of heat flow $B \to D$. (As, steady state is given)

$$\Rightarrow \quad \frac{90 - \theta}{R} + \frac{90 - \theta}{R} = \frac{\theta - 0}{R}$$

where, R = thermal resistance of each rod (which is same for all rods).

Solving, we get $3\theta = 180$

$$\Rightarrow \quad \theta = 60°C$$

8. (c) $\frac{dQ}{dt} = KA\left(\frac{\Delta\theta}{\Delta x}\right) \quad \Rightarrow \quad X = \left(\frac{\Delta\theta}{\Delta x}\right) = \frac{1}{K} \cdot \left(\frac{\frac{dQ}{dt}}{A}\right)$

$$\Rightarrow \quad X \propto \frac{1}{K} \quad \Rightarrow \quad X_A < X_B < X_C$$

9. (c) Let thermal resistance of each rod is R.

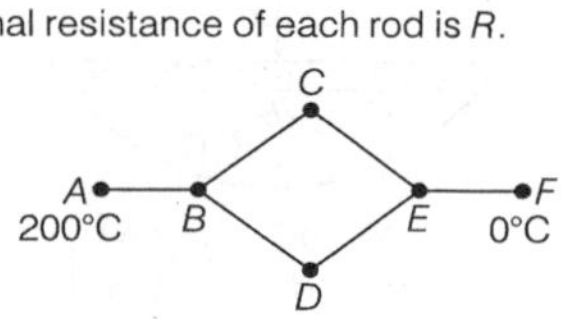

Equivalent combination is

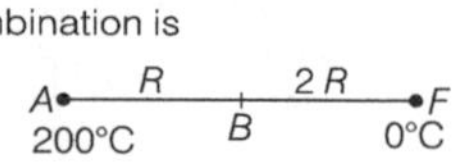

Effective resistance between B and F is $2R$.

If T is temperature of B, then

$$R\,T_2 + 2RT_1 = (R + 2R)\,T$$

$$\Rightarrow \quad \frac{T_1 + T_2}{3} = T$$

$$\Rightarrow \quad \frac{200 + 0}{3} = T$$

$$\Rightarrow \quad T = \frac{200}{3} \quad \Rightarrow \quad T = 66.67°C$$

10. (c) To increase thickness from y_1 to y_2, time taken

$$t = \frac{\rho L}{2K\theta}(y_2^2 - y_1^2)$$

$$\therefore \quad \Delta t_1 : \Delta t_2 : \Delta t_3 :: (1^2 - 0^2) : (2^2 - 1^2) : (3^2 - 2^2)$$

$$\Delta t_1 : \Delta t_2 : \Delta t_3 :: 1 : 3 : 5$$

11. (d) Rate of heat generated = Rate of heat outflow

$$\frac{1}{J} \cdot \frac{V^2}{R} \cdot t = KA\frac{\Delta\theta}{l} \cdot t$$

where, $J = 4.2\ \text{J cal}^{-1}$

$$\Rightarrow \quad \frac{1}{4.2} \times \frac{(200)^2}{20} = \frac{20 \times 1 \times (25 - \theta)}{0.2 \times 10^{-2}}$$

$$\Rightarrow \quad \frac{200 \times 200}{4.2 \times 20 \times 100} = 25 - \theta$$

$$\Rightarrow \quad \theta = 25 - 4.8 = 21.2\ °C$$

12. (b) Igen-Hausz experiment shows that, $K \propto l^2$

$$\Rightarrow \quad \frac{K_1}{K_2} = \frac{l_1^2}{l_2^2}$$

$$\Rightarrow \quad \frac{l_1}{l_2} = \sqrt{\frac{K_1}{K_2}} = \sqrt{\frac{10}{9}} = \frac{\sqrt{10}}{3}$$

13. (d) Heat current, $H = \frac{\Delta\theta}{R}$

***Case* I** When rods are in parallel,

$$\therefore \quad R_P = \frac{R_1 R_2}{R_1 + R_2} = \frac{\frac{1}{5KA} \times \frac{1}{KA}}{\frac{1}{5KA} + \frac{1}{KA}} = \frac{1}{6KA}$$

***Case* II** When rods are in series,

$$\therefore \quad R_S = R_1 + R_2 = \frac{1}{5KA} + \frac{1}{KA} = \frac{6}{5KA}$$

So, ratio of heat currents is

$$\frac{H_P}{H_S} = \frac{R_S}{R_P} \text{ (or } \Delta\theta \text{ is same)} = \frac{\frac{6}{5KA}}{\frac{1}{6KA}} = \frac{36}{5}$$

14. (d) As, material is same and sides are insulated.

$\therefore$ For any section in steady state,

Rate of heat in flow = Rate of heat out flow.

15. (c) $R_{AB} = \frac{l}{KA} = R_{CD} \quad \Rightarrow \quad R_{eq(BC)} = \frac{\pi l}{2KA}$

So, $$\frac{100 - 0}{\frac{2l}{KA} - \frac{\pi l}{2KA}} = \frac{T_C - 0}{\frac{l}{KA}}$$

$$\Rightarrow \quad T_C = \frac{100}{2 + \frac{\pi}{2}} = \frac{100}{2 + 1.57} \quad \Rightarrow \quad T \approx 28°C$$

16. (a,b,c,d) $R_{eq} = R_1 + R_2 = \frac{l_1}{AK_1} + \frac{l_2}{AK_2}$

$$H = \frac{dQ}{dt} = \frac{A\,\Delta\theta}{R} = \frac{(T_1 - T_2)A}{R_1 + R_2} = \frac{(T_1 - T_2)A}{\left(\frac{l_1}{K_1} + \frac{l_2}{K_2}\right)}$$

Also, $$\frac{T_1 - T_2}{R_1 + R_2} = \frac{T_1 - T}{R_1} \quad \Rightarrow \quad T = \frac{T_1\frac{l_2}{K_2} + T_2\frac{l_1}{K_1}}{\left(\frac{l_1}{K_1} + \frac{l_2}{K_2}\right)}$$

And $R = R_1 + R_2$ so,

$$K_{eq} = \frac{l_1 + l_2}{A(R_1 + R_2)} = \frac{l_1 + l_2}{\left(\frac{l_1}{K_1} + \frac{l_2}{K_2}\right)}$$

17. (a,c) $H = \frac{dQ}{dt} = \frac{4\pi K r_1 r_2 \Delta T}{r_2 - r_1}$ and $H = L\frac{dm}{dt}$

Equating we get, $\frac{dm}{dt} = \frac{\pi}{2}\ \text{gs}^{-1}$

Also, $t = \frac{m}{dm/dt} = 589\ \text{s} < 10\ \text{min}$

18. (b,d) Let θ is the temperature of point C.

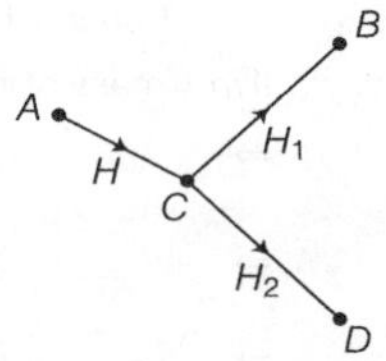

Then, $H = H_1 + H_2$, in steady state.

$\Rightarrow \frac{KA}{l}(T_A - \theta) = \frac{KA}{l}(\theta - T_B) + \frac{KA}{l}(\theta - T_D)$

$\Rightarrow T_A - \theta = 2\theta - T_B - T_D$

$\Rightarrow 3\theta = T_A + T_B + T_D$

$\Rightarrow \theta = \frac{40 + 30 + 20}{3}$

$= 30\ °\text{C}$

So, no heat flow occurs across BC and heat flows form C to D.

19. (d) **Hint** Use $\frac{dQ}{dt} = \frac{KA}{l}(\theta_1 - \theta_2)$

and then put condition given in each option.

20. (b,c) Rods 2, 3, 4 are in parallel and rods 5, 6, 7, 8 and 9 are also in parallel.

$\therefore$ In terms of thermal resistance, given combination of rods is equivalent to

$T_A = 100°\text{C}$ 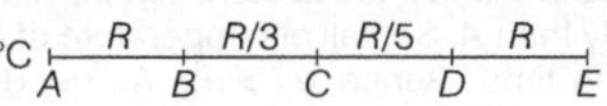 $T_E = 0°\text{C}$

Using $\frac{1}{R_P} = \Sigma \frac{1}{R_i} i$

or the given combination is

$T_A = 100°\text{C}$ 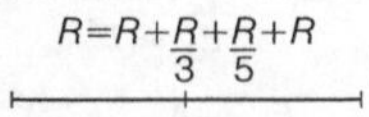 $T_E = 0°\text{C}$

$R = R + \frac{R}{3} + \frac{R}{5} + R = \frac{38}{15}R = \frac{38}{15}\left(\frac{L}{KA}\right)$

So, heat flow rate

$H = \frac{dQ}{dt} = \frac{T_A \cdot T_E}{R_{eq}} = \frac{100}{\frac{38}{15}\left(\frac{L}{KA}\right)} = \frac{750}{19}\left(\frac{KA}{L}\right)$

21. (a) Electrons also participate in conduction process and number of available free electrons increases with temperature.

22. (c) Statement I is correct but Statement II is incorrect.

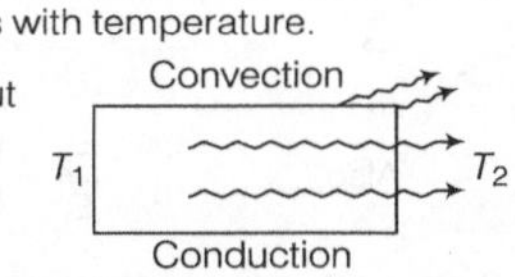

When there is no insulation on side,

Heat loss per second,

$$\frac{dQ}{dt} = \underset{\text{(convection loss)}}{hA(T - T_0)} + \underset{\text{(conduction loss)}}{KA\left(\frac{\Delta T}{\Delta x}\right)}$$

So, temperature gradient falls along the rods length.

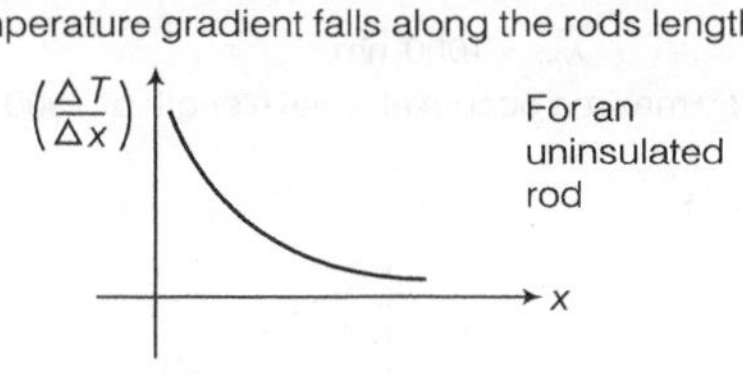

23. (b) In steady state, j_E = constant

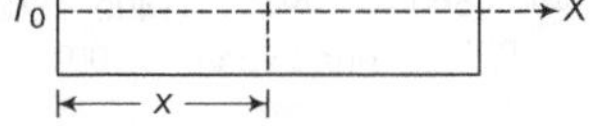

$\therefore \quad j_E = \frac{K(T_0 - T)}{x} \quad \Rightarrow T = \frac{-j_E}{K} \cdot x + T_0$

24. (c) We start with finding temperature of interface. Let temperature of interface = T_0

Then, $T_0 = \frac{K_1 \theta_1 L_2 + K_2 \theta_2 L_1}{K_1 L_2 + K_2 L_1}$

$\Rightarrow T_0 = \frac{K \times 0 \times 2 + 3K \times 100 \times 1}{K \times 2 + 3K \times 1} = \frac{300K}{5K} = 60°\text{C}$

Now, temperature at distance y from this interface, then

$T = \frac{-j_E y}{K} + T_0$, where $j_E = \frac{K(0 - 60)}{1}$

$= \frac{-j_E \cdot y}{3K} + 60 = \frac{-K(0 - 60) \times 1}{3K \times 1} + 60$

$= \frac{60}{3} + 60 = 20 + 60 = 80°\text{C}$

25. (c) Electrons do diffuse but at same time, same number of electrons with less energies fall back to hot region. So, net exchange of electrons across any section is zero and there is no potential drop across a single conductor.

26. (b) Conduction in solids at low temperatures, energy is transmitted *via* phonon vibration i.e. random (erratic) vibration of lattice structure. At moderately high and very high temperatures, electrons carry energy from one part of solid to another. High energy electrons diffuse into older regions.

Conduction in gases depends on pressure and temperature due to molecular collisions.

Conduction in liquids depends on only temperature due to molecular collision and electron drift.

Hence, (i) → (p), (ii) → (p,q), (iii) → (r), (iv) → (q,r)

27. (4) In steady state, thermal current $C_1 = \frac{dQ}{dt} = -K 4\pi r^2 \frac{dT}{dr}$

$\Rightarrow \frac{dT}{dr} = \frac{-C_1}{4\pi K} \cdot \frac{1}{r^2}$

$\Rightarrow dT = \frac{-C_1}{4\pi K} \cdot \frac{1}{r^2} \cdot dr$

Integrating, we have

$T = \frac{C_1}{4\pi K r} + C_2$

At $r = r_0$, $T = 2T_0$

and at $r = 2r_0$, $T = T_0$

$\Rightarrow C_2 = 0$

and $C_1 = 8\pi r_0 K T_0$

So, $T = \frac{2r_0 T_0}{r}$

$\therefore T\left(r = \frac{3r_0}{2}\right) = \frac{4T_0}{3}$

$\therefore p = 4$

28. (2) Temperature of water at the time $t = \theta$, heat supplied in the next dt duration is

$dQ = \frac{KA}{L}(\theta_0 - \theta)dt$

Let dQ raises temperature of m mass of water from θ to $\theta + d\theta$.

Then, $dQ = msd\theta$

Hence, $\int_0^t dt = \frac{Lms}{KA}\int_{\theta_1}^{\theta_2} \frac{d\theta}{\theta_0 - \theta}$

$\Rightarrow t = \frac{Lms}{AK} \cdot \log\left(\frac{\theta_0 - \theta_1}{\theta_0 - \theta_2}\right)$

Substituting the values, we have

$t = \frac{2 \times 10 \times 4200}{0.5 \times 20} \cdot \log\left(\frac{90 - 10}{90 - 30}\right) = 8400 \cdot \log\left(\frac{80}{60}\right)$

$= 8400 \log\left(\frac{4}{3}\right) = 2 \times 4200 \log\frac{4}{3}$

$\therefore N = 2$

DPP-5 **Heat Transfer-II** (Convective and Radiative Heat Transfer)

1. (i) Convection involves the transfer of heat between a surface at a given temperature T_s and a fluid at a bulk temperature T_b. For flow adjacent to a hot or cold surface, T_b is the temperature of fluid far from the surface. For boiling or condensation, T_b is the saturation temperature of the fluid.

For flow in a tube, T_b is mean temperature measured at a particular cross-section of the tube.

Basic relationship for heat transfer by convection is of same form as that of conduction.

$$\frac{dQ}{dt} = hA\Delta T$$

where, $\frac{dQ}{dt}$ = rate of heat transfer (Js^{-1})

h = connective heat transfer coefficient ($Jm^{-1}{}^\circ C^{-1}s^{-1}$)

A = Area of surface (m^2)

ΔT = temperature difference $(T_s - T_b)$ in °C

(ii) $\frac{dQ}{dt} = hA\Delta T = h(2\pi rl)\Delta T$

$= 2 \times 10^{-3} \times 2 \times 3.14 \times 15 \times 10^{-2} \times 6 \times (110 - 20)$

$= 2 \times 10^{-3} \times 2 \times 3.14 \times 15 \times 10^{-2} \times 6 \times 90$

$= 101736 \times 10^{-5} Js^{-1} \approx 2.8 \times 10^{-4} Jh^{-1}$

2. We assume that, metal wall of boiler is thin and conductivity of metal is high. So, we take temperature of inner surface of insulation is nearly 600°C.

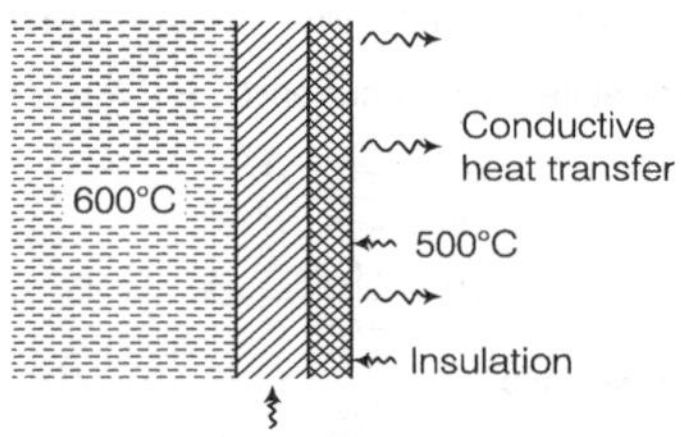

Now, heat flux obtained through the insulation is

$$\frac{dQ}{dt} = KA\left(\frac{\Delta T}{\Delta x}\right)$$

so, $\frac{dQ}{Adt} = \frac{K\Delta T}{\Delta x}$ = heat flow rate per square metre

$$\frac{1}{A}\cdot\frac{dQ}{dt} = \frac{0.08 \times (600 - 500)}{2 \times 10^{-2}}$$

$$= 400 \ Js^{-1}m^{-2} \text{ or } Wm^{-2}$$

Now, this heat flux is removed by convection current of medium outside the wall so that temperature of wall does not increase.

$\therefore \quad \frac{dQ}{dt} = hA(T_s - T_b)$

$\Rightarrow \quad T_s - T_b = \left(\frac{1}{A}\cdot\frac{dQ}{dt}\right) \times \frac{1}{h}$

$\Rightarrow \quad T_s - T_b = \frac{400}{2} = 200$

$\Rightarrow \quad 500 - T_b = 200 \Rightarrow 500 - 200 = T_b$

$\Rightarrow \quad T_b = 300°C$

3. Net rate of energy lost by radiation mode is

$$P = \sigma Ae(T^4 - T_0^4) \qquad \ldots(i)$$

Given, $\sigma = 5.67 \times 10^{-8} Wm^{-2}K^{-4}$

$e = 0.9, \quad A = 1.5 \ m^2$

$T = 35°C = 273 + 35 = 308 \ K$

$T_0 = 20°C = 293K$

So, substituting the values in Eq. (i), we have

$P = 5.67 \times 10^{-8} \times 1.5 \times 0.9 \times [(308)^4 - (293)^4]$

$\Rightarrow \quad P = 1.25 \ W$

Total energy lost in

1 hour = 125W × 3600 s = 450000 J = 4.5×10^5 J

If m = mass of ice melted, then $mL = P \times \Delta t$

$\Rightarrow \quad m = \frac{P \times \Delta t}{L}$

$\Rightarrow \quad m = \frac{4.5 \times 10^5}{3.33 \times 10^5} \quad (\because L = 3.33 \times 10^5 Jkg^{-1})$

$\Rightarrow \quad m \approx 1.4 \ kg$

4. Earth is taken as sphere of radius R absorbing sunlight in a projected area of πR^2 and emitting radiations in all directions perpendicular to surface, over an area of $4\pi R^2$. In steady state,

$$P_{in} = P'_{out}$$

$$e\,(1340 Wm^{-2})\,\pi R^2 = e\sigma(4\pi R^2)T^4$$

$\Rightarrow \quad T = \left(\frac{1340}{4 \times 5.67 \times 10^{-8}}\right)^{1/4}$

$= 277K = 4°C$

5. (c) Initially, both A and B are at same temperature but B loses heat much rapidly than A. So, fall of temperature of B is more than that of A in same time. Hence, $e_A > e_B$. As good emitters are also good absorbers so, $a_A > a_B$.

6. (d) Area under graph $= \int_0^\infty E_\lambda \cdot d\lambda$ = Total amount of thermal energy emitted per unit time, per unit area for all possible wavelengths.

And $\quad E_\lambda \propto T^4$

So, $\quad \frac{E_1}{E_2} = \frac{A_1}{A_2} = \left(\frac{T_1}{T_2}\right)^4$

$\Rightarrow \quad \frac{E_{T_1}}{E_{2000}} = \frac{16}{1} = \left(\frac{T}{2000}\right)^4$

$\Rightarrow \quad T = 4000 \ K$

7. (a) By Wien's displacement law, ΔT = constant

$\therefore \quad \lambda_{m_1}T_1 = \lambda_{m_2}T_2$

$\Rightarrow \quad \frac{T_2}{T_1} = \frac{\lambda_{m_1}}{\lambda_{m_2}} = \frac{\lambda_0}{3\lambda_0/4} = \frac{4}{3}$

Also, $\quad \frac{E_2}{E_1} = \left(\frac{T_2}{T_1}\right)^4 = \left(\frac{4}{3}\right)^4 = \frac{256}{81}$

8. (c) By Wien's displacement law, $\lambda_m T = b$

$\Rightarrow \quad \lambda_m = \frac{b}{T} = \frac{3 \times 10^6 nmK}{3000 \ K}$

$\Rightarrow \quad \lambda_m = 1000 \ nm$

So, energy maxima occurs at a wavelength of 1000 nm.

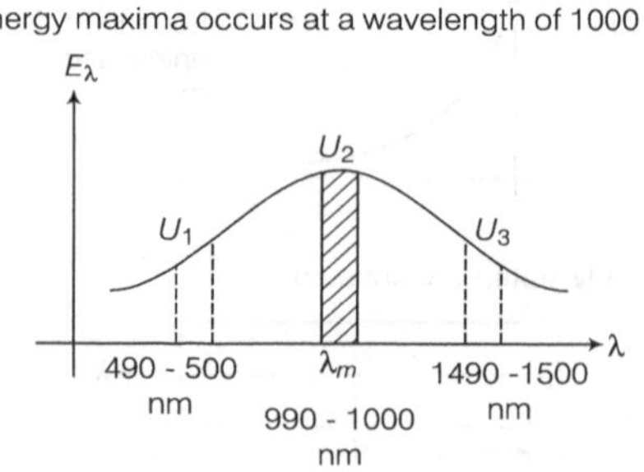

$\therefore \quad U_2 > U_1$ and $U_2 > U_3$

9. (b) Energy radiated in t second,

$$Q = P\cdot t = \sigma AT^4\cdot t$$

$$\therefore \quad \frac{Q_1}{Q_2} = \frac{\sigma A_1T_1^4\cdot t}{\sigma A_2T_2^4\cdot t} = \frac{A_1}{A_2}\cdot\left(\frac{T_1}{T_2}\right)^4$$

$$= \frac{4\pi R_1^2}{4\pi R_2^2}\cdot\left(\frac{T_1}{T_2}\right)^4 = \left(\frac{R_1}{R_2}\right)^2\cdot\left(\frac{T_1}{T_2}\right)^4$$

$$= \left(\frac{R}{2R}\right)^2\cdot\left(\frac{2T}{T}\right)^4 = \frac{2^4}{2^2} = 2^2 = 4:1$$

10. (c) By Stefan's law rate of cooling, $\frac{dT}{dt} = \frac{eA\sigma}{mC}\cdot(T^4 - T_0^4)$

Now, as both cuboids are of same material and same surface finish so;

$$\frac{\text{Rate of cooling of first cuboid}}{\text{Rate of cooling of second cuboid}} = \frac{A_1/V_1}{A_2/V_2}$$

$$= \frac{2(ab + bc + ac)\times(2a\times 2b\times 2c)}{abc\times 2(2a\cdot 2b + 2b\cdot 2c + 2a\cdot 2c)}$$

$$= \frac{8}{4} = \frac{1}{2}$$

So, $$\frac{\frac{dT}{dt_1}}{\frac{dT}{dt_2}} = \frac{8}{4} = \frac{1}{2}$$

$$\Rightarrow \quad \frac{dt_2}{dt_1} = \frac{1}{2} \qquad \text{(As, } \Delta T \text{ is same for both)}$$

$$\Rightarrow \quad dt_2 = \frac{1}{2}dt_1 \Rightarrow t_2 = \frac{1}{2}\cdot t_1 = \frac{1}{2}\times 100 \Rightarrow t_2 = 50\text{ s}$$

11. (a) By Newton's law of cooling,

$$\frac{\theta_1 - \theta_2}{t} \propto \left(\frac{\theta_1 + \theta_2}{2} - \theta\right)$$

so, $$\frac{80 - 60}{60} \propto \frac{80 + 60}{2} - 30 \qquad \ldots\text{(i)}$$

and $$\frac{60 - 50}{t} \propto \frac{60 + 50}{2} - 30 \qquad \ldots\text{(ii)}$$

Solving Eqs. (i) and (ii), we get

$$t = 48\text{ s} \approx 50\text{ s}$$

which is less than one minute.

12. (b) By Newton's law of cooling, if temperature of body decreases by $d\theta$ in time dt, then

$$\frac{d\theta}{dt} = \frac{-K}{mC}(\theta - \theta_0) \Rightarrow 2.3026\times\log_{10}\left(\frac{T_1 - T_0}{T_2 - T_0}\right) = Kt$$

Case I

$$2.3026\log_{10}\left(\frac{64 - 16}{52 - 16}\right) = K\times 10$$

$$\Rightarrow \quad 2.3026\log_{10}\left(\frac{48}{36}\right) = 10K \qquad \ldots\text{(i)}$$

Case II

$$2.306\log_{10}\left(\frac{52 - 16}{T_2 - 16}\right) = K\times 5 \qquad \ldots\text{(ii)}$$

From Eqs. (i) and (ii), we have

$$\frac{48}{36} = \frac{36}{T_2 - 16} \Rightarrow T_2 = 43°\text{C}$$

13. (c) $\frac{\sigma T^4\cdot 4\pi R^2}{4\pi r^2} = 1400$

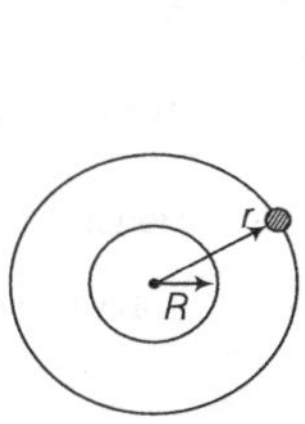

$$\Rightarrow \quad T^4 = \frac{1400r^2}{\sigma R^2}$$

Solving, we get

$$T = \left(\frac{1400r^2}{\sigma R^2}\right)^{\frac{1}{4}} = 10^{\frac{15}{4}}\text{K}$$

14. (a) Rate of cooling = Cooling by radiation per second + Cooling by conduction per second.

$$\therefore \quad -\frac{dT}{dt} = K_1(T - T_A) + \frac{KA}{LC}(T - T_A)$$

$$\Rightarrow \quad \int_{T_1}^{T_2}\frac{dT}{T - T_A} = -\left(K_1 + \frac{KA}{LC}\right)\int_0^t dt$$

$$\Rightarrow \quad \log\left(\frac{T_2 - T_A}{T_1 - T_A}\right) = -\left(K_1 + \frac{KA}{LC}\right)\cdot t$$

Note *In conduction,*

$$\frac{dQ}{dt} = \frac{KA(T - T_A)}{L} = C\left(-\frac{dT}{dt}\right)$$

$$\therefore \quad -\frac{dT}{dt} = \frac{KA}{LC}(T - T_A)$$

15. (a,b) Power radiated and surface area of A and B is same.

$$\Rightarrow \quad e_A\sigma T_A^4\cdot A = e_B\sigma T_B^4\cdot A$$

$$\Rightarrow \quad \frac{T_A}{T_B} = \left(\frac{e_B}{e_A}\right)^{\frac{1}{4}} = \left(\frac{0.81}{0.01}\right)^{\frac{1}{4}} = 3$$

$$\Rightarrow \quad T_B = \frac{T_A}{3} = \frac{5802}{3} = 1934\text{ K}$$

Also, $\lambda_m T$ = constant

$$\therefore \quad \lambda_A T_A = \lambda_B T_B$$

so, $$\lambda_A = \lambda_B\cdot\left(\frac{T_B}{T_A}\right) = \frac{\lambda_B}{3}$$

Also, $$\lambda_B - \lambda_A = 1\,\mu\text{m} \Rightarrow \lambda_B - \frac{\lambda_B}{3} = 1\,\mu\text{m}$$

$$\frac{2}{3}\lambda_B = 1\,\mu\text{m}$$

$$\Rightarrow \quad \lambda_B = \frac{3}{2} = 1.5\,\mu\text{m}$$

16. (a,d) $\frac{dQ}{dt}$ = Rate of change in energy of room.

$$= E_{\text{in}} - E_{\text{out}}$$

As, the room temperature raises due to fan motor, rate of heat loss also increases. At some point, rate of heat loss = rate of heat produced and

$$\therefore \quad \frac{dQ}{dt} = 0 \Rightarrow E_{\text{in}} = E_{\text{out}}$$

And $E_{\text{in}} = E_{\text{out}} = UA(T_i - T_0)$,

where, U = overall heat transfer coefficient

$$\therefore \quad Q = 200\text{ W} = \frac{6\text{ W}}{\text{m}^2\text{°C}}\times 30\text{m}^2(T_i - 25)$$

$$\Rightarrow \quad T_i = 26.1°\text{C}$$

So, temperature of room is steady after reaching 26.1°C. Fan motor converts electrical energy into magnetic and heat energy. As, part of energy is converted finally in mechanical form so equivalent heater voltage is not 200 W.

17. (a,b,c) $\frac{dQ}{dt}$ convection = Heat loss rate due to convection

$$= hA(T_1 - T_2)$$

$$= \frac{6\text{W}}{\text{m}^2\text{°C}}\times 1.6\text{ m}^2\times(29 - 20)\text{°C}$$

$$= 86.4\text{W}$$

$\frac{dQ}{dt}$ radiation = Heat loss rate due to radiation = $\sigma\varepsilon A(T_1^4 - T_2^4)$

$\frac{dQ}{dt}$ radiation is maximum when $t = 1$

$$\therefore \quad \frac{dQ}{dt}\text{ radiation (maximum)} = 5.67\times 10^{-8}\frac{\text{W}}{\text{m}^2\text{K}^4}\times 1.6\text{ m}^2$$

$$\times\{(29 + 273)^4 - (20 + 273)^4\}\text{K}^4$$

$$= 86\text{W}$$

18. (a,d) Heat transfer from a hot surface by conduction to air.

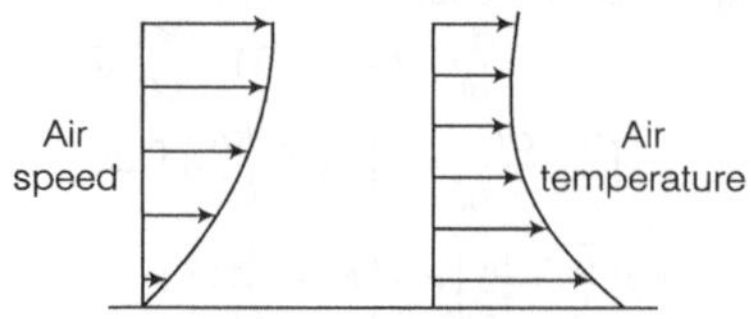

$\frac{dT}{dx}$ = negative as we move across a slab width x so to make $\frac{dQ}{dt}$ positive or negative sign is put in the equation. Amount of energy transferred by radiation does not depend on temperature of medium. Conversion of the sun's energy into plant's energy does not have any adverse effect. Only conversions made by humans are adverse.

19. (a,b,c,d) From 7.8×10^3 Å to 3×10^5 Å, infrared region of electromagnetic waves lies which generally carries heat. Ideal black body acts like a good absorber initially and it appears black.

Without gravity a heated and less dense portion is not experiencing an upward force.

Green and red are complementary colours, this flows from Kirchhoff's law.

20. (d) $\lambda_m \propto \frac{1}{T}$ and $A \propto T^4$.

21. (b) Curve at higher temperature :

(i) Lies above curve at lower temperature.

(ii) Never intersects curve at lower temperature.

(iii) The maxima occurs at a lower wavelength.

22. (c) At 50°C, temperature of ball is constant.

So, rate of heat generation = Rate of heat loss

$\therefore$ Heat loss rate is 20 W.

23. (b) At 30°C, rate of heat loss will be

$$\frac{dQ}{dt} = \frac{20}{3}\frac{J}{s} = \frac{20}{3} W$$

24. (d) In 5 min $= 5 \times 60$ s, heat generated

$$= P \times t = 20 W \times 5 \times 60 s = 6000 J$$

Heat absorbed by metal $= ms\Delta T = 1 \times (500) \times 10 = 5000 J$

So, heat lost by ball $= 6000 - 5000 = 1000 J$

25. (b) Both Statement I and Statement II are correct but Statement II is not the correct explanation of Statement I.

Frequency increases with amount of energy radiated per second. So, a blue star will be hotter.

26. (d) Statement I is incorrect as Jupiter is more far from Sun, solar constant is much less for Jupiter. Statement II is correct.

27. (a) Both Statement I and Statement II are correct and Statement II correctly explains Statement I.

$$\frac{dQ}{dt} = eA\sigma(T^4 - T_0^4)$$

$$\therefore \quad mC\frac{dT}{dt} = eA\sigma(T^4 - T_0^4) \Rightarrow \frac{dT}{dt} \propto \frac{A}{m} = \frac{4\pi R^2}{\frac{4}{3}\pi R^3 \cdot \rho}$$

$$\therefore \quad \frac{dT}{dt} \propto \frac{1}{\rho R}$$

28. (c) By definition,

$$r = \text{Reflectance} = \frac{\text{Heat reflected}}{\text{Heat falling over surface}} = \frac{Q_r}{Q}$$

Similarly, $a = \frac{Q_a}{Q}$ and $t = \frac{Q_t}{Q}$

Now, $r + a + t = \frac{Q_r}{Q} + \frac{Q_a}{Q} + \frac{Q_t}{Q} = \frac{Q_r + Q_a + Q_t}{Q} = 1$

$\Rightarrow$ $r + a + t = 1$ – for every type of surface

For a perfect reflector,

$$r = 1, a = 0 \text{ and } t = 0$$

For a perfect absorber,

$$r = 0, a = 1, t = 0$$

For a perfect transmitter,

$$r = 0, a = 0, t = 1$$

Hence, (i) → (p,q,s), (ii) → (p,r,s), (iii) → (p,q,r), (iv) → (p)

29. (4) As, $\frac{95 - 90}{30} = K\left(\frac{95 + 90}{2} - \theta_0\right)$

and $\frac{55 - 50}{70} = K\left(\frac{55 - 50}{2} - \theta_0\right)$

$\therefore$ $\frac{7}{3} = \frac{92.5 - \theta_0}{52.5 - \theta_0}$

so, $\theta_0 = 22.5$

If liquid cools from 50°C to 45° C in t second, then

$$\frac{55 - 45}{t} = K\left(\frac{50 - 45}{2} - \theta_0\right)$$

Using $\theta_0 = 22.5$ and dividing, we get

$$\frac{t}{30} = \frac{92.5 - 22.5}{47.5 - 22.5} \Rightarrow t = 84 \text{ s} = 4 \times (21) \text{ s}$$

$\therefore$ $k = 4$

30. (3)

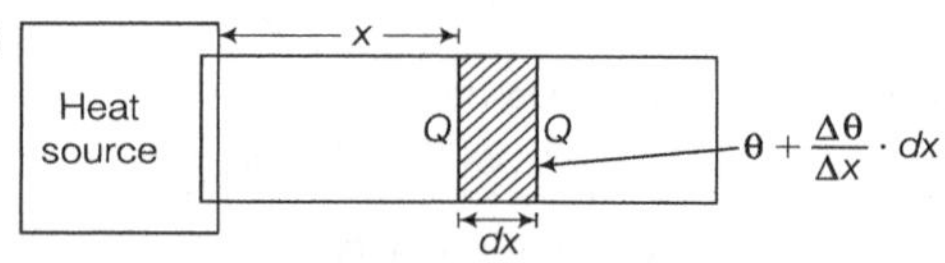

For a section of length dx distance x from hot end,

Heat entering this section is

$$Q_1 = -KA\frac{d\theta}{dt}$$

Heat going out from opposite end of section by conduction,

$$Q_2 = -KA\frac{d}{dx}\left(\theta + \frac{d\theta}{dx} \cdot \delta x\right)$$

Heat remains with this section,

$$Q = Q_1 - Q_2 = KA\frac{d^2\theta}{dx^2} \cdot \delta x$$

Before steady state, a part of this heat is radiated to atmosphere.

Amount of heat required to heat the element

$$= A \cdot \delta x \cdot \rho \cdot s \cdot \frac{d\theta}{dt},$$

where, ρ = density, s = specific heat

Amount of heat radiated to atmosphere $= e \cdot p \cdot \delta x \cdot \theta$

where, e = emissivity, p = perimeter

Thus, $KA\frac{d^2\theta}{dx^2} \cdot \delta x = A \cdot \delta x \cdot \sigma \cdot \rho \cdot s \cdot \frac{d\theta}{dt} + e \cdot p \cdot \delta x \cdot \theta$

or $\frac{d^2\theta}{dx^2} = \frac{\rho s}{K} \cdot \frac{d\theta}{dt} + \frac{e\rho}{KA} \cdot \theta$

At steady state, $\frac{d\theta}{dt} = 0$

Hence, $\frac{d^2\theta}{dx^2} = \frac{ep}{KA} \cdot \theta$

Substituting the values, we get

$$\frac{d^2\theta}{dx^2} = \frac{0.9 \times 2\pi r}{100 \times \pi r^2} \cdot \theta = \frac{0.9 \times 2}{100 \times r} \cdot \theta \quad (\because r = 0.1)$$

$$= \frac{18}{100} \cdot \theta = 2 \times 9 \times \theta \times 10^{-2}$$

So, $N^2 = 9 \Rightarrow N = 3$

Revisal Problems (JEE Main)

1. (c) Momentum is conserved,

$$\because \quad m_b v_b + m_{Al} \times 0 = (m_b + m_{Al})v'$$

$$\Rightarrow \quad v' = \frac{50 \times 10}{50 + 950} = 0.5 \text{ ms}^{-1}$$

Loss in KE = Initial KE − Final KE

$$= \frac{1}{2}m_b v_b^2 - \frac{1}{2}(m_b + m_{Al})v'^2$$

$$= \frac{1}{2} \times \frac{50}{1000} \times (10)^2 - \frac{1}{2} \times \frac{1000}{1000} \times (0.5)^2$$

$$= \frac{1}{2} \times \frac{4750}{1000} = 2.375$$

We assume no loss of heat energy to atmosphere as it is not given in question.

If ΔT = rise in temperature of bullet + Aluminium slab, then

$$m_b C_b \Delta T + m_{Al} C_{Al} \Delta T = 2.375 \quad \text{...(i)}$$

Also, $$\frac{m_b C_b \Delta T}{m_{Al} C_{Al} \Delta T} = \frac{m_b C_b}{m_{Al} C_{Al}} = \frac{50 \times 150}{950 \times 900}$$

$$= \frac{1}{114} \quad \text{...(ii)}$$

Substituting in Eq. (i), we get

$$115\, m_b C_b \Delta T = 2.375$$

So, $$m_b C_b \Delta T = \frac{2.375}{115}$$

and $$m_{Al} C_{Al} \Delta T = \frac{114 \times 2.375}{115}$$

Difference in amount of heat absorbed

$$= \frac{114 \times 2.375}{115} - \frac{2.375}{115} = \frac{113}{115} \times 2.375$$

$$= \frac{113}{115} \times 95\% \text{ of initial KE}$$

2. (d) ΔV_{Hg} = Change in volume of mercury = $\gamma V_0 \Delta T$

$$= 1.8 \times 10^{-4} \times 25 \times (100 - 20) = 0.36 \text{ cm}^3$$

$$\Delta V_g = \gamma V \cdot \Delta T = 0.054 \text{ cm}^3$$

So, tube now can hold 0.054 cm^3 extra amount of Hg and remaining flows out.

$\therefore$ Overflow amount = 0.36 − 0.054 = 0.306 cm^3

3. (d) $dp = -\rho g dH = -\frac{pM}{RT} \cdot g dH$

$$\Rightarrow \quad \int_{p_0}^{p} \frac{dp}{p} = -\int_0^H \frac{Mg}{RT} \cdot dH \quad \Rightarrow \quad p = p_0 e^{-\left(\frac{Mg}{RT}\right)H}$$

4. (c) Displaced volume = Upthrust

At temperature of 20°C, volume of displaced fluid

$$= V_0 (1 + \gamma \times 20) = \frac{1}{2} \times V_0 \quad \text{...(i)}$$

At temperature of 80°C, volume of displaced fluid

$$= V_0 (1 + \gamma \times 80) = \frac{3}{4} \times V_0 \quad \text{...(ii)}$$

Dividing Eq. (i) by Eq. (ii), we get

$$\frac{\frac{1}{2}}{\frac{3}{4}} = \frac{1 + 20\gamma}{1 + 80\gamma} \quad \Rightarrow \quad \gamma = 10^{-2}{}^\circ\text{C}^{-1}$$

5. (c) Let x number of moles transfer from hot bulb to cold bulb.

$$pV = nRT$$

$$\Rightarrow \quad 76 \times V = nR \times 273 \quad \text{...(A and B)}$$

$$p'V = (n + x)R \times 273 \quad \text{...(A)}$$

$$p'V = (n - x)R \times 335 \quad \text{...(B)}$$

Solving these equations, we get

$$\frac{p'}{76} = \frac{n + x}{n} = 1 + \frac{x}{n} = 1 + \frac{62}{602} = 1.103$$

$$\Rightarrow \quad p' = 83.83 \text{ cm of Hg}$$

6. (a) Contraction in rod,

$$\Delta l = l \alpha \Delta T = 0.5 \times 10^{-5} \times 100$$

$$= 5.0 \times 10^{-4} \text{ m}$$

Rod is extended by mass m,

$$\Delta l = \frac{Fl}{YA}$$

$\Rightarrow$ Stress developed in rod

$$= \frac{F}{A} = \frac{\Delta l \times Y}{l} = \frac{5 \times 10^{-4} \times 10^{11}}{5 \times 10^{11}} = 10^6 \text{ Nm}^{-2}$$

Potential energy stored in the rod is

$$U = \frac{1}{2} \times \text{Stress} \times \text{Strain}$$

$$= \frac{1}{2}\left(\frac{F}{A}\right) \times \left(\frac{\Delta l}{l}\right)$$

$$= \frac{1}{2} \times 10^6 \times \frac{5 \times 10^{-4}}{0.5} = 500 \text{ J}$$

7. (b) Heat lost by water is gained by the thermometer and finally temperature is 44.4°C.

$\therefore$ He at lost = Heat gained

$$m_w C_w (\Delta T)_w = m_T \cdot C_T (\Delta T)_T$$

$$300 \times 1 \times (T - 44.4) = 55 \times 0.2 \times 29.5$$

$$\Rightarrow \quad T - 44.4 = \frac{55 \times 0.2 \times 29.5}{300}$$

$$\Rightarrow \quad T - 44.4 \approx 1.1$$

$$\Rightarrow \quad T \approx 45.5^\circ\text{C}$$

8. (c) Given condition is

Initially, $$L = \sqrt{l_2^2 - \left(\frac{l_1}{2}\right)^2}$$

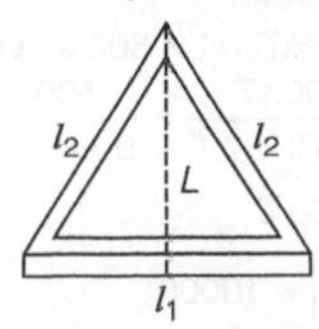

After temperature change,

$$L = \sqrt{l_2'^2 - \left(\frac{l_1'}{2}\right)^2}$$

Equating $$l_2^2 - \frac{l_1^2}{4} = l_2'^2 - \frac{l_1'^2}{4}$$

$$\Rightarrow \quad l_2^2 - \frac{l_1^2}{4} = l_2^2(1 + \alpha_2 \Delta T)^2 - \frac{l_1^2}{4}(1 + \alpha_1 \Delta T)^2$$

$$\Rightarrow \quad \frac{l_1^2}{4}[(1 + \alpha_1 \Delta T)^2 - 1] = l_2^2[(1 + \alpha_2 \Delta T)^2 - 1]$$

$$\Rightarrow \quad \frac{l_1^2}{4}(\alpha_1^2 \Delta T^2 + 2\Delta T \cdot \alpha_1) = l_2^2(\alpha_2^2 \Delta T^2 + 2\alpha_2 \Delta T)$$

Neglecting $\alpha_1^2 \Delta T^2$, $\alpha_2^2 \Delta T^2$ as it is very small.

$$\Rightarrow \quad \frac{l_1^2}{4} \cdot 2\Delta T \cdot \alpha_1 = l_2^2 \cdot 2\alpha_2 \Delta T$$

$$\Rightarrow \quad \frac{l_1^2}{l_2^2} = 4\frac{\alpha_2}{\alpha_1}$$

$$\Rightarrow \quad \frac{l_1}{l_2} = 2\sqrt{\frac{\alpha_2}{\alpha_1}}$$

9. (d) Angular momentum remains conserved.

$\therefore \quad I_0\omega_0 = I_1\omega_1$

$$\frac{2}{5}MR_0^2\omega_0 = \frac{2}{5}MR_1^2\omega_1$$

$$\Rightarrow \quad \omega_1 = \frac{\omega_0 \cdot R_0^2}{R_1^2} = \frac{\omega_0 \cdot R_0^2}{R_0^2(1+\alpha\Delta T)^2}$$

$$= \frac{\omega_0}{1+2\alpha\Delta T} = \frac{\omega_0}{1+0.004}$$

10. (b) For ΔT increase of temperature change in length ΔL of ring is

$$\Delta L = L\,\alpha\,\Delta T$$

or $\quad \frac{\Delta L}{L} = \alpha\,\Delta T$

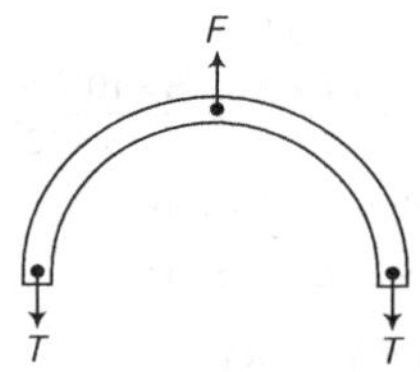

Let T = tension in ring, then stress in ring will be

$$\frac{T}{A} = \frac{Y\Delta L}{L} = Y\,\alpha\,\Delta T$$

or $\quad T = \text{Tension} = YA\,\alpha\,\Delta T$

As force, $\quad F = 2T = 2YA\,\alpha\,\Delta T$

11. (d) Change in volume of metal piece,

$$\Delta V = V_1 \cdot \gamma \cdot \Delta T$$

$$= \frac{m}{\rho} \cdot 3\alpha \cdot \Delta T$$

$$= \frac{1}{1000} \times 3 \times 3 \times 10^{-5} \times \Delta T$$

Work done against pushing the atmosphere,

$$E_1 = p\Delta V = \frac{10^5 \times 3 \times 3 \times 10^{-5}}{1000} \times \Delta T$$

$$= 9 \times 10^{-3}\Delta T$$

Heat absorbed by metal,

$$E_2 = mC\,\Delta T = 1 \times 500 \times \Delta T = 500\,\Delta T$$

Now, $\quad \frac{E_2}{E_1} = \frac{500\Delta T}{9 \times 10^{-3}\,\Delta T} = \frac{500}{9 \times 10^{-3}}$

and $\quad E_1 + E_2 = 10000$ J

$$\Rightarrow \quad E_1\left(1 + \frac{500}{9 \times 10^{-3}}\right) = 10000$$

$$\Rightarrow \quad E_1 = \frac{10000 \times 9 \times 10^{-3}}{(500 + 9 \times 10^{-3})}$$

$$\approx \frac{90}{500} = \frac{9}{50}$$

So, $\quad \frac{E_1}{E} = \frac{E_1}{E_1 + E_2} = \frac{9/50}{10000}$

$$= \frac{9}{5 \times 10^5} \text{ part}$$

12. (b) When temperature is increased, same volume remains submerged. Volume of cube increases and density of fluid decreases, so upthrust remains same.

$$\Rightarrow \quad F = F'$$

$$\Rightarrow \quad V\rho_L g = V'\rho_L' g$$

$$\Rightarrow \quad Ah_1\,\rho_L = A(1 + 2\,\alpha_s\Delta T)h_1 \cdot \frac{\rho_L}{(1+\gamma_L\Delta T)}$$

$$\Rightarrow \quad \gamma_L = 2\alpha_s = 2 \times 10^{-5}\,{}^\circ\text{C}^{-1}$$

13. (d) Equating pressures at p_1 and p_2

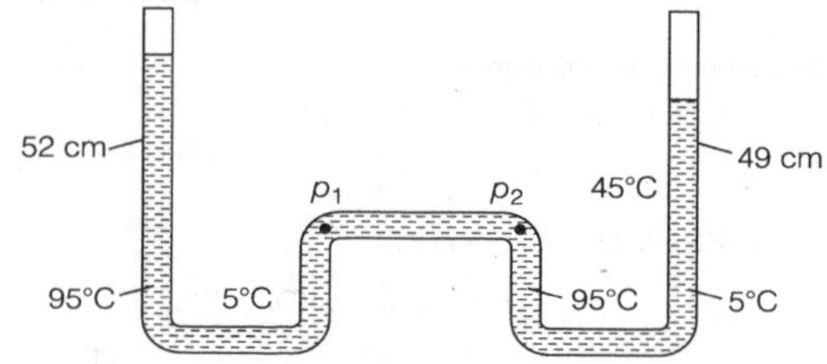

$$p_0 + \gamma_1\rho_{95°}g - \gamma\rho_{5°}g = p_0 - \gamma_2\rho_{5°}g - \gamma\rho_{95°}g$$

$$\Rightarrow \quad \frac{\rho_{95°}}{\rho_{5°}} = \frac{\gamma_2 + \gamma}{\gamma_1 + \gamma} = \frac{\dfrac{\rho_0}{1+95\gamma}}{\dfrac{\rho_0}{1+5\gamma}}$$

$$\Rightarrow \quad \frac{1+5\gamma}{1+95\gamma} = \frac{51+49}{52+49} = \frac{100}{101}$$

$$\Rightarrow \quad 101(1+5\gamma) = 100(1+95\gamma)$$

$$\Rightarrow \quad 101 + 505\gamma = 100 + 9500\,\gamma$$

$$\Rightarrow \quad 1 = 8995\,\gamma$$

$$\Rightarrow \quad \gamma = \frac{1}{8995}$$

$$\therefore \quad \alpha = \frac{\gamma}{3} = \frac{1}{26985}\,{}^\circ\text{C}^{-1}$$

14. (b) Loss of weight at 27°C = 46 − 30 = 16 = $V_1 \times 1.24\rho_l \times g$

Loss of weight at 42°C = 46 − 30.5 = 15.5 = $V_2 \times 1.2\rho_l \times g$

$$\Rightarrow \quad \frac{16}{15.5} = \frac{V_1}{V_2} \times \frac{1.24}{1.20}$$

As, $\quad V_2 = V_1\,[1 + 3\alpha\,(t_2 - t_1)]$

We have, $3\,\alpha\,(42 - 27) = 0.001042$

$$\Rightarrow \quad \alpha = 2.316 \times 10^{-5}\,{}^\circ\text{C}^{-1}$$

15. (c) $P\Delta t = mC\Delta T$

$$\frac{\Delta T}{\Delta t} = \frac{P}{mC} \quad \text{or} \quad \frac{\Delta t}{\Delta T} = \frac{mC}{P}$$

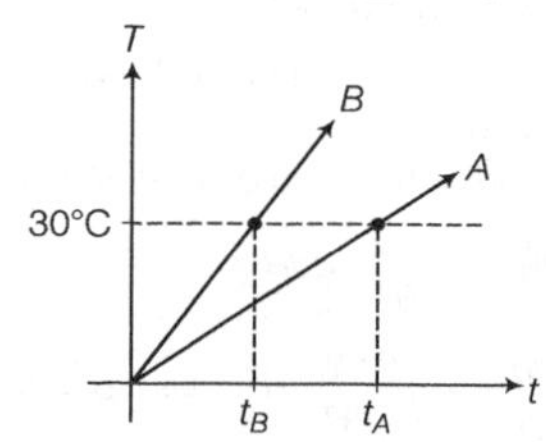

So, $\quad \frac{\Delta t}{\Delta T} = \frac{\text{Perpendicular}}{\text{Base}} = \tan\theta$

$$\therefore \quad \frac{\Delta t_A}{\Delta t_B} = \frac{\tan\theta_A}{\tan\theta_B} = \frac{\tan 30°}{\tan 60°} = \frac{1}{3}$$

16. (b) $V = V_0(1 + r\Delta\theta) \quad \Rightarrow \quad L^3 = L_x \cdot L_y \cdot L_z$

$$= L_0(1+\alpha_1\Delta\theta) \cdot L_0(1+\alpha_2\Delta\theta) \cdot L_0(1+\alpha_2\Delta\theta)$$

$$= L_0(1+\alpha_1\Delta\theta) \cdot L_0^2(1+\alpha_2\Delta\theta)^2$$

$$= L_0^3(1 + \alpha_1\Delta\theta + \alpha_2^2\Delta\theta^2 + 2\alpha_2\Delta\theta + \alpha_1\alpha_2^2\Delta\theta^2 + 2\alpha_1\alpha_2\Delta\theta^2)$$

$$\approx L_0^3(1 + \alpha_1\Delta\theta + 2\alpha_2\Delta\theta)$$

or $\quad 1 + r\Delta\theta = 1 + \alpha_1\Delta\theta + 2\alpha_2\Delta\theta$

$$\Rightarrow \quad r = \alpha_1 + 2\alpha_2$$

$$= 1.2 \times 10^{-5} + 2 \times 2.4 \times 10^{-5}$$

$$= 6.0 \times 10^{-5}\,{}^\circ\text{C}^{-1}$$

17. (a)
$$\frac{dQ}{dt} = H = KA\frac{\Delta T}{\Delta x}$$

$$\Rightarrow \quad \Delta T = \frac{1}{K}\left(\frac{H\Delta x}{A}\right)$$

which remarkable, $\quad y = mx$

$$\text{Slope} = \frac{1}{K}\left(\frac{H}{A}\right) \text{ or slope} \propto \frac{1}{K}$$

18. (d) Each layer forms a series of thermal resistances.

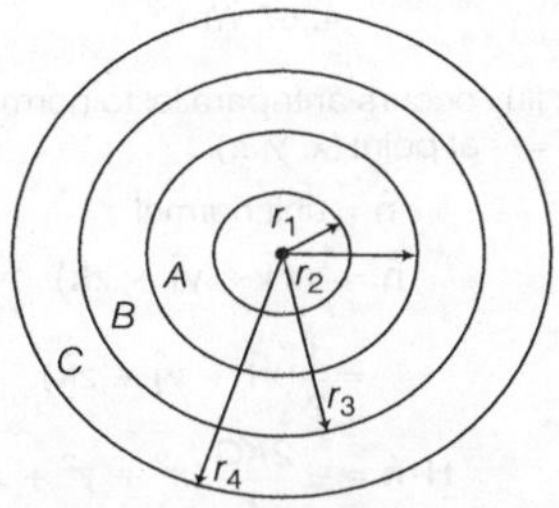

Equivalent circuit,
$$\frac{\log(r_2/r_1)}{2\pi K_A L} = \frac{\log(r_3/r_2)}{2\pi K_B L} = \frac{\log(r_4/r_3)}{2\pi K_C L}$$

Heat flow rate through a cylinder is
$$\frac{dQ}{dt} = H = KA\frac{\Delta T}{\Delta r}$$

$$= K2\pi L\left(\frac{r_o - r_i}{\log\left(\frac{r_o}{r_i}\right)}\right)\cdot\frac{T_o - T_i}{(r_o - r_i)} = \frac{2\pi KL\Delta T}{\log(r_o/r_i)}$$

and thermal resistance is $R_{th} = \dfrac{\ln\left(\frac{r_o}{r_i}\right)}{2\pi KL_1}$

So, for the assembly, heat flow rate per metre length of tube will be
$$\frac{dQ}{dt} = \frac{2\pi L(T_1 - T_4)}{\frac{\ln(r_2/r_1)}{K_A} + \frac{\ln(r_3/r_2)}{K_B} + \frac{\ln(r_4/r_3)}{K_C}}$$

$$= \frac{2\pi \times 1 \times (500 - 40)}{\frac{\ln\left(\frac{2}{1}\right)}{1} + \frac{\ln\left(\frac{8}{2}\right)}{2} + \frac{\ln\left(\frac{64}{8}\right)}{3}}$$

$$= \frac{2\pi \times 1 \times 460}{\ln(2) + 2\cdot\frac{\ln(2)}{2} + \frac{3\ln(2)}{3}}$$

$$= \frac{2\pi}{\ln(2)} \times 460 \text{ W}$$

19. (c) Equivalent electrical analog of wall is

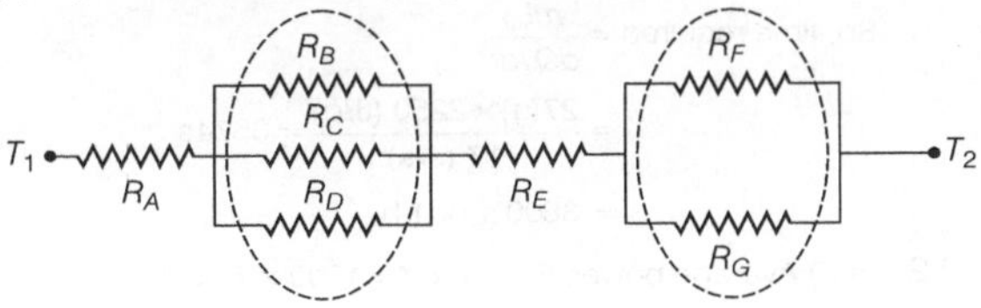

$$\frac{1}{R_{eq}} = \frac{1}{1.0} + \frac{1}{2.0} = \frac{2+1}{2}$$

$$\therefore \quad R_{eq} = \frac{2}{3}$$

$$\frac{1}{R_{eq}} = \frac{1}{0.2} + \frac{1}{0.4} + \frac{1}{0.8}$$

$$= \frac{4+2+1}{0.8}$$

$$R_{eq} = \frac{0.8}{7}$$

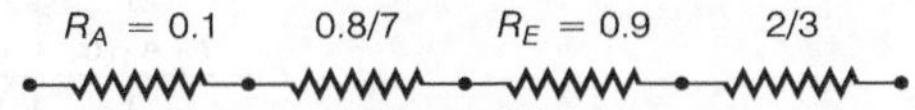

$$\therefore \quad R_{total} = 0.1 + \frac{0.8}{7} + 0.9 + \frac{2}{3}$$
$$\approx 1.78 \text{ m}^2\text{°C W}^{-1}$$

20. (a) p = Momentum transferred to surface

= Momentum from absorption + momentum from reflection

$$= \frac{U}{C} + \frac{2U}{C}$$

$$= \frac{3U}{C}$$

$$\Rightarrow \quad \frac{3U}{C} = 3 \times 10^{-4} \text{ kg - ms}^{-1}$$

$$\Rightarrow \quad U = 10^{-4} \times 3 \times 10^{8} \frac{\text{kgm}}{\text{s}} \times \frac{\text{m}}{\text{s}}$$

$$\Rightarrow \quad U = 3 \times 10^{4} \text{ kgm}^2\text{s}^{-2}$$

$$= 3 \times 10^{4} \text{ kg}\cdot\frac{\text{m}}{\text{s}^2}\cdot\text{m}$$

$$= 3 \times 10^{4} \text{ J} = 30 \text{ kJ}$$

So, emissive power of source $= \dfrac{\text{Energy}}{\text{Area} \times \text{time}}$

$$= \frac{30 \times 10^{4}}{30 \times 10^{-4} \times 1}\ \frac{\text{J}}{\text{m}^2\cdot\text{s}} \text{ or Wm}^{-2}$$

$$= 10^{8} \text{ Wm}^{-2}$$

21. (a) As, $Q \to 0$ as $T \to 0$

$\therefore$ As, Q cannot be negative T must be zero or positive. Both Statement I and Statement II are correct and Statement II correctly explains Statement I.

22. (b) Both Statement I and Statement II are correct but Statement II is not correct explanation for Statement I.

Normal boiling point of substance is point of intersection of normal intersection of normal pressure line and liquid vapour line.

23. (d) Statement I is incorrect only when we consider no radiative and convective losses from sides of the rod.

Statement II is correct.

24. (c) Statement I is correct but Statement II is incorrect.

Body enclosed remains at same temperature as energy received = energy radiated.

25. (c) Statement I is correct but Statement II is incorrect.

For natural convection currents, there must be an gravitational pull.

For an orbiting satellite, gravitational pull provides necessary centripetal force.

Revisal Problems (JEE Advanced)

1. (a) Walls are rigid so, $\delta_I + \delta_{II} = 0$

$$\Rightarrow \left(\alpha_1 \Delta TL - \frac{FL}{Y_1 A}\right)_I + \left(\alpha_2 \Delta TL - \frac{FL}{Y_2 A}\right)_{II} = 0$$

$$\Rightarrow (\alpha_1 + \alpha_2) L \Delta T = \frac{FL}{A}\left(\frac{1}{Y_1} + \frac{1}{Y_2}\right)$$

$$\Rightarrow F = \frac{T \cdot A \cdot (\alpha_1 + \alpha_2)}{\left(\frac{1}{Y_1} + \frac{1}{Y_2}\right)}$$

2. (c) New length of rod = Original length + Thermal expansion – Compression due to force

$$= L\left(1 + \alpha_1 \Delta T - \frac{F}{AY_1}\right)$$

3. (d) Similarly for second rod,

$$L' = L\left(1 + \alpha_2 \Delta T - \frac{F}{AY_2}\right)$$

4. (b) Heat supplied expands block and so work is done in pushing atmosphere, also heat is absorbed by the block. Let ΔT = rise in temperature, then change in volume,

$$\Delta V = V_0 \gamma \Delta T$$

$$\therefore \quad V_2 = V_1(1 + \gamma \Delta T)$$

Work done in pushing atmosphere = $p \Delta V = pV_0 \gamma \Delta T$

Heat absorbed by the block = $mC\Delta T$

So, $mC\Delta T + pV_0\gamma\Delta T = 20000$

$$\Rightarrow \Delta T = \frac{20000}{mC + pV_0\gamma}$$

$$\Rightarrow \Delta T = \frac{20000}{0.5 \times 400 + 10^5 \times \frac{0.5}{8 \times 10^3} \times 8 \times 10^{-5}}$$

$$= \frac{20000}{200 + \frac{1}{2000}} \approx 100°C$$

So, $T_{final} \approx 100 + 30 = 130°C$

5. (a) Work done against atmosphere

$$= p\Delta V = pV_0\gamma\Delta T$$

$$= 10^5 \times \frac{0.5}{8 \times 10^3} \times 8 \times 10^{-5} \times 100 = 0.05 \text{ J}$$

6. (c) Remaining heat goes into block increasing its internal energy, which is slightly less than 20000 J.

7. (a) Taking centre of ball at origin, $T \propto r^2$

or $T = Cr^2$

As, $r^2 = x^2 + y^2 + z^2$

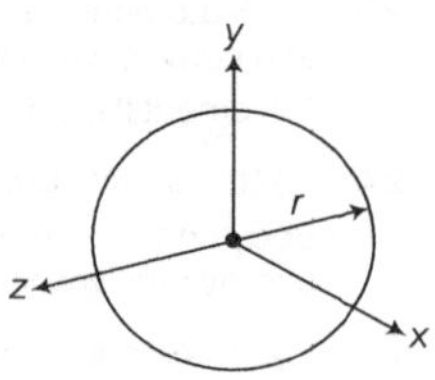

$$T(1, 1, 1) = 400 \text{ K}$$

$$\therefore \quad T = 400 = C \cdot (1^2 + 1^2 + 1^2)$$

or $C = \frac{400}{3}\left(\frac{\text{K}}{\text{m}^2}\right)$

or $T = \frac{400}{3}(x^2 + y^2 + z^2)$

This is temperature function of given space.

Now, heat flux is given by

$$\mathbf{H} = -K\Delta T = -\left(\hat{\mathbf{i}}\frac{\partial T}{\partial x} + \hat{\mathbf{j}}\frac{\partial T}{\partial y} + \hat{\mathbf{k}}\frac{\partial T}{\partial z}\right)$$

$$\therefore \quad \mathbf{H} = \frac{400}{3}K(\hat{\mathbf{i}}\, 2x + \hat{\mathbf{j}} 2y + \hat{\mathbf{k}} 2z)$$

$$= \frac{-80 \times 400 \times 2}{3}(x\hat{\mathbf{i}} + y\hat{\mathbf{j}} + z\hat{\mathbf{k}})$$

So, magnitude of heat flux is

$$|\mathbf{H}| = \frac{80 \times 800}{3} \times \sqrt{x^2 + y^2 + z^2} = 80 \times 800 \times \frac{2}{3}$$

(As for a sphere of radius 2, $r^2 = x^2 + y^2 + z^2 = 4$

$$\therefore \quad |\mathbf{H}| = \frac{12800}{3} \text{ Wm}^{-2} \text{ into the sphere.}$$

$$\approx 4267 \text{ Wm}^{-2}$$

8. (a) Now, heat flux occurs anti-parallel to normal to $x^2 + y^2 + z^2 = r^2$ at point (x, y, z).

$$\therefore \quad \hat{\mathbf{n}} = \text{unit normal}$$

$$\Rightarrow \quad \hat{\mathbf{n}} = \frac{1}{r}(x\hat{\mathbf{i}} + y\hat{\mathbf{j}} + z\hat{\mathbf{k}})$$

$$= \frac{1}{2}(x\hat{\mathbf{i}} + y\hat{\mathbf{j}} + z\hat{\mathbf{k}})$$

So, $\mathbf{H} \cdot \hat{\mathbf{n}} = -\frac{2KC}{r}(x^2 + y^2 + z^2)$

For sphere of radius r, $x^2 + y^2 + z^2 = r^2$

$$\therefore \quad \mathbf{H} \cdot \hat{\mathbf{n}} = -2KCr$$

$\therefore$ Heat flow rate across the sphere is

$$\frac{dQ}{dt} = \mathbf{H} \cdot \hat{\mathbf{n}} = -2KCr$$

$$= -2 \times 80 \times \frac{400}{3} \times 2$$

$$= \frac{-128000}{3} \text{ W}$$

$$= 42666.66 = 42667 \text{ W}$$

9. (a) $a \to b$, solid to liquid phase conversion : melting,
$a \to d$, solid to gas or vapour : sublimation,
$d \to a$, gaseous to solid state : deposition,
$d \to c$, gas to liquid phase transformation : condensation.
Hence, (i) $\to$ (t), (ii) $\to$ (r), (iii) $\to$ (q), (iv) $\to$ (p)

10. (d) In (a), (b) and (c) bonds breaks and absorbs energy in b to a transformation is from liquid to solid, so bonds are formed.

11. (a) Time = $\frac{\text{Heat required}}{\text{Rate of heat transfer}}$

$$= \frac{mL}{(dQ/dt)}, \; n = \frac{m}{M} = \frac{nML}{\frac{dQ}{dt}}$$

$$n = \frac{m}{M} = \frac{\text{Sample mass}}{\text{Molar mass}}$$

Also, 1 mL = 1 g of water.

Now, molar mass of water = 18 g

$\therefore$ 1.5 mole of H_2O = 27 g

So, time required = $\frac{mL_V}{dQ/dt}$

$$= \frac{27(\text{g}) \times 2250 \text{ (J/g)}}{17 \text{ (J/s)}} = 3574\text{s}$$

$$\approx 3600 \text{ s or 1 h}$$

12. (a,d) Average power, $P_{av} = I \times A = 1000 \times 8 \times 20$

$$= 1.6 \times 10^5 \text{ W}$$

Average radiation pressure,

$$p = \frac{I}{C} = \frac{1000}{3 \times 10^8} \text{ Nm}^{-2}$$

$$= 3.33 \times 10^{-6} \text{ Nm}^{-2}$$

Force = Pressure × Area = $p \times A$

$$= 3.33 \times 10^{-6} \times 8 \times 20$$

$$= 5.33 \times 10^{-4} \text{ N}$$

13. (b,c,d) Following figure shows distribution of energy.

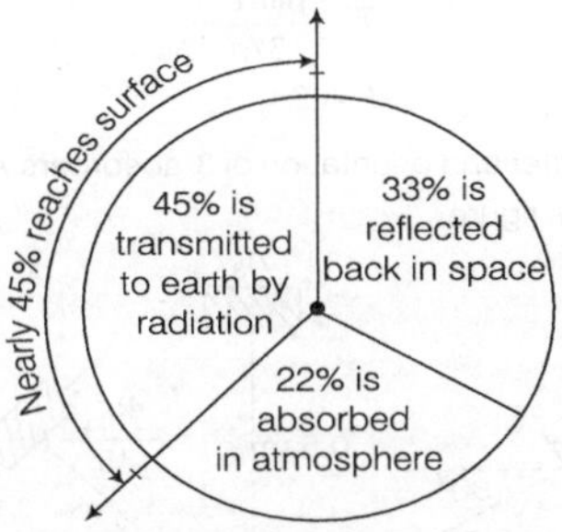

14. (a,b) For gases and transparent solids like glass and salt crystals at elevated temperature, emission is a volumetric phenomena.

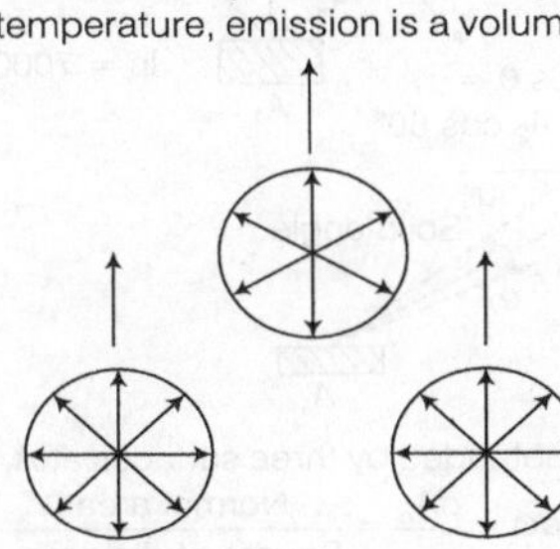

In most solids and liquids, radiation emitted from interior molecules (or atoms) is strongly absorbed by adjoining molecules. So, radiation emission is a surface phenomena in solids and liquids.

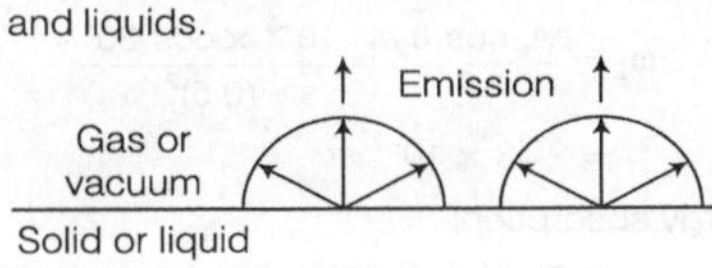

15. (a,d) Rods are connected as :

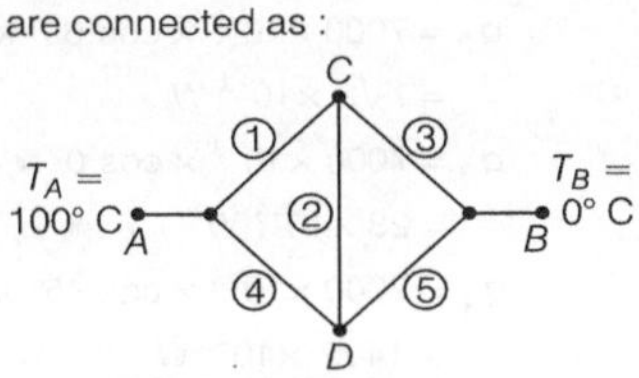

Following electrical analogy, $T_C = T_D$.

No heat flow through rod 2.

$\Rightarrow$ Equivalent network of rods is

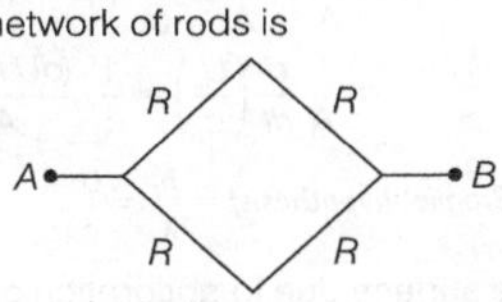

$\Rightarrow$ Equivalent resistance is R, i.e.

$$R = \frac{l}{KA}$$

16. (a,b) $H = KA\frac{d\theta}{dr} = K(2\pi r l)\frac{d\theta}{dr} \Rightarrow \int_{r_1}^{r_2}\frac{dr}{r} = \frac{2\pi K l}{H}\int_{\theta_1}^{\theta_2} d\theta$

$$\Rightarrow \frac{dQ}{dt} = H = \frac{2\pi K l(\theta_2 - \theta_1)}{\log\left(\frac{r_2}{r_1}\right)}$$

Substituting the values, we have

$$H = 80\,\pi$$

So, $\frac{dm}{dt}\cdot L = 80\,\pi$

$$\frac{dm}{dt} = \frac{80\pi}{L} = \frac{80\pi}{80 \times 4200}\ \text{kg s}^{-1} = \frac{\pi}{4200}\text{kg s}^{-1}$$

17. (d) Heat loss is due to convection as well as radiation.

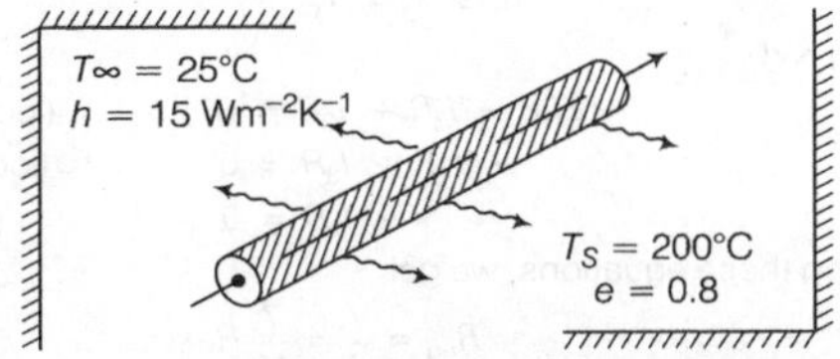

$$q = q_{\text{convection}} + q_{\text{radiation}}$$

$$\therefore \quad q = \lambda\pi DL(T_S - T_\infty) + e\,\pi DL\,\sigma(T_S^4 - T_{\text{Surr}}^4)$$

$$\Rightarrow \frac{q}{L} = 15 \times \pi \times 0.07\,(200 - 25) + 0.8 \times \pi \times 0.07 \times 5.67 \times 10^{-8} \times (473^4 - 298^4)$$

$$\Rightarrow \frac{q}{L} = 577 + 421 = 998\ \text{Wm}^{-1}$$

18. (d) Iodine melts at 1 atm pressure at 113.6°C which its normal melting point. It is also triple point of iodine.

19. (b) Wood piece just sinks, when density of wood = density of benzene.

$$\therefore \quad \rho_{t\,(\text{wood})} = \rho_{t\,(\text{benzene})}$$

$$\frac{\rho_w}{1 + \gamma_w \Delta T} = \frac{\rho_b}{1 + \gamma_b \Delta T}$$

$$\Rightarrow \rho_w + \rho_w\gamma_b\,\Delta T = \rho_b + \rho_b\gamma_w\Delta T \Rightarrow \frac{\rho_w - \rho_b}{\rho_b\gamma_w - \rho_w\gamma_b} = \Delta T$$

$$\Rightarrow \Delta T = \frac{880 - 900}{900 \times 1.5 \times 10^{-4} - 880 \times 1.2 \times 10^{-3}} = 21.7°\text{C}$$

20. (c) As, $\lambda_{m_1}T_1 = \lambda_{m_2}T_2$ and $\frac{P_0}{P} \propto \left(\frac{T_0}{T}\right)^4$

$$\Rightarrow \frac{P}{P_0} = \frac{256}{81} = \left(\frac{4}{3}\right)^4$$

$$\therefore \frac{T_0}{T} = \frac{3}{4} = \frac{\lambda_{m_2}}{\lambda_{m_1}} \Rightarrow \lambda_{m_2} = \frac{3}{4}\lambda_{m_1}$$

Shift of wavelength,

$$\Delta\lambda = \lambda_{m_2} - \lambda_{m_1} = \frac{3}{4}\lambda_m - \lambda_m = -\frac{1}{4}\lambda_m$$

Wavelength shifts to a smaller value.

21. (a) If θ = temperature of junction C, then

$$\frac{2KA\,(\theta - 100)}{l} + \frac{(K/2)A(\theta - 0)}{l} = 200$$

$$\Rightarrow \theta = 880°\text{C}$$

Also, $\frac{KA(880 - 0)}{2l} \times t = 440 \times 80$

$$\Rightarrow t = 800\ \text{s}$$

22. (c) We make use of electrical analogy and Kirchhoff's laws to solve this problem.

Case I Hot junction is A and cold junction is D.

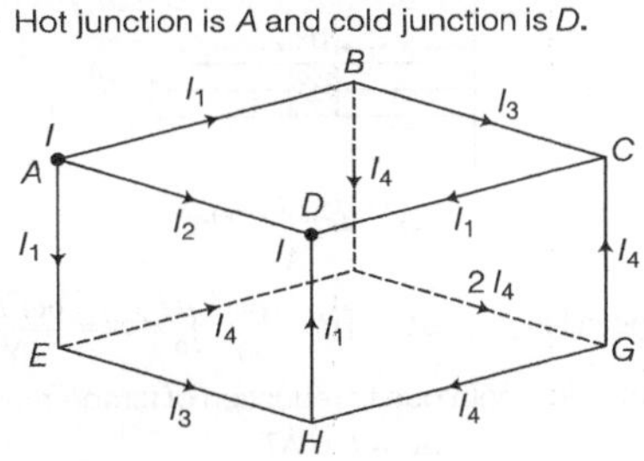

Current entring from A is $I = 2I_1 + I_2$

Current leaving from D is $I = 2I_1 + I_2$

also, $I_1 = I_3 + I_4$

Using KVL,

$$-2I_1R - I_3R + I_2R = 0 \quad \text{(loop } ABCDA\text{)}$$

$$-4I_4R + I_3R = 0 \quad \text{(loop } EFGHE\text{)}$$

Also, $$V - I_2R = 0 \quad \text{(loop } ADA\text{)}$$

Solving these equations, we get

$$R_{eq} = \frac{V}{I} = \frac{7}{12}R$$

As, thermal resistance in this resistance

$$R = \frac{l}{KA} = \frac{1}{100 \times 10 \times 10^{-4}}$$

$$= 0.1 \text{ cal}^{-1}\text{s}^\circ\text{C}$$

So, $$R_{eq} \text{ or } R_{AD} = \frac{7}{12} \times 0.1 = \frac{0.7}{12} \text{ cal}^{-1}\text{s}^\circ\text{C}$$

***Case* II** Hot junction is A and cold junction is H.

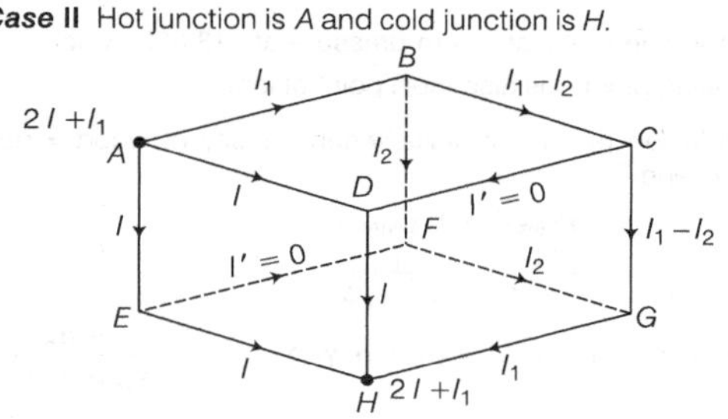

$$(2I + I_1)R_{eq} = IR + IR \quad \text{(loop } AHA\text{)}$$

$$I_2R + (I_1 - I_2)R - IR = 0 \quad \text{(loop } ABCDA\text{)}$$

$$-(I_2R + I_1R) + IR = 0 \quad \text{(loop } EFGHE\text{)}$$

Solving these, we get

$$R_{eq} = R_{AH} = \frac{3}{4}R = \frac{0.3}{4} \text{ cal}^{-1}\text{s}^\circ\text{C}$$

***Case* III** Hot junction is A and cold junction is G.

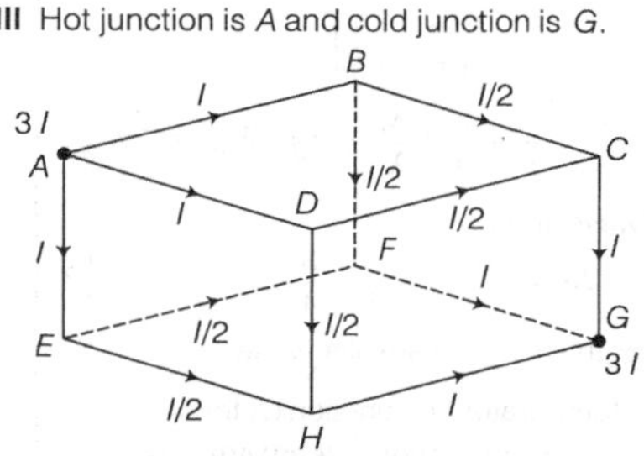

$$R_{eq} = R_{AG} = \frac{5}{6}R$$

$$= \frac{0.5}{6} \text{ cal}^{-1}\text{s}^\circ\text{C}$$

Hence, (i) → (r), (ii) → (p), (iii) → (q)

23. (3) Elongation dl in an elemental length dx at a distance x from axis of rotation is given by

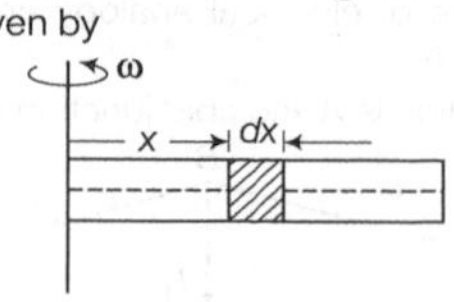

$$dl = \frac{(\rho A\omega^2 x dx)x}{A \cdot Y}$$

$\therefore$ Change in length $= \Delta L = \int_0^L dl = \frac{\rho\omega^2}{Y}\int_0^L x^2 dx = \frac{\rho\omega^2 L^3}{3Y}$ ...(i)

and change in length due to reduction of temperature will be

$$\Delta L = L\,\alpha\,\Delta T \quad \text{...(ii)}$$

From Eqs. (i) and (ii), we get

$$\Delta T = \frac{\rho\omega^2 L^2}{3Y\alpha}$$

$$\therefore \quad k = 3$$

24. (3) A_1 is emitter and orientation of 3-absorbers A_2, A_3 and A_4 is as shown in the figure.

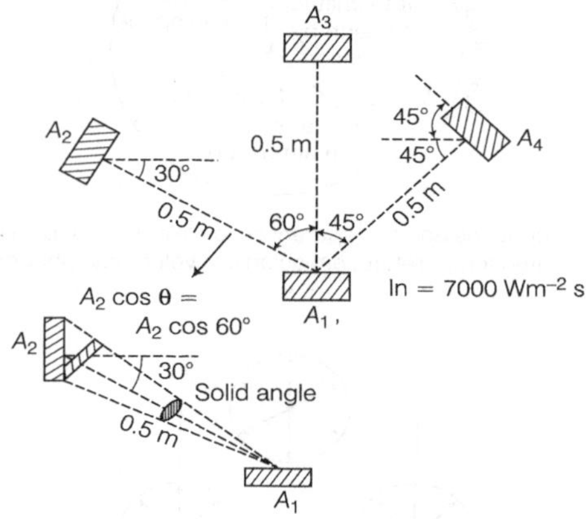

Solid angles subtended by three surfaces at A_1 are

$$d\omega = \frac{dA_n}{r^2} = \frac{\text{Normal area}}{\text{Square of distance}}$$

So, $$\omega_3 = \omega_4 = \frac{A}{r^2} = \frac{10^{-3}}{(0.5)^2} = 4 \times 10^{-3} \text{ sr}$$

and $$\omega_2 = \frac{A_2 \cos\theta_2}{r_2^2} = \frac{10^{-3} \times \cos 30^\circ}{(0.5)^2}$$

$$= 2\sqrt{3} \times 10^{-3} \text{ sr}$$

Rate of energy absorption,

$$q = I \times A_n \times \omega$$

$$\therefore \quad q_2 = 7000 \times 10^{-3} \times \cos 60^\circ \times 2\sqrt{3} \times 10^{-3}$$

$$= 7\sqrt{3} \times 10^{-3} \text{ W}$$

$$q_3 = 7000 \times 10^{-3} \times \cos 0^\circ \times 4 \times 10^{-3}$$

$$= 28 \times 10^{-3} \text{ W}$$

$$q_4 = 7000 \times 10^{-3} \times \cos 45^\circ \times 4 \times 10^{-3}$$

$$= 14\sqrt{2} \times 10^{-3} \text{ W}$$

Now, pressure exerted on the surface,

$$p = \frac{F}{A} = \frac{1}{A} \times \frac{dp'}{dt}, \; p' = \text{momentum}$$

$$= \frac{1}{A}\frac{d}{dt}\left(\frac{U}{c}\right) = \frac{1}{c} \cdot \frac{(dU/dt)}{A}$$

Note $p = \frac{h}{\lambda}$ *(de-Broglie' hypothesis)* $= \frac{h_f}{\lambda_f} = \frac{U}{c}$

So, force on the surface due to absorption of radiation is

$$F = p \times A = \frac{dU/dt}{c}$$

As, $$F = m \times a$$

Acceleration of surfaces, $a = \frac{dU/dt}{m \times c} = \frac{q}{m \times c}$

$$\Rightarrow \quad a_2 : a_3 : a_4 :: q_2 : q_3 : q_4$$

$$\Rightarrow \quad \frac{a_2}{a_4} = \frac{q_2}{q_4} = \frac{7\sqrt{3} \times 10^{-3}}{14\sqrt{2} \times 10^{-3}}$$

$$\Rightarrow \quad \frac{a_2}{a_4} = \frac{\sqrt{3}}{2\sqrt{2}} \Rightarrow \frac{a_2^2}{a_4^2} = \frac{3}{8}$$

Hence, $$k = 3$$

JEE Main & AIEEE Archive

1. (c) $p = \frac{1}{3}\left(\frac{U}{V}\right) \propto T^4$

For an ideal gas $pV = nRT$

$\therefore \quad p = \frac{nR'T}{V}$ (R' – molar gas constant)

$\therefore \quad \frac{nR'T}{V} \propto T^4$

$\frac{nR'}{V} \propto T^3$

$\therefore \quad \frac{1}{V} \propto T^3$

$\frac{1}{\frac{4\pi}{3}R^3} \propto T^3$

$T^3 \propto \frac{1}{R^3}$

$T \propto \frac{1}{R}$

2. (a) Cubical expansion,

$\Delta V = \gamma V \Delta T = 3\,\alpha\, V \Delta T$ (as $\gamma = 3\alpha$)

$= 3 \times 23 \times 10^{-6} \times \left(\frac{4}{3}\pi (10)^3\right) \times 100 \quad \left(\because r = \frac{d}{2} = 10\text{ cm}\right)$

$= 28.9$ cc

3. (a) As, work done = 0 (as, $W = p\Delta V$ and $\Delta V = 0$, so $W = 0$)

$\Delta U = mC\Delta T = 100 \times 10^{-3} \times 4184 \times (50 - 30) \approx 8.4$ kJ

4. (b) We know that,

Rate of flow of heat, $\frac{dQ}{dt} = KA\frac{d\theta}{dx}$

In steady state, flow of heat,

$d\theta = \frac{dQ}{dt} \cdot \frac{1}{KA}\,dx$

$\Rightarrow \quad \theta_H - \theta = K'x \quad \Rightarrow \quad \theta = \theta_H - K'x$

Equation $\theta = \theta_H - K'x$ represents a straight line.

JEE Advanced & IIT JEE Archive

1. (2) $\frac{p_A}{p_B} = \frac{A_A T_A^4}{A_B T_B^4} = \frac{R_A^2 T_A^4}{R_B^2 T_B^4}$

$\Rightarrow \quad \frac{T_B}{T_A} = \sqrt{\frac{R_A}{R_B}} \cdot \left(\frac{p_B}{p_A}\right)^{1/4}$

According to Wien's displacement law.

$\frac{\lambda_A}{\lambda_B} = \frac{T_B}{T_A} = \sqrt{400}\left(\frac{1}{10^4}\right)^{1/4} = 2$

2. (a) $R_1 = R_1 + R_2 = \left(\frac{l}{KA}\right) + \left(\frac{l}{2KA}\right) = \frac{3}{2}\left(\frac{l}{KA}\right)$

$\frac{1}{R_{II}} = \frac{1}{R_1} + \frac{1}{R_2} = \frac{KA}{l} + \frac{2KA}{l}$ or $R_{II} = \frac{l}{3KA} = \frac{R_1}{4.5}$

Since, thermal resistance R_{II} is 4.5 times less than thermal resistance R_1.

$\therefore \quad t_{II} = \frac{t_1}{4.5} = \frac{9}{4.5}\text{ s} = 2\text{ s}$

3. (d) The process may be assumed to be isobaric.

$\therefore \quad Q = n\,C_p\,\Delta T = (2)\left(\frac{5}{2}R\right)(5)$

$= 5 \times 8.31 \times 5$
$= 207.75\text{ J} \approx 208\text{ J}$

4. (c) Let temperature of middle plate in steady state is T_0.

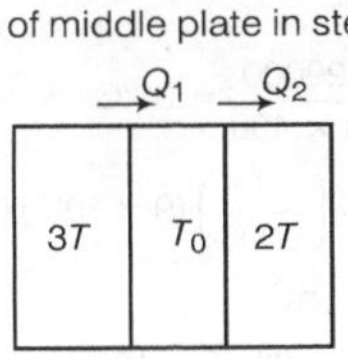

$Q_1 = Q_2$

$Q =$ net rate of heat flow

$\therefore \quad \sigma A(3T)^4 - \sigma A T_0^4 = \sigma A T_0^4 - \sigma A(2T)^4$

Solving this equation, we get

$T_0 = \left(\frac{97}{2}\right)^{1/4} T$

5. (a,c,d) Thermal resistance, $R = \frac{l}{KA}$

$\therefore \quad R_A = \frac{L}{(2K)(4Lw)}$ (Here, w = width)

$= \frac{1}{8Kw}$

$R_B = \frac{4L}{3K\,(Lw)} = \frac{4}{3Kw}$

$R_C = \frac{4L}{(4K)(2Lw)} = \frac{1}{2Kw}$

$R_D = \frac{4L}{(5K)(Lw)} = \frac{4}{5Kw}$

$R_E = \frac{L}{(6K)(Lw)} = \frac{1}{6Kw}$

$R_A : R_B : R_C : R_D : R_E = 15 : 160 : 60 : 96 : 12$

So, let us write, $R_A = 15\,R$, $R_B = 160\,R$ etc and draw a simple electrical circuit as shown in the figure.

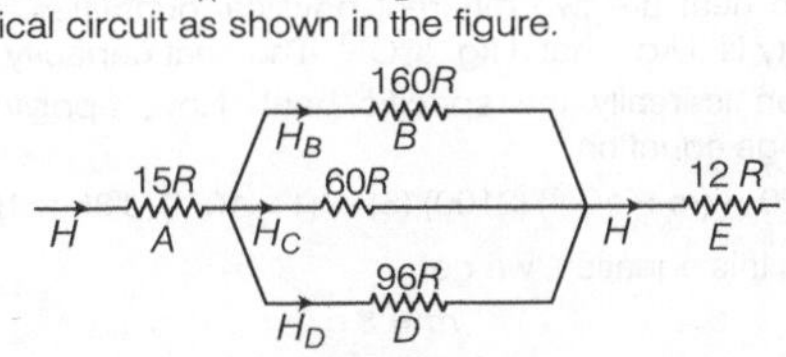

H = Heat current = Rate of heat flow

$H_A = H_E = H$ (let)

$\therefore$ Option (a) is correct.

In parallel current distributes in inverse ratio of resistance.

$\therefore \quad H_B : H_C : H_D = \frac{1}{R_B} : \frac{1}{R_C} : \frac{1}{R_D}$

$= \frac{1}{160} : \frac{1}{60} : \frac{1}{96} = 9 : 24 : 15$

$\therefore \quad H_B = \left(\frac{9}{9 + 24 + 15}\right) H = \frac{3}{16} H$

$H_C = \left(\frac{24}{9 + 24 + 15}\right) H = \frac{1}{2} H$

and $$H_D = \left(\frac{15}{9+24+15}\right)H = \frac{5}{16}H$$

$$H_C = H_B + H_D$$

∴ Option (d) is correct.

Temperature difference (let us call it T)

$$= \text{(Heat current)} \times \text{(Thermal resistance)}$$

$$T_A = H_A R_A = (H)(15R) = 15\,HR$$

$$T_B = H_B R_B = \left(\frac{3}{16}H\right)(160\,R) = 30\,HR$$

$$T_C = H_C R_C = \left(\frac{1}{2}H\right)(60\,R) = 30HR$$

$$T_D = H_D R_D = \left(\frac{5}{16}H\right)(96\,R) = 30\,HR$$

$$T_E = H_E R_E = (H)(12\,R) = 12\,HR$$

Here, T_E is minimum. Therefore option (c) is also correct.

6. (3) $\Delta l_1 = \frac{FL}{AY} = \frac{mgL}{\pi r^2 Y}$ = Increase in length

$\Delta l_2 = L\,\alpha\,\Delta\theta$ = Decrease in length

To regain its original length, $\Delta l_1 = \Delta l_2$

$$\therefore \quad \frac{mgL}{\pi r^2 Y} = L\,\alpha\,\Delta\theta$$

$$\therefore \quad m = \left(\frac{r^2 Y\,\alpha\,\Delta\theta}{g}\right)$$

Substituting the values, we get

$$m \approx 3\text{ kg}$$

7. (9) $$\lambda_m \propto \frac{1}{T}$$

$$\therefore \quad \frac{\lambda_A}{\lambda_B} = \frac{T_B}{T_A} = \frac{500}{1500} = \frac{1}{3}$$

$$E \propto T^4 A$$

(where, A = surface area = $4\pi R^2$)

$$\therefore \quad E \propto T^4 R^2$$

$$\frac{E_A}{E_B} = \left(\frac{T_A}{T_B}\right)^4 \left(\frac{R_A}{R_B}\right)^2 = (3)^4 \left(\frac{6}{18}\right)^2 = 9$$

8. (8) Language of question is slightly wrong. As, heat capacity and specific heat are two different physical quantities. Unit of heat capacity is J kg^{-1} not J kg^{-1}°C^{-1}. The heat capacity given in the question is really the specific heat. Now, applying the heat exchange equation

$$420 = (m \times 10^{-3})(2100)(5) + (1 \times 10^{-3})(3.36 \times 10^5)$$

Solving this equation, we get

$$m = 8\text{ g}$$

9. (9) Heat will flow both sides from point P.

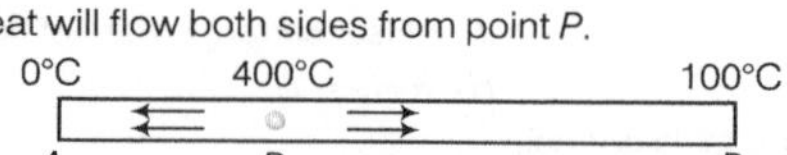

$$L_1 \frac{dm_1}{dt} = \left(\frac{\text{Temperature difference}}{\text{Thermal resistance}}\right)_1 = \frac{400}{(\lambda x)/KA} \quad \text{...(i)}$$

$$L_1 \frac{dm_2}{dt} = \frac{400 - 100}{(100-\lambda)x/KA} \quad \text{...(ii)}$$

In about two equations,

$$\frac{dm_1}{dt} = \frac{dm_2}{dt} \text{ (given)}$$

Here, $L_1 = 80\ \text{calg}^{-1}$

and $L_2 = 540\ \text{calg}^{-1}$

Solving these two equations, we get $\lambda = 9$.

10. (c) (i) → (s), (ii) → (q), (iii) → (p,q), (iv) → (q,r)

11. (d) Since, the temperature of black body is constant, total heat absorbed = total heat radiated.

12. 0.05 kg steam at 373 K $\xrightarrow{Q_1}$ 0.05 kg water at 373 K

0.05 kg water at 373 K $\xrightarrow{Q_2}$ 0.05 kg water at 273 K

0.45 kg ice at 253 K $\xrightarrow{Q_3}$ 0.45 kg ice at 273 K

0.45 kg ice at 273 K $\xrightarrow{Q_4}$ 0.45 kg water at 273 K

$Q_1 = (50)(540) = 27000\text{ cal} = 27\text{ kcal}$

$Q_2 = (50)(1)(100) = 5000\text{ cal} = 5\text{ kcal}$

$Q_3 = (450)(0.5)(20) = 4500\text{ cal} = 4.5\text{ kcal}$

$Q_4 = (450)(80) = 36000\text{ cal} = 36\text{ kcal}$

Now, since $Q_1 + Q_2 > Q_3$ but $Q_1 + Q_2 < Q_3 + Q_4$ ice will come to 273 K from 253 K but whole ice will not melt. Therefore, temperature of the mixture is 273 K.

13. (a) Energy gained by water (in 1 s)

= Energy supplied – Energy lost

= 1000 J – 160 J = 840 J

Total heat required to raise the temperature of water from 27°C to 77°C is $ms\Delta\theta$.

Hence, the required time

$$t = \frac{ms\Delta\theta}{\text{Rate by which energy is gained by water}}$$

$$= \frac{(2)(4.2 \times 10^3)(50)}{840} = 500\text{ s} = 8\text{ min } 20\text{ s}$$

∴ Correct answer is (a).

14. (c) Glass of bulb heats due to filament by radiation. Hence, correct answer is (c).

15. (c) $\lambda_m T$ = constant

From the graph $T_3 > T_2 > T_1$

Temperature of Sun will be maximum.

Therefore, (c) is the correct option.

Note *Although graphs are not very clear.*

16. (a) 1 calorie is the heat required to raise the temperature of 1 g of water from 14.5°C to 15.5°C at 760 mm of Hg.

17. (i) From $\Delta Q = ms\Delta T$

$$\Delta T = \frac{\Delta Q}{ms} = \frac{20000}{1 \times 400} = 50°\text{C}$$

(ii) $$\Delta V = V\gamma\,\Delta T = \left(\frac{1}{9000}\right)(9 \times 10^{-5})(50)$$

$$= 5 \times 10^{-7}\ \text{m}^3$$

$$\therefore\ W = p \cdot \Delta V = (10^5)(5 \times 10^{-7}) = 0.05\text{ J}$$

(iii) $$\Delta U = \Delta Q - W = (20000 - 0.05)\text{ J}$$

$$= 19999.95\text{ J}$$

2. Kinetic Theory of Gases and Thermodynamics

DPP-1 Kinetic Theory of Gases-I (Real and Ideal Gases)

1. (i) Mean speed is

$$v_{av} = \frac{v_1 + v_2 + \cdots + v_{12}}{12} = \frac{6.0 + 2.0 + \cdots + 8.0}{12}$$

$$= \frac{60}{12} = 5\ \text{cms}^{-1}$$

(ii) v_{rms} = rms speed

$$= \sqrt{\left(\frac{v_1^2 + v_2^2 + \cdots + v_{12}^2}{12}\right)}$$

$$= \sqrt{\frac{(6.0)^2 + (2.0)^2 + \cdots + (8.0)^2}{12}} = \sqrt{29.67} \approx 5.45\ \text{cms}^{-1}$$

(iii) As, particle has no preferred direction,

$$v_{rms}^2 = v_x^2 + v_y^2 + v_r^2 = 3v_x^2$$

$$\Rightarrow \quad v_x = \frac{v_{rms}}{\sqrt{3}}$$

Time taken for 1 crossing of the room is

$$t = \frac{d}{v} = \frac{d}{v_x} = \frac{\sqrt{3}\,d}{v_{rms}}$$

and so, time for one round trip is $2t = \dfrac{2\sqrt{3}\,d}{v_{rms}}$

Hence, number of back and forth round trips per second

$$= \frac{1}{2t} = \frac{v_{rms}}{2\sqrt{3}d} = \frac{5.45\ \text{cms}^{-1}}{2 \times 1.7 \times 1\ \text{cm}}$$

$= 1.59 \approx$ one and half trip nearly in 1 second.

(iv) Most probable speed is value occuring most of the time in list.

$\therefore \quad v_{mp} = 6\ \text{cms}^{-1}$

2. (i) $v_{rms} = \sqrt{\left(\dfrac{3kT}{m}\right)}$; $m = 4\ \text{amu} = 4 \times 1.66 \times 10^{-27}\ \text{kg}$

$k = 1.38 \times 10^{-23}\ \text{JK}^{-1}$, $T_{Sun} = 6000\ \text{K}$

$$\therefore \quad v_{rms} = \sqrt{\frac{3 \times 1.38 \times 10^{-23} \times 6000}{4 \times 1.66 \times 10^{-27}}}$$

$$= 6116\ \text{ms}^{-1} \approx 6 \times 10^3\ \text{ms}^{-1}$$

(ii) $\dfrac{v_{rms_2}}{v_{rms_1}} = \sqrt{\dfrac{T_2}{T_1}} = \dfrac{2}{1}$

$\Rightarrow \quad T_2 = 4T_1 = 4 \times 6000 = 24000\ \text{K}$

(iii) From gas equation, $pV = nRT$, if volume and amount of gas are held constant, then the temperature is proportional to the pressure.

$$pV = nRT \Rightarrow p = \frac{nR}{V} \cdot T$$

$$\Rightarrow \quad p \propto T$$

Thus, when pressure is tripled, temperature is also tripled.

$\therefore$ v_{rms} will be multiplied by a factor of $\sqrt{3}$.

3. (i) 2×10^{-3} kg of H_2 have 6.02×10^{23} molecules.

$\therefore$ Number of molecules in one gram $= \dfrac{6.02 \times 10^{23}}{2}$

$= 3.01 \times 10^{23}$ molecules

So, molecular energy is

$$E = N \times \text{KE of 1 molecule} = N \times \frac{3}{2} k_B T$$

$$= \frac{1}{2} N_A \times \frac{3}{2} k_B T = \frac{3}{4} RT$$

$$= \frac{3}{4} \times 8.3 \times 323 = 2010.675\ \text{J} \approx 2011\ \text{J}$$

(ii) Boltzmann distribution law is

$$n_V(E) = n_0 e^{-E/k_B T}$$

so, $\quad n_V(E_2) = n_0 e^{-E_2/k_B T}$

$$n_V(E_1) = n_0 e^{-E_1/k_B T}$$

$$\Rightarrow \quad \frac{n_V(E_2)}{n_V(E_1)} = e^{-(E_2 - E_1)/k_B T}$$

In given problem, $E_2 - E_1 = 0.216$ eV and

$$k_B T = 1.38 \times 10^{-23} \times 2500 \times \frac{1}{1.6 \times 10^{-19}}\ \text{eV}$$

$$= 0.216\ \text{eV}$$

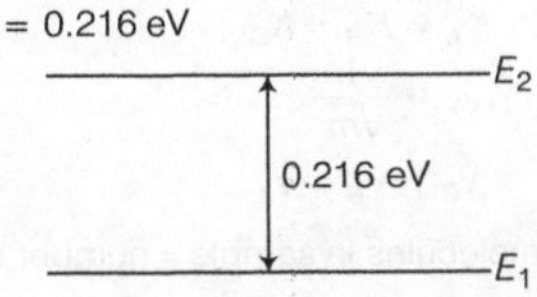

$\therefore$ Required ratio

$$= \frac{n_V(E_2)}{n_V(E_1)} = e^{-1} \approx \frac{1}{3}$$

So, nearly $\frac{1}{3}$rd atoms are in higher energy level.

(iii) Ratio of the number of molecules at higher energy to lower energy is $e^{-\Delta E/k_B T}$ where, ΔE = difference of energy levels.

$$\Delta E = E_2 - E_1$$

$$= \frac{-13.6}{4} - (-13.6) = 10.2\ \text{eV}$$

$$= 10.2 \times 1.6 \times 10^{-19}\ \text{J} = 1.63 \times 10^{-18}\ \text{J}$$

$$k_B T = 1.38 \times 10^{-23} \times 273 = 3.77 \times 10^{-21}\ \text{J}$$

As, this is much less than excitation energy, nearly all the atoms will be in the ground state.

$$(2.70 \times 10^{25})\, e^{\left(-\frac{1.63 \times 10^{-18}}{3.77 \times 10^{-27}}\right)} \approx (2.70 \times 10^{25})\, e^{-433}$$

This number is less than 1 so, there is not a single atom in excited state.

4. When a molecule with speed v makes a head on collision with a block moving with speed v_0.

Momentum transferred to block (per collision) $= 2m(v + v_0)$

Number of collisions in time,

$$\Delta t = \frac{1}{2}(v + v_0)\, n \Delta t\, A$$

$\therefore$ Momentum transferred in time, $\Delta t = m(v + v_0)^2 n A \Delta t$ from front surface and momentum transferred in time $\Delta t = m(v - v_0)^2\, nA\Delta t$ from back surface.

$\therefore$ Net momentum transferred to block

$$= mnA((v + v_0)^2 - (v - v_0)^2$$

$$= mnA\,(4v v_0) = 4(mnAv)v_0 = 4\rho A v v_0$$

$$= 4 \times 4 \times 10^{-3} \times 1 \times 10 \times v$$

$$= 16 \times 10^{-2} v$$

where, v = speed of gas molecule

$$= \sqrt{\frac{kT}{m}} = \sqrt{\frac{1.38 \times 10^{-23} \times 300}{4.14 \times 10^{-23}}} = \sqrt{100} = 10\ \text{ms}^{-1}$$

$\therefore$ Drag force on block $= 4\rho A \sqrt{\dfrac{kT}{m}} \cdot v_0$

$$= 16 \times 10^{-2} v$$

$$= 16 \times 10^{-2} \times 10 = 160\ \text{N}$$

5. (d) $\frac{p_1V_1}{T_1} = \frac{p_2V_2}{T_2}$

$$\Rightarrow \quad \frac{V_1}{V_2} = \frac{p_2T_1}{p_1T_2} = \frac{2 \times 300}{1 \times 400} = \frac{3}{2}$$

As, $p_1 = \frac{1}{3}\frac{M}{V_1}C_1^2$ (where, C = rms speed)

and $p_2 = \frac{1}{3}\frac{M}{V_2}C_2^2$

$$\Rightarrow \quad C_2^2 = C_1^2 \times \frac{V_2}{V_1} \times \frac{p_2}{p_1} = (100)^2 \times \frac{2}{3} \times 2$$

$$\Rightarrow \quad C_2 = \frac{200}{\sqrt{3}} \text{ ms}^{-1}$$

6. (c) In mixture, temperature is same for every type of molecule A, B and C, then

$$\Rightarrow \quad K_A = K_B = K_C$$

Also, $v_{rms} \propto \frac{1}{\sqrt{m}}$

$$\therefore \quad v_C > v_B > v_A$$

7. (c) Number of molecules in sample = number of moles × N_A

$$= \frac{0.5}{2} \times 6 \times 10^{23}$$

So, total molecular volume

$$= \frac{0.5}{2} \times 6 \times 10^{23} \times 10^{-30} \approx 2.5 \times 10^{-7} \text{ m}^3$$

Final volume of gas using ideal gas laws

$$= \frac{V}{100} = \frac{3^3 \times 10^{-6}}{100} \approx 2.7 \times 10^{-7} \text{m}^3$$

As, molecular volume occupies a very large part of total volume initially as well as finally.

∴ ideal gas condition does not hold.

8. (d) Initially,

$$(v_{rms})^2 = \frac{\Sigma n_i v_i^2}{\Sigma n_i}$$

$$= \frac{10 \times (200)^2 + 20 \times (400)^2 + \cdots}{100}$$

$$(v_{rms})^2 = (408 \times 1000) \text{ m}^2\text{s}^{-2}$$

$$\Rightarrow \quad v_{rms} = 639 \text{ m s}^{-1}$$

So, temperature $T = \frac{1}{3}\left(\frac{mv_{rms}^2}{k_B}\right)$

$$= \frac{3 \times 10^{-26} \times 4.08 \times 10^5}{3 \times 1.38 \times 10^{-23}} = 296 \text{ K}$$

As, $v_{escape} = 900 \text{ ms}^{-1}$, so, all the particles with $v_{rms} > v_{escape}$ or escape speed > 900 m s^{-1} will be lost from the planet producing a cooling effect by evaporation.

After a long time when all molecules with $v_{rms} > v_{escape}$ are lost, the rms speed of remaining molecules is

$$v_{rms}^2 = \frac{10 \times (200)^2 + 20 \times (400)^2 + 40 \times (600)^2 + 20 \times (800)^2}{90}$$

$$= 342 \times 1000 \text{ m}^2\text{s}^{-2}$$

$$v_{rms} = 584 \text{ ms}^{-1}$$

and so, temperature of planet $T = \frac{1}{3}\frac{mv_{rms}^2}{k} = 248 \text{ K}$

Hence, fall of temperature estimated = 296 − 248 = 48 K

9. (d) Mean square velocity of molecule, $v^2 = \frac{3kT}{m}$

For gas A, x-component of mean square velocity

$$v_{Ax}^2 = \frac{1}{3}v^2 = \frac{kT}{m}$$

For gas B,

$$v_{Bx}^2 = \frac{1}{3} \cdot \frac{3kT}{2m} = \frac{kT}{2m}$$

$$\therefore \quad \frac{v_{Ax}^2}{v_{Bx}^2} = \frac{2}{1} \Rightarrow \frac{v_{Ax}}{v_{Bx}} = \frac{\sqrt{2}}{1}$$

10. (c) $C_{rms} \geq 11.2 \text{ k ms}^{-1}$

$$\Rightarrow \quad \sqrt{\frac{3RT}{M}} \geq 11.2 \times 10^3$$

$$\Rightarrow \quad \sqrt{\frac{3 \times 8.3 \times T}{0.032}} \geq 11.2 \times 10^3$$

$$\Rightarrow \quad T = \frac{(11.2 \times 10^3)^2 \times 0.032}{3 \times 8.3} = 1.61 \times 10^5 \text{K}$$

11. (d) Force = $Nm\frac{\Delta v}{\Delta t}$

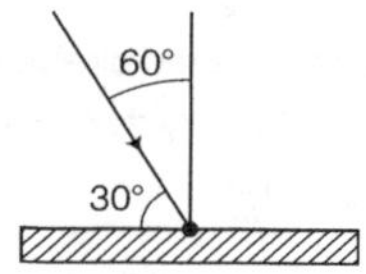

$$\text{Pressure} = \frac{\text{Force}}{\text{Area}} = \frac{Nm\Delta v}{A\Delta t}$$

$$= \frac{200 \times 5 \times 10^{-3} \times 6\,[(\sin 30° - (-\sin 30°)]}{0.5 \times 30}$$

$$= \frac{1000 \times 10^{-3}\left(2 \times 6 \times \frac{1}{2}\right)}{0.5 \times 30} = \frac{0.2}{0.5}\left(\frac{\text{N}}{\text{m}^2}\right) = \frac{2}{5} \text{ Pa}$$

12. (a) $pV = \frac{1}{3}Nm\,v_{av}^2$

$$= \frac{1}{3} \times 20 \times m \times \frac{(2 \times v^2 + 3 \times (2v)^2 + \cdots)}{2 + 3 + 5 + 4 + 3 + 2 + 1} = 106\,mv^2$$

13. (b) For n moles of gas, van der Waals' equation for a real gas is

$$\left(p + \frac{n^2a}{V^2}\right)(V - nb) = nRT \Rightarrow p = \frac{nRT}{V - nb} - \frac{n^2a}{V^2}$$

Comparing this with,

$$p = \frac{RT}{2V - b} - \frac{a}{4V^2} \quad \text{or} \quad p = \frac{\frac{1}{2}RT}{V - \frac{1}{2}b} - \frac{\left(\frac{1}{2}\right)^2 \times a}{V^2}$$

we get, number of moles

$$n = \frac{1}{2} = \frac{m}{M}$$

$$\Rightarrow \quad m = n \times M = \frac{1}{2} \times 44 = 22\text{g}$$

14. (b) $pV = nRT = \frac{m}{M}RT$

V and T = constant $\Rightarrow$ $Vdp = \frac{dm}{M}RT$

$$\Rightarrow \quad dm = \frac{MV}{RT} \cdot dp \quad \ldots\text{(i)}$$

Also, $p = \frac{m}{V} \cdot \frac{R}{M} \cdot T$

$$\Rightarrow \quad p = \rho \cdot \frac{RT}{M} \Rightarrow \frac{M}{RT} = \frac{\rho}{p} \quad \ldots\text{(ii)}$$

Substituting in Eq. (i), we have

$$dm = \frac{\rho \cdot V \cdot dp}{p}$$

$$\therefore \quad \Delta m = \frac{\rho V \Delta p}{p} = \frac{1.3\frac{\text{kg}}{\text{m}^3} \times 30 \times 10^{-3}\text{m}^3 \times 0.5 \times 10^5\frac{\text{N}}{\text{m}^2}}{1 \times 10^5\left(\frac{\text{N}}{\text{m}^2}\right)}$$

$$= 19.5 \times 10^{-3}\text{kg} \approx 20 \text{ g}$$

15. (a) $\left(p + \dfrac{aT^2}{V}\right)V^c = RT + b$

$$p + aT^2V^{-1} = RTV^{-c} + bV^{-c}$$

$$p = (RT + b)V^{-c} - (aT^2)V^{-1}$$

Comparing with

$$p = AV^m - BV^n$$

we have, $m = -c$ and $n = -1$

16. (a,c,d) Mean KE $= \dfrac{1}{2} mv_{rms}^2 = \dfrac{1}{2} m \cdot \dfrac{3kT}{m}$

$$= \frac{3}{4} m \cdot \left(\frac{2kT}{m}\right) = \frac{3}{4} mv_p^2$$

$$v_{rms} = \sqrt{\frac{3kT}{m}},\ v_p = \sqrt{\frac{2kT}{m}},\ \bar{v} = \sqrt{\frac{8}{\pi} \cdot \frac{kT}{m}}$$

$\therefore$ $v_{rms} > \bar{v} > v_p$

Also, $v_{rms} : \bar{v} : v_p :: \sqrt{3} : \sqrt{\dfrac{8}{\pi}} : \sqrt{2}$

$\Rightarrow$ $v_{rms} : \bar{v} : v_p :: \sqrt{3} : \sqrt{2.5} : \sqrt{2}$

Also, KE of 1g of a gas $= \dfrac{3}{2}\dfrac{R}{M} \cdot T = \dfrac{3}{2}\dfrac{k}{m} T$

17. (a,c,d) At same pressure, volumetric expansion coefficient is same. Then,

$\gamma = \dfrac{1}{273}$ for every ideal gas.

Mean translational KE $= \dfrac{3}{2} kT$

Pressure decrease causes a decrease in number of molecules, so gap between molecules (mean free path) increases.

18. (a,b,c,d) For an ideal gas sample, Maxwell's speed distribution curve at two different temperatures are as shown in the figure.

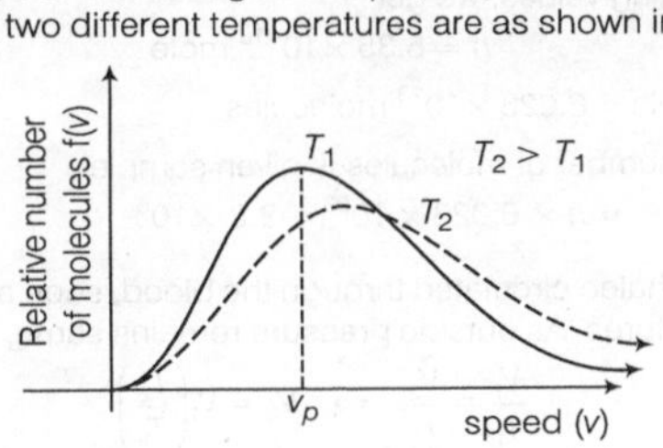

So, all options are correct.

19. (b,c,d) Average height $\bar{h}$ of a molecule of atmosphere is

$$\bar{h} = \frac{\int_0^\infty h n_V(h)\, dh}{\int_0^\infty n_V(h)\, dh}$$

$$\Rightarrow \quad \bar{h} = \frac{\int_0^\infty h e^{-mgh/k_B T} \cdot dh}{\int_0^\infty e^{-mgh/k_B T} dh}$$

Integrating by using 'by parts' procedure, we get

$$\bar{h} = \frac{(k_B T/mg)^2}{(k_B T/mg)} = \frac{k_B T}{mg}$$

So, average height increases with temperature.

Also, $\quad$ GPE $= U_g = mg\bar{h}$

$$= mg \cdot \frac{k_B T}{mg} = k_B T$$

$\therefore$ GPE is a function of temperature only.

20. (c,d) **Hint** Refer to postulates and assumptions of kinetic theory.

21. (d) $\left[p + a\left(\dfrac{N}{V}\right)^2\right](V - Nb) = kNT$

$$\Rightarrow \quad \left[p + a\left(\frac{N}{V}\right)^2\right]\left(1 - \frac{N}{V}b\right) = k\left(\frac{N}{V}\right) \cdot T$$

$$\Rightarrow \quad p + a\left(\frac{N}{V}\right)^2 = \frac{kNT}{V} \cdot \left(1 - \frac{N}{V}b\right)^{-1}$$

Using binomial expansion of $(1 - x)^{-1}$ and rearranging, we get

$$p = \frac{kNT}{V}\left[1 + \left(b - \frac{a}{kT}\right)\frac{N}{V} + b^2\left(\frac{N}{V}\right)^2 + \cdots\right]$$

22. (a) Boyle's temperature is that temperature at and above which a real gas behaves like an ideal gas.

Approximating we have,

$$p = \frac{kNT}{V}\left[1 + \left(b - \frac{a}{kT}\right)\frac{N}{V}\right]$$

Above equation becomes

$$p = \frac{kNT}{V}$$

When $\left(b - \dfrac{a}{kT}\right) = 0$

or $\quad T = \dfrac{a}{bk}$

This is the Boyle's temperature.

23. (d) Using $T_c = \dfrac{8a}{27bR}$, $p_c = \dfrac{a}{27b^2}$

and $\quad V_c = 3Nb$

and eliminating a and b, we get

$$\left(\frac{p}{p_c} + \frac{3V_c^2}{V^2}\right)\left(\frac{V}{V_c} - \frac{1}{3}\right) = \frac{8}{3}\left(\frac{T}{T_c}\right)$$

24. (d) Statement I is incorrect but Statement II is correct.

Average KE is a function of temperature.

$E_{av} = \dfrac{3}{2} k_B T$ (KE of one molecule)

As, density of both gases is not same, so their average speeds must be different.

25. (d) Statement I is incorrect as temperature change also depends on "how the volume changes". In an quanistatic isothermal process, pressure can increase and volume can decrease at a constant temperature.

26. (a) V_{av} at 27°C, 300 K for helium atom is 1.4 km s^{-1}. So, a significant fraction of He molecules have speeds greater than 11.2 km s^{-1}.

27. (a) Statement I and Satement II are correct and Statement II is a correct explanation of Statement I.

KE of 1 molecule $= \dfrac{3}{2} k_B T$

$\therefore$ KE of 1 mole $= N_A \times 10^{-3} \times \dfrac{3}{2} k_B T = N_A \times 10^{-3} \times \dfrac{3}{2} k_B T$

(As, 1 gram mole $= N_A \times 10^{-3}$ molecules)

$\therefore$ $\quad$ KE per mole $= 3 \times 10^{-3} \dfrac{RT}{2}$ $\quad$ $(\because k_B N_A = R)$

$= 3.4$ kJ $\quad$ $(\because T = 273$ K$)$

DPP-2 **Kinetic Theory of Gases-II** (Gas Laws)

1. One mole or 44.01 g of CO_2 gas has a volume of

$$V = \frac{44.01\text{g}}{1.977\ \text{gL}^{-1}} = 22.26\ \text{L} = 22260\ \text{cm}^3$$

Volume per molecule is

$$V = \frac{22260}{6.022 \times 10^{23}} \times \frac{10^{24}\ \text{Å}^3}{1\ \text{cm}^3}$$

$$= 36900\ \text{Å}^3 \text{ per molecule}$$

So, here's the comparison.

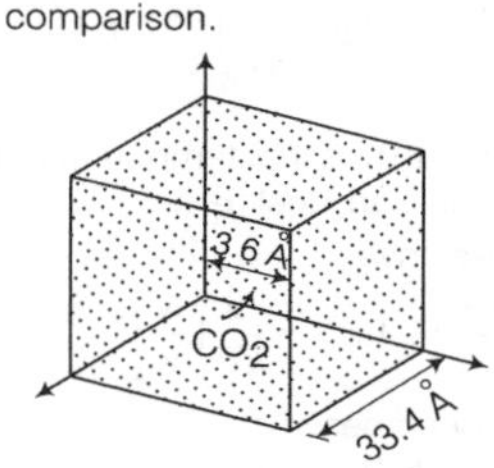

Relative size of a CO_2 molecule and the volume per molecule available to it.

Also, notice the small deviation from ideal gas behaviour.

2. First we find number of moles of Helium and Xenon in any given sample. A suitable choice is a 100 g sample.

$$\eta_{He} = \frac{50\text{g}}{4\text{g (mole)}^{-1}} = 12.5 \text{ moles}$$

$$\eta_{Xe} = \frac{50\text{g}}{131.3\ \text{g(mole)}^{-1}} = 0.381 \text{ moels}$$

Mole fraction of each component is

$$\mu_{He} = \frac{12.5}{12.5 + 0.381} = 0.970$$

$$\mu_{Xe} = \frac{0.381}{12.5 + 0.381} = 0.030$$

Partial pressure α mole fraction or $p_j = \mu p$

$$p_{He} = 0.970\ p = 0.97 \times 0.8 = 0.776 \text{ atm}$$

$$p_{Xe} = 0.030\ p = 0.030 \times 0.8 = 0.024 \text{ atm}$$

3. Cylinder before removal of membrane

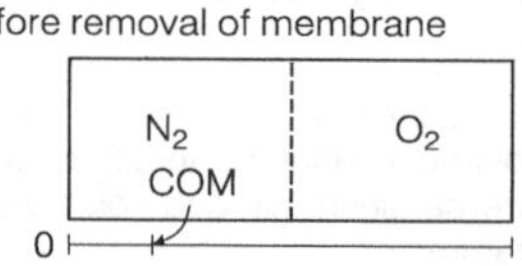

We assume temperature is same. COM of system must be on N_2 side (or $p_{N_2} > p_{O_2}$).

As, there is no external force is involved.

$$\Delta X = X_{\underset{\text{(final)}}{COM}} - X_{\underset{\text{(Initial)}}{COM}}$$

Now, $p_{N_2} V_{N_2} = n_{N_2} kT$

and $p_{O_2} V_{O_2} = n_{O_2} kT \Rightarrow \dfrac{p_{N_2}}{p_{O_2}} = \dfrac{n_{N_2}}{n_{O_2}}$

$\Rightarrow \dfrac{n_{N_2}}{n_{O_2}} = \dfrac{2}{1};$

As, $p_{N_2} = 2p_{O_2}$

And mass of given volumes of gases is

$$m_{N_2} = n_{N_2} M_{N_2} \Rightarrow m_{O_2} = n_{O_2} M_{O_2}$$

Now, $X_{\underset{\text{(Initial)}}{COM}} = \dfrac{\Sigma X_i m_i}{\Sigma m_i}$

$$= \frac{n_{N_2} \cdot M_{N_2} \cdot X_{N_2} + n_{O_2} M_{O_2} X_{O_2}}{(n_{N_2} M_{N_2} + n_{O_2} M_{O_2})}$$

$$= \frac{2 \times 28.01 \times 10 + 1 \times 32 \times 30}{2 \times 28.01 + 32} = 17.27 \text{ cm}$$

and $X_{\underset{\text{(final)}}{COM}} = \dfrac{2 \times 28.01 \times 20 + 1 \times 32 \times 20}{2 \times 28.01 + 32} = 20 \text{ cm}$

$\therefore$ Δx = distance moved by cylinder

$= X_{final} - X_{initial} = 20 - 17.27 = 2.7$ cm

As, com moves to right, the cylinder moves to the left.

4. (i) $1 \text{ pascal} = 1\dfrac{\text{N}}{\text{m}^2} = 1\dfrac{\text{N}\cdot\text{m}}{\text{m}^3} = 1\ \text{Jm}^{-3}$

(ii) For any monatomic ideal gas,

$$E_{int} = \frac{3}{2} nRT$$

For any ideal gas, energy of translation is same.

$$\therefore \quad E_{trans} = \frac{3}{2} nRT = \frac{3}{2} pV$$

$$\Rightarrow \quad \frac{E}{V} = \text{energy per unit volume} = \frac{3}{2} p$$

5. (c) $p_1 V_1 = p_2 V_2$, T = constant

$$\therefore \quad p_2 = p_1 \left(\frac{V_1}{V_2}\right) = 2 \times \left(\frac{1000}{800}\right) = 2.5 \text{ atm}$$

6. (c) $n = \dfrac{pV}{RT}$

$$p = 10^{-13}(\text{m}) \times (13.6 \times 10^3) \frac{\text{kg}}{\text{m}^3} \times 9.8 \frac{\text{N}}{\text{kg}}$$

$$= 133.3 \times 10^{-10} \frac{\text{N}}{\text{m}^2}$$

$$V = 1\ \text{cm}^3 = 10^{-6}\ \text{m}^3$$

$$T = 27°\text{C} = 300\ \text{K}$$

Substituting values, we get

$$n = 5.35 \times 10^{-18} \text{ mole}$$

As, 1 mole $= 6.023 \times 10^{23}$ molecules

$\therefore$ Total number of molecules is given sample

$= n \times 6.023 \times 10^{23} = 3.2 \times 10^6$

7. (d) Air inhaled circulated through the blood, soon accquires body temperatures. As outside pressure remains same,

$$\Rightarrow \quad \frac{V_1}{T_1} = \frac{V_2}{T_2} \Rightarrow V_2 = V_1 \left(\frac{T_2}{T_1}\right)$$

$$\Rightarrow \quad V_2 = 500 \times \left(\frac{270}{300}\right) = 450 \text{ mL}$$

8. (d) $pV = nRT = \dfrac{N}{N_A} \cdot R \cdot T = Nk_B T$

So, $\dfrac{pV}{k_B T} = N$ = Number of molecules.

9. (a) $pV^{3/2}$ = constant, $pV = nRT$

Combining $pV^{3/2} = k_1$

$\Rightarrow pV \cdot \sqrt{V} = k_1 \Rightarrow nRT \cdot \sqrt{V} = k_1$

$\Rightarrow T\sqrt{V}$ = constant $\Rightarrow T_1 V_1^{1/2} = T_2 V_2^{1/2}$

when $V_2 = \left(\dfrac{V_1}{2}\right)$,

$$T_2 = T_1 \left(\frac{V_1}{V_2}\right)^{1/2} = T_1 (2^{1/2})$$

So, final temperature of gas is $T_2 = \sqrt{2}\ T$

10. (a) $p = \dfrac{a}{1 + \left(\dfrac{V}{b}\right)^2} = \dfrac{RT}{V}$

When $V = b$, $\dfrac{a}{1 + (1)^2} = \dfrac{RT}{b} \Rightarrow T = \dfrac{ab}{2R}$

11. (b) Number of moles are conserved as containers are sealed.

$\therefore \quad n_1 + n_2 = n$

$\Rightarrow \quad \frac{p_1V_1}{RT_1} + \frac{p_2V_2}{RT_2} = \frac{p_3V_3}{RT_3}$

where, $\quad V_3 = V_1 + V_2 \Rightarrow \frac{p_1V_1 + p_2V_2}{(V_2 + V_1)} = p_3$

or $\quad T_1 = T_2 = T_3$

$\Rightarrow \quad p_3 = \frac{1.4 \times 0.1 + 0.7 \times 0.15}{0.10 + 0.15} = 0.98$ MPa

12. (d) $\frac{p_1V_1}{T_1} = \frac{p_2V_2}{T_2} \Rightarrow T_2 = T_1\left(\frac{p_2}{p_1}\right)\left(\frac{V_2}{V_1}\right)$

$p_1 = 305 + 101 = 406$ kPa

$p_2 = 360 + 101 = 461$ kPa

$\therefore \quad T_2 = (273 + 15)\left(\frac{461}{406}\right)(1.00) = 327$ K

$= 327 - 273 = 54°C$

13. (c) $p_N \cdot V = n_N RT$ and $p_H V = n_H RT$

$\Rightarrow \quad \frac{n_H}{n_N} = \frac{p_H}{p_N} = \frac{3.5}{4.5} = 0.778$

But $\quad n_N = \frac{m}{M} = \frac{18}{28} = 0.643$ k mol

$\therefore \quad n_H = n_N\,(0.778) = 0.5$ k mol

As, $\quad n_H = \frac{m}{M} \Rightarrow m = M \cdot n_H = 0.5 \times 2 = 1$ kg

14. (d) $p \propto \frac{1}{V^2}$ and $V \propto T^{2/3}$
$(T = \text{constant}) \quad (p = \text{constant})$

Also, $V \propto n$ (for constant p and T) combining there, we have

$pV^2 = n^2RT^{4/3}$

15. (a,d) 1 mole of gas at 0°C occupies a volume of 22.4 L.

Volume of room $= 5 \times 3 \times 2.5$ m^3

So, $\quad n$ = number of moles

$= \frac{5 \times 3 \times 2.5}{22.4 \times 10^{-3}} \approx 1700$

Air contains about 20% O_2 and 80% N_2. The molecular masses are $2 \times 16u = 32u$ and $2 \times 14u = 28u$

$\therefore$ Average molecular mass $= 29u$

$\therefore$ 1 mole of air has mass about 29 g = 0.029 kg

$\therefore$ Mass of air is room

$= 1700 \times 0.029 \approx 50$ kg

16. (b,d) KE for an ideal gas $= \frac{f}{2}nRT = \frac{5}{2}nRT$

Also, KE of gas w.r.t. ground = KE of COM w.r.t. ground + KE of gas w.r.t. COM of system

$= \frac{1}{2}mv^2 + \frac{5}{2}nRT$

17. (a,b) Mass of $H_2 = m_1$ and mass of He $= m_2$

$m_1 + m_2 = 10^{-2}$ kg $= 10 \times 10^{-3}$ kg

$p_1 + p_2 = 4.15 \times 10^5$ Nm^{-2}

$(p_1 + p_2)V = \left(\frac{m_1}{n_1} + \frac{m_2}{n_2}\right)RT$

$\Rightarrow \quad 4.15 \times 10^5 \times 2 \times 10^{-2} = \left(\frac{m_1}{2 \times 10^{-3}} + \frac{m_2}{4 \times 10^{-3}}\right)RT$

$\Rightarrow \quad 2m_1 + m_2 = 12.48 \times 10^{-3}$

So, $\quad m_1 = m_{H_2} = 2.5 \times 10^{-3}$ kg

$m_2 = m_{He} = 7.5 \times 10^{-3}$ kg

18. (a,b) Reaction is

$C_2H_6 + \frac{7}{2}O_2 \longrightarrow 2CO_2 + 3H_2O$

and $\quad C_2H_4 + 3O_2 \longrightarrow 2CO_2 + 2H_2O$

Let there are x moles of Ethane (C_2H_6) and y moles of Ethene (C_2H_4), then, from above reactions,

Moles of O_2 used $= \frac{7}{2}x + 3y$

Moles of O_2 in reaction $= \frac{130}{32}$

$\therefore \quad \frac{7}{2}x + 3y = \frac{130}{32}$ ...(i)

Also, $\quad x + y = \frac{pV}{RT}$

$\Rightarrow \quad x + y = \frac{1 \times 40}{0.0821 \times 400} = 1.218$ ...(ii)

From Eqs. (i) and (ii) we have

$x = 0.812$ and $y = 0.406$

$\therefore$ Moles fraction are $\frac{0.812}{1.218}$ and $\frac{0.406}{1.218}$ or 0.67 and 0.33.

19. (b,d) Partial pressure of helium

$= \frac{4}{5} \times 20 = 16$ atm

$\therefore$ Partial pressure of methane $= 20 - 16 = 4$ atm

Ratio of rates of diffusion is

$\frac{r_{He}}{r_{CH_4}} = \frac{p_{He}}{p_{CH_4}} \times \left(\frac{M_{CH_4}}{M_{He}}\right)^{1/2}$

$= \frac{16}{4} \times \left(\frac{16}{4}\right)^{1/2} = 8:1$

20. (a) In container A, H_2 has lower molecular mass.

$\therefore$ Rms speed is more and thermal energy is same.

In container B the pressure is more, thermal energy is same and internal energy is more.

Hence, (i) $\rightarrow$ (q, r), (ii) $\rightarrow$ (p, r, s), (iii) $\rightarrow$ (p, s), (iv) $\rightarrow$ (q)

21. (d) Number of molecules

$= N$ = number of moles $\times 6.023 \times 10^{23}$

$= \frac{10000\text{g}}{18\text{g}} \times 6.02 \times 10^{23} = 3.34 \times 10^{26}$

22. (c) After one day, 10^{-1} of original molecules remain. After two days, fraction of original remained $= 10^{-2}$

and so on ...

$\therefore$ After 26 days, 10^{-26} th fraction i.e. only 3 molecules of original batch remains. After 27 days probably none.

23. (c) Pot water is $\left(\frac{10}{1.32 \times 10^{21}}\right)$ fraction of hydrosphere.

$\therefore$ Molecules remain $= \frac{10}{1.32 \times 10^{21}} \times 3.34 \times 10^{26}$

$= 2.53 \times 10^6$ molecules

DPP-3 **Kinetic Theory of Gases-III**
(Degree of Freedom and Energy of Gas)

1. (i) Average KE of translation per molecule, if an ideal gas is $\frac{3}{2}kT$.

where, k = Boltzmann's constant and T = temperature of gas in kelvins.

Now, $k = 1.38 \times 10^{-23}$ J/molecule · K, T = 0°C ≈273 K

∴ Average KE of 1 molecule

$$= \frac{3}{2} \times 1.38 \times 10^{-23} \times 273$$

$$= 5.65 \times 10^{-21} \text{ J/molecule}$$

(ii) 1 mole of gas has a total of $N_A = 6.02 \times 10^{23}$ molecules

∴ KE of translation per mole of gas

= KE of 1 molecule × Avagadro's number

∴ KE (1 mole)

$$= \frac{3}{2} k_B T \times N_A = 5.65 \times 10^{-21} \times 6.02 \times 10^{23}$$

= 3400 J/molecule

(iii) Molar mass of gas = 4 g

∴ Mass of 6.023×10^{23} number of molecules is 4g.

∴ 1 gram of gas contains N = number of molecules

$$= \frac{N_A}{M} = \frac{6.023 \times 10^{23}}{4}$$

So, average translational energy of 1 gram of gas = $N \times$ KE of 1 molecule

$$= \frac{N_A}{M} \times \frac{3}{2} k_B T = 851 \text{ J}$$

2. (i) $T = 27°C = 27 + 273 = 300$ K

By law of equipartition of enegy , average KE of gas molecule associated with each degree of freedom is $\frac{1}{2}k_B T$.

As, an oxygen molecule has 3-degree of freedom with respect to translation, the average KE of translation of the molecule is

$$3 \times \frac{1}{2} k_B T = \frac{3}{2} k_B T$$

$$= \frac{3}{2} \times 1.38 \times 10^{-23} \times 300$$

$$= 6.21 \times 10^{-2} \text{ J/molecule}$$

(ii) An oxygen molecule, being diatomic, has a total of 5 degree of freedom.

∴ Total energy of a molecule $= 5 \times \frac{1}{2} k_B T = \frac{5}{2} k_B T$

$$= \frac{5}{2} \times (1.38 \times 10^{-23}) \times 300$$

$$= 10.35 \times 10^{-21} \text{ J/molecule}$$

(iii) As oxygen molecule has 3 translational degree of freedom and 2 rotational degree of freedom.

∴ $\frac{3}{5}$th part of total energy is translational and $\frac{2}{5}$th part of total energy is rotational.

3. (i) $1 \text{ eV} = 1.6 \times 10^{-19}$ J ⇒ $10.2 \text{ eV} = 16.32 \times 10^{-19}$ J

Let at T temperature, a hydrogen atom will have an average KE of 16.32×10^{-19} J.

Then, $\frac{3}{2} k_B T = 16.32 \times 10^{-19}$ J

$$\Rightarrow \quad T = \frac{2}{3} \times \frac{16.32 \times 10^{-19}}{1.38 \times 10^{-23}} = 8 \times 10^4 \text{ K}$$

(ii) As, a very high temperature results to liberate such energy by thermal methods, de-excitation of atoms occurs in form of radiative transitions.

4. (i) Let $\beta = \frac{\text{Increase in total energy}}{\text{Increase in translational energy}}$

$$\Rightarrow \beta = \frac{C_V \cdot dT}{\frac{3}{2} R dT} = \frac{2}{3}\frac{C_V}{R} = \frac{2}{3}\left(\frac{C_V}{C_p - C_V}\right) = \frac{2}{3}\left(\frac{1}{\gamma - 1}\right)$$

For gases like argon, helium or mercury vapour, $\gamma = \frac{5}{3}$

$$\therefore \quad \beta = \frac{2}{3}\left(\frac{1}{\frac{5}{3} - 1}\right) = 1$$

so, total energy is same as translational energy. For gases like oxygen, value of γ is $\frac{7}{5}$.

$$\text{So,} \quad \beta = \frac{2}{3}\left(\frac{1}{\frac{7}{5} - 1}\right) = \frac{5}{3}$$

∴ Rotational and vibrational energy is $\frac{2}{3}$ of translational energy.

Hence, we can say that, smaller the value of γ, greater is the atomicity of the gas.

(ii) Total number of modes (ways) in which a system can passes energy is called degree of freedom (f). Independent modes can be translational, rotational or vibrational.

General expression for number of degree of freedom is

$$f = 3A - B$$

where, A = number of independent particles and

B = numbner of independent restrictions.

For a monatomic gas, $A = 1$, $B = 0$

$$\therefore \quad f = 3$$

For a diatomic gas, $A = 2$, $B = 1$

$$\therefore \quad f = 3 \times 2 - 1 = 6 - 1 = 5$$

For a triatomic non-linear molecule, $A = 3$, $B = 3$

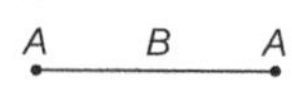

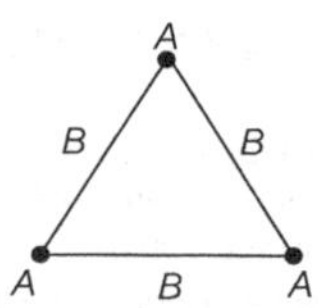

$$\therefore \quad f = 3 \times 3 - 3 = 6$$

For a triatomic linear molecule, $A = 3$, $B = 2$

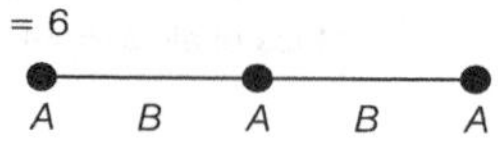

$$\therefore \quad f = 3 \times 3 - 2 = 7$$

Above degree of freedom are at room temperature range. At high temperatures, in case of di and polyatomic molecules, the atoms with in the molecule may vibrate.

∴ The molecule will have an additional '2 degree of freedom. One for PE change and one for KE change.

(iii) As, KE of 1 mole of gas $E = \frac{f}{2} \cdot RT$, f = degree of freedom.

$$\therefore \quad \frac{dE}{dT} = nC_V \text{ or } C_V = \frac{dE}{dT}$$

$$\text{So,} \quad C_V = \frac{d}{dT}\left(\frac{f}{2}RT\right) \Rightarrow C_V = \frac{f}{2}R$$

From $C_p - C_V = R, C_p = R + C_V$

$$\Rightarrow \quad C_p = \left(1 + \frac{f}{2}\right)R$$

So, ratio $\frac{C_p}{C_V} = \gamma = 1 + \frac{2}{f}$

$$\therefore \quad \gamma = 1 + \frac{2}{f}$$

5. (d) For a solid molecule there is no degree of freedom associated with translation and rotation but due to vibration along X, Y or Z axes, it has $3 \times 2 = 6$ degree of freedom.

6. (d) Energy per mole of a gas $= \frac{f}{2}RT = \frac{f}{2}pV$, f = degree of freedom.

For a monatomic and a diatomic gas, degree of freedom for translational motion are same; $f = 3$.

$\therefore$ Translational energy for a diatomic gas $= \frac{3}{2}RT = \frac{3}{2}pV$

and total energy for a diatomic gas $= \frac{5}{2}RT = \frac{5}{2}pV$

Hence, ratio $= \frac{3/2\ pV}{5/2\ pV} = \frac{3}{5}$

7. (c) Argon is a monatomic gas, so there are only 3 possible translatory modes of motion. Hence, the argon molecule can posses only translatory KE.

8. (c) Speed of sound in a gas column is $v_s = \sqrt{\frac{\gamma p}{\rho}}$

$$\Rightarrow \quad (330)^2 = \frac{\gamma(1.01 \times 10^5)}{1.3} \Rightarrow \gamma = 1.41$$

But $$\gamma = 1 + \frac{2}{f} \Rightarrow f = \frac{2}{\gamma - 1} = \frac{2}{1.4 - 1} = 5$$

9. (d) Given, radius $b = 3 \times 10^{-10}$ m

At STP 1 kmol of a gas occupies 22.4 m^3.

$$\therefore \quad \frac{N}{V} = \frac{6.02 \times 10^{26} \text{ molecules}}{22.4 \text{ m}^3}$$

so, mean free path,

$$\lambda = \frac{1}{4\pi\sqrt{2}b^2(N/V)} = \frac{22.4}{4 \times \pi \times \sqrt{2} \times (3 \times 10^{-10})^2 \times 6.02 \times 10^{26}}$$

$$= 2.4 \times 10^{-8} \text{ m}$$

10. (c) $\frac{N}{V} = \frac{1}{4\pi\sqrt{2}b^2\lambda}$

where, λ = mean free path,

$$p = \frac{k_B T \cdot N}{V} = \frac{k_B T}{4\pi\sqrt{2}b^2 \times \lambda} = \frac{1.38 \times 10^{-23} \times 293}{4\pi\sqrt{2} \times (3 \times 10^{-10}) \times 0.5}$$

$$= 5.1 \text{ MPa}$$

11. (d) $\frac{1}{2}mv_{rms}^2 = \frac{3}{2}k_B T = mgh$

$$\Rightarrow \quad h = \left(\frac{1}{m} \cdot \frac{3k_B T}{2g}\right) = \frac{3 \times 1.38 \times 10^{-23} \times 300}{2 \times 9.81 \times 4.65 \times 10^{-26}} \approx 14 \text{ km}$$

12. (a) Monatomic gas possess only translatory degree of freedom whereas a diatomic gas possess both translatory as well as rotatory degree of freedom.

$\therefore$ Specific heat of monatomic gas is small and so, it shows a large temperature change.

13. (c) λ = mean free path $= \frac{1}{\sqrt{2}\ \pi n d^2}$

where, $n = \frac{N}{V} = \frac{p}{k_B T}$

$$\therefore \quad \lambda = \frac{k_B \cdot T}{\sqrt{2} \cdot \pi \cdot p \cdot d^2}$$

So, $$\frac{\lambda_2}{\lambda_1} = \frac{k_B T_2}{\sqrt{2}\ \pi p_2 d^2} \cdot \frac{\sqrt{2}\pi p_1 d^2}{k_B T_1}$$

$$\Rightarrow \quad \frac{\lambda_2}{\lambda_1} = \frac{p_1}{p_2}$$

$$\Rightarrow \quad \lambda_2 = \lambda_1 \cdot \frac{p_1}{p_2} = \frac{9.5 \times 10^{-8} \times p_1}{\left(\frac{p_1}{100}\right)} = 9.5 \times 10^{-6} \text{ m}$$

14. (c) Given, $v_{rms} = \sqrt{\frac{3kT}{m}} = 4.53 \times 10^2 \text{ ms}^{-1}$

So, $$v_{av} = \sqrt{\frac{8\ kT}{\pi\ m}} = 4.53 \times 10^2 \times \sqrt{\frac{3\pi}{8}} \text{ ms}^{-1}$$

Also, collision frequency = number of collisions is 1 second

$$= \frac{v_{av}}{\lambda} = \frac{4.53 \times 10^2 \sqrt{\frac{3\pi}{8}}}{6.85 \times 10^{-8}}$$

$$= 6.6 \times 10^9 \times \sqrt{\frac{3\pi}{8}} \text{ per second}$$

15. (a,b,d) C is incorrect or atomic vibration results in two more degree of freedom.

16. (b,c,d) Law of equipartition of energy is applicable to all substances like solids, liquids and gases.

By Dulong and Petit law, molar specific heat for any solids approaches to $3R$ with rise in temperature.

Energy is stored in form of vibrational energy of atoms in molecules. They have 3 degree of freedom for KE and 3 degree of freedom associated with PE of vibration along X, Y and Z axes.

17. (b,c,d) λ = mean free path $= \frac{kT}{4\pi\sqrt{2} \cdot r^2 \cdot p}$

So, $$\lambda = \text{mean free path} = \frac{kT}{4\pi\sqrt{2} \cdot r^2 \cdot p}$$

$$= \frac{1.38 \times 10^{-23} \times 300}{4\pi\sqrt{2}\left(\frac{1}{\sqrt{\pi}} \times 10^{-10}\right)^2 \times (\sqrt{2} \times 10^{-3})}$$

$$= \frac{1.38 \times 300 \times 10^{-23}}{8 \times 10^{-23}} = 51.75 \text{ m}$$

Also, side length of box = 10.35 m

$\therefore$ Collision frequency is given by

$$\bar{\nu} = \frac{\lambda}{t_{\text{molecular collisions}}} = \frac{l_{box}}{t_{\text{wall collisions}}}$$

$$\therefore \quad \bar{\nu} = \frac{\lambda}{l_{box}} = \frac{51.75}{10.35} = 5$$

So, wall collisions are nearly 5 times more frequent than the intermolecular collisions. Also, option (c) is true in every condition.

18. (c,d) Option (a) is true but cannot be concluded from observed value of 'C_v' for SO_2.

As, few molecules have vibrating atoms even at ordinary temperatures.

so, SO_2 gas behaves like a mixture of two different type of molecules one with 6 degree of freedom and other with 8 degreee of freedom (both occupying $\frac{1}{2}$ part of SO_2 gas).

19. (a,c,d) Oxygen has 3 translational and 2 rotational degree of freedom.

$\therefore$ $\frac{3}{5}$th part of total energy is translational and $\frac{2}{5}$ th part of total energy is rotational.

Now, $$\frac{3}{5} \times E = 2202 \quad \text{(given)}$$

So, rotational energy is $\frac{2}{5} \times E = 2202 \times \frac{2}{3} = 1468$

Also, $$E = 3670 \text{ J}$$

$$\Rightarrow \quad 3670 = \frac{5}{2}k_B T \times \frac{N_A}{M} \times m$$

$$\Rightarrow \quad 3670 = \frac{5}{2} \times 1.38 \times 10^{-23} \times \frac{6.02 \times 10^{23}}{32} \times 20 \times T$$

$$\Rightarrow \quad T = \frac{3670 \times 2 \times 32}{20 \times 5 \times 1.38 \times 6.02} = 283 \text{ K}$$

$$\therefore \quad T = 283 - 273 = 10°\text{C}$$

20. (b) Both Statement I and Statement II are correct but Statement II is not correct explanation for Statement I.

A gas column can transmit on those sound waves for which wavelength of sound wave is greater than mean free path. So,

mean free path $\propto \dfrac{1}{\text{pressure of gas}} \propto$ wavelength $\propto \dfrac{1}{\text{frequency}}$

21. (d) Statement I is incorrect but Statement II is correct.

H_2 gas can have $C_v = \frac{3}{2}R$ at low temperature range which steadily increases with the temperature to $\frac{5}{2}R$ and then to $\frac{7}{2}R$.

That means all values of C_v between $\frac{3}{2}R$ and $\frac{7}{2}R$ exists at different temperatures.

22. (d) Statement I is incorrect but Statement II is correct.

Each vibration mode accounts for increase of degree of freedom by 2.

23. (a) Rotational KE $= \frac{1}{2}I\omega^2$

$$I = 2(mr^2) = 2 \times 35 \times 1.67 \times 10^{-27}) \times (10^{-10})^2$$
$$= 1.17 \times 10^{-45} \text{ kg-m}^2$$
$$\omega = 2 \times 10^{12} \text{ s}^{-1}$$

$\therefore$ $$K_{rot} = \frac{1}{2}I\omega^2 = 2.33 \times 10^{-21} \text{ J}$$

24. (b) Total energy = Rotational energy + Translational energy

$$= f \cdot \left(\frac{1}{2}k_B T\right)$$

As, for a diatomic molecule there are 3 translational and 2 rotational degree of freedom,

$\therefore$ Energy associated with rotation is $\frac{2}{5}$th of total energy.

So, total energy of a Cl_2 molecule = $\frac{5}{2}$ × rotational energy

$$= \frac{5}{2} \times 2.33 \times 10^{-2} \text{ J} = 5.8 \times 10^{-21} \text{ J}$$

25. (c) $\frac{5}{2}k_B T = 5.8 \times 10^{-21} \Rightarrow T = \dfrac{2 \times 5.8 \times 10^{-21}}{5 \times 1.38 \times 10^{-23}}$

$$= 1.68 \times 10^2 = 168 \text{ K}$$
$$= 168 - 273 = -105°\text{C}$$

26. (4) $\lambda = \dfrac{1}{\sqrt{2}\pi n d^2}$, where d = molecular diameter

$$n = \frac{p}{kT}; \quad \lambda = \frac{kT}{\sqrt{2}\pi p d^2}$$

Now given that, $\lambda = \dfrac{k(200)}{\sqrt{2}\pi \cdot p \cdot d^2} = 2.5 \times 10^{-7}$ m

At pressure $\frac{p}{4}$ and temperature of 800 K, mean free path

$$\lambda' = \frac{k(800)}{\sqrt{2} \cdot \pi \cdot \frac{p}{4} \cdot d^2} = 16 \times 2.5 \times 10^{-7} \text{ m}$$
$$= 40 \times 10^{-7} \text{ m} = 4 \times 10^{-6} \text{ m}$$

27. (4) Mean relaxation time or average time between collisions is the mean free path divided by the mean speed of molecules. The highest frequency for a sound wave is the inverse of this average collision time.

$$pV = NkT \Rightarrow \frac{N}{V} = \frac{p}{kT}$$

Also, mean free path $= \lambda = \dfrac{1}{4\pi\sqrt{2}r^2(N/V)}$

$$= \frac{kT}{4\pi\sqrt{2}r^2 p}$$

Average molecular speed,

$$v_{av} = \sqrt{\frac{8kT}{\pi m}}$$

$\therefore$ $$f_{max} = \frac{v_{av}}{\lambda} = \frac{\sqrt{\frac{8}{\pi} \cdot \frac{kT}{m}}}{\left(\frac{kT}{4\pi\sqrt{2}r^2 p}\right)}$$
$$= 16pr^2\sqrt{\frac{\pi}{mkT}}$$

$\therefore$ $$N^2 = 16 \Rightarrow N = 4$$

28. (c) KE of 1 mole $= \frac{f}{2}RT$

$f = 3$ for monatomic

$f = 5$ for diatomic

$f = 6$ for triatomic non-linear

$f = 7$ for triatomic linear molecules.

Hence, (i) → (s), (ii) → (r), (iii) → (q), (iv) → (p)

DPP-4 Specific Heat of Solids, Liquids and Gases

1. (c) $$dQ = mCdT$$

$\Rightarrow$ $$C = \frac{dQ}{mdT}$$

2. (d) For a real gas, value of specific heat depends on the process, it ranges from 0 to ∞.

3. (d) $U_2 - U_1 = mC_V(T_2 - T_1)$

4. (d) All other are correct as in MKS system, J = mechanical equivalent of heat is used to convert work units into heat units.

5. (d) Enthalpy, $h = u + p\Delta V$

$$= C_p \Delta T$$
$$= C_V \Delta T + R\Delta T$$

6. (b) $Q = \Delta U + W = U_2 - U_1 + W$

$$= mC_V(T_2 - T_1) + W$$
$$= 0.3 \times 0.75 (433 - 133) - 30$$
$$= -3 \text{ kJ}$$

7. (a) $$\Delta Q = nC_V\Delta T$$

$\Rightarrow$ $$C_V = \frac{\Delta Q}{n\Delta T} = \frac{3}{2}R$$

As, $$C_V = \frac{f}{2}R \Rightarrow \frac{3}{2}R = \frac{f}{2}R$$

$\Rightarrow$ $$f = 3$$

So, gas is monatomic.

8. (a) $\Delta Q = mC_V\Delta T$

$$20750 = 2.5\, C_V \cdot \Delta T \quad \text{...(i)}$$

Also, $C_p - C_V = 4150$ and $\dfrac{C_p}{C_V} = 1.4$

$\Rightarrow$ $$C_V = \frac{4150}{0.5} = 10375 \frac{\text{J}}{\text{kg-K}}$$

So, from Eq. (i), $\Delta T = \dfrac{20750}{2.5 \times 10375} = \dfrac{2}{2.5} = \dfrac{4}{5}$ K

9. (a) Molar specific heat = Molecular weight × Gram specific heat

$$\Rightarrow \quad 2.98 = M \times \frac{0.075 \times 10^3}{10^3}\left(\frac{\text{calorie}}{\text{mole-K}}\right)$$

$$\Rightarrow \quad M = \frac{2.98}{0.075} = 3.97\,\text{g}$$

Now, mass of 1 mole of argon atoms = 39.7 g

$\therefore$ Mass of 1 atom × N_A = 39.7

$$\Rightarrow \quad m(1\text{ atom}) = \frac{39.7}{6.023 \times 10^{23}} = 6.60 \times 10^{-23}\,\text{g}$$

10. (b) Fraction of heat that rises internal energy

$$= \frac{\Delta U}{\Delta Q} = \frac{nC_V\Delta T}{nC_p\Delta T} = \frac{1}{\gamma}$$

Now, $\gamma(\text{diatomic}) = \frac{7}{5}$

So, heat which rises internal energy

$$= \frac{1}{\gamma} \times \Delta Q = \frac{5}{7} \times 105 = 75\,\text{cal}$$

11. (b) Fraction of heat that increases internal energy

$$= \frac{1}{\gamma}\text{ of total heat}$$

$\therefore$ Fraction which does work $= 1 - \frac{1}{\gamma}$

12. (a) Work done $= \left(1 - \frac{1}{\gamma}\right) \times \Delta Q_{\text{total}}$

$$\Rightarrow \quad 25 = \left(1 - \frac{1}{\gamma}\right) \times \Delta Q \qquad \ldots\text{(i)}$$

Now, $\gamma = 1 + \frac{2}{f} = 1 + \frac{2}{6} = 1 + \frac{1}{3} = \frac{4}{3}$

$\therefore$ From Eq. (i), $25 = \left(1 - \frac{3}{4}\right) \times \Delta Q \Rightarrow \frac{1}{4} \times \Delta Q = 25\,\text{J}$

$$\Rightarrow \quad \Delta Q = 100\,\text{J}$$

Heat absorbed by gas

$$= 100\,\text{J} - 25\,\text{J} = 75\,\text{J}$$

13. (b) KE = Heat absorbed by gas in the adiabatic process.

$$\Rightarrow \quad \frac{1}{2}mv^2 = nC_V\Delta T = \frac{m}{M}C_V\Delta T$$

$$\Rightarrow \quad \Delta T = \frac{Mv^2(\gamma - 1)}{2R}; \quad \left(C_V = \frac{R}{\gamma - 1}\right)$$

14. (c) For state A, $C_p - C_V = R$, so gas acts like an ideal gas.

For state B, $C_p - C_V \neq R$, So gas acts like a real gas.

$\therefore p_A < p_B$ and $T_A > T_B$

15. (c) For mixture, $\gamma_{(\text{mixture})} = \dfrac{\dfrac{n_1\gamma_1}{\gamma_1 - 1} + \dfrac{n_2\gamma_2}{\gamma_2 - 1}}{\dfrac{n_1}{\gamma - 1} + \dfrac{n_2}{\gamma - 1}}$

$$\therefore \quad \gamma_{\text{mixture}} = \left(\frac{\dfrac{1 \times \frac{5}{3}}{\frac{5}{3} - 1} + \dfrac{2 \times \frac{7}{5}}{\frac{7}{5} - 1}}{\dfrac{1}{\frac{5}{3} - 1} + \dfrac{2}{\frac{7}{5} - 1}}\right) = \frac{19}{13}$$

As $\frac{1}{\gamma}$ part is used to raise internal energy, $1 - \frac{1}{\gamma}$ part is used to do external work.

Hence, part of heat utilised in doing external work.

$$1 - \frac{1}{19/13} = 1 - \frac{13}{19} = \frac{6}{19}\text{ part}$$

16. (a,c) Total energy of an atom $= f \times \frac{1}{2}kT = 6 \times \frac{1}{2}kT = 3kT$

Atom possesses both KE and PE during vibration.

17. (b,c,d) Molar specific heat of any metal at high temperatures is $3R$ by Dulong-Petit law.

18. (a,c) $m = 0.44\,\text{kg}$, $T_1 = 180 + 273 = 453\,\text{K}$

$$\frac{V_2}{V_1} = 3, \; T_2 = 15 + 273 = 288\,\text{K}$$

$$W_{12} = 52.5\,\text{kJ}, \quad \frac{C_p}{C_V} = \gamma = 1.41\text{ (air)}$$

Now, 1–2 is an adiabatic process

$$\therefore \quad W_{12} = \frac{mR(T_1 - T_2)}{\gamma - 1}$$

$$52.5 = \frac{0.44 \times R\,(453 - 288)}{1.41 - 1}$$

$$\Rightarrow \quad R = 0.296\text{ units}$$

$$\therefore \quad C_p - C_V = 0.296\text{ units}$$

$$\frac{C_p}{C_V} = 1.41$$

$$\Rightarrow \quad (1.41 - 1)C_V = 0.296$$

$$\Rightarrow \quad C_V = 0.722\,\text{kJkg}^{-1}\text{K}^{-1}$$

$$C_p = 1.018\,\text{kJkg}^{-1}\text{k}^{-1}$$

19. (a,b) For $n = 1$, $C_n = \infty$

$\therefore$ Process is isothermal as $\Delta T = 0$

We have $C_n = \dfrac{Q}{m(\Delta T)} = \infty \Rightarrow \Delta T = 0$

As process is isothermal, $\Delta U = 0$, then $\Delta Q = \Delta U + \Delta W$

$\Rightarrow \quad \Delta Q = \Delta W$; isothermal process

In isothermal process entropy does not remains constant.

Also, isothermal process is quaristatic i.e. reversible.

20. (b,d) ΔS_1 (constant volume process) $= mC_V \ln\left(\frac{T_2}{T_1}\right)$

ΔS_2 (polytropic process, pV^n = constant)

$$= mC_V \times \left(\frac{\gamma - n}{n - 1}\right)\ln\left(\frac{T_2}{T_1}\right)$$

As change in entropy is same for both process,

$$\frac{\Delta S_2}{\Delta S_1} = 1 \Rightarrow \frac{\gamma - n}{n - 1} = 1$$

For monatomic gas, $\gamma = \frac{5}{3}$

$$\Rightarrow \quad \frac{\frac{5}{3} - n}{n - 1} = 1 \Rightarrow \frac{5}{3} - n = n - 1$$

$$\Rightarrow \quad 2n = \frac{5}{3} + 1 = \frac{8}{3}; \quad n = \frac{4}{3}$$

For monatomic ideal gas,

$$C_V = \text{molar specific heat at constant volume} = \frac{3}{2}R$$

$\therefore$ Molar specific heat for polytropic process is

$$C_n = C_V\left|\left(\frac{\gamma - n}{1 - n}\right)\right| = C_V\left|\left(\frac{\frac{5}{3} - \frac{4}{3}}{1 - \frac{4}{3}}\right)\right| = C_V\left|\left(\frac{\frac{1}{3}}{\frac{-1}{3}}\right)\right| = C_V = \frac{3}{2}R$$

21. (a) (i) $C = \dfrac{\Delta Q}{m\Delta T} = \dfrac{0}{m(\Delta T)} = 0$; (ii) $C = \dfrac{\Delta Q}{m(\Delta T)} = \dfrac{\Delta Q}{m(0)} = \infty$

(iii) $C = \dfrac{\Delta Q}{m(\Delta T)} > 0, \Delta T > 0$; (iv) $C = \dfrac{\Delta Q}{m\Delta T} < 0, \Delta T < 0$

Hence, (i) → (s), (ii) → (r), (iii) → (q), (iv) → (p)

DPP-5 **Thermodynamics-I** (Work, Heat and Internal Energy)

1. (d) In case of $\int pdV$ and $\int Vdp$ integral can be evaluated, when p can be expressed in terms of V (or V in terms of p). It is an exact differential and hence not a property.

In case of $\int pdV + \int Vdp$,

$$\int pdV + \int Vdp = \int (pdV + Vdp)$$
$$= \int d(pV) = pV$$

So, it can be evaluated without knowing a relation between p and V. It is an exact differential and hence it is a property.

2. (b) Displacement work

$$= \int_{\text{cylinder}} pdV + \int_{\text{balloon}} pdV = 0 + \int_{\text{balloon}} pdV$$
$$= 101.3 \text{ kPa} \times 0.6 \text{ m}^3 = 60.795 \text{ kJ} = 60.8 \text{ kJ}$$

As, balloon pushes the atmosphere

$\therefore$ Work done by the atmosphere

$$= -60.8 \text{ kJ}$$

3. (d) Work done by stirring device upon the system (negative work for system)

$$= W_1 = -2\pi NT = -2\pi \times 9500 \times \frac{1}{2\pi} \text{ (N-m)}$$
$$= -9500 \text{ N-m}$$

Work done by the system on surroundings

$$= \text{piston work (positive work for system)}$$
$$= (p \cdot A) \times l = p \times \frac{\pi d^2}{4} \times l$$
$$= 101.3 \times \frac{\pi}{4} \times \frac{1}{\pi} \times 10 \times 10^{-2}$$
$$= \frac{1013 \times 10^{-2}}{4} = 2.53 \text{ kJ}$$

Hence, net work transfer for the system

$$W_{net} = -9500 + 2.53 = -9497.47 \text{ kJ}$$

4. (b) $W = \int_{V_1}^{V_2} pdV = \int_{V_1}^{V_2} \frac{3}{V} dV$

$$V_1 = \frac{3}{p_1} = \frac{3}{1.5 \times 10^5} = 2 \times 10^{-5} \text{ m}^3$$
$$V_2 = \frac{3}{7.5 \times 10^5} = 0.4 \times 10^{-5} \text{ m}^3$$

$\therefore$
$$W = \int_{2 \times 10^{-5}}^{0.4 \times 10^{-5}} \frac{3}{V} dV$$
$$= 3 [\log (0.4 \times 10^{-5}) - \log (2 \times 10^{-5})]$$
$$= 3 \log\left(\frac{0.4}{2}\right) = -3 \log\left(\frac{2}{0.4}\right)$$
$$= -3 \log (5) = -3 \times 1.61 = -4.83 \text{ J}$$

5. (c) Internal energy is a state function and heat is a path function. Also, Q = area below curve

$\therefore$ $Q_1 < Q_2 < Q_3$ and $U_1 = U_2 = U_3$

6. (c) $\Delta Q = \Delta U + \Delta W$

$$150 = \Delta U + 110 \Rightarrow \Delta U = 40 \text{ J}$$

7. (a) For a van der Waals' gas, $p = \frac{nRT}{V - nb} - \frac{an^2}{V^2}$

So, work done $= \int_{V_1}^{V_2} pdV = \int_{V_1}^{V_2} \left(\frac{nRT}{V - nb} - \frac{an^2}{V^2}\right) dV$

$$= nRT \left[(\log(V - nb))_{V_1}^{V_2} - an^2 \left(-\frac{1}{V}\right)_{V_1}^{V_2} \right]$$
$$= nRT \log_e \left(\frac{V_2 - nb}{V_1 - nb}\right) + an^2 \left(\frac{V_1 - V_2}{V_1 V_2}\right)$$

8. (c) According to work-energy theorem,

$$\Delta W = \Delta K$$

But $\Delta K = 0 \Rightarrow \Delta W = 0$

$\Rightarrow$ $W_{gas} + W_{atm} + W_{ext} = 0$

$\Rightarrow$
$$W_{ext} = -(W_{gas} + W_{atm})$$
$$= -\left(nRT \ln \frac{V_f}{V_i} + nRT\right) = nR(1 - \ln 2)$$

9. (d) Process 1 is clockwise (positive work), so work done = Area = πr^2

Process 2 is anticlockwise (negative work)

So, work done = Area = $-\pi(1.5r)^2$

$$\text{Ratio} = \frac{\pi r^2}{-\pi \frac{9}{4} r^2} = -\frac{4}{9}$$

10. (c) $\Delta Q = \Delta U + \Delta W$

$$\Rightarrow \Delta U = \Delta Q - \Delta W = 540 - \frac{p\Delta V}{J}$$
$$\Rightarrow \Delta U = 540 - \frac{1.013 \times 10^5 (1671 - 1) \times 10^{-6}}{4.2}$$
$$= 540 - 39.7 \approx 500 \text{ cal}$$

11. (a,c) Volume decreases during melting of ice.

$\therefore$ Positive work is done on the ice-water system by the atmosphere (expansion of atmosphere).

Heat is absorbed by ice-water system.

So, internal energy increases.

12. (b,c) $pV = nRT$

$$\left(p_{atm} + \frac{mg}{A}\right)Ah = nRT \Rightarrow n = \frac{\left(p_{atm} + \frac{mg}{A}\right)Ah}{RT}$$

Also, gas on heating lifts piston against gravity, spring and atm pressure.

$\therefore$
$$W_{gas} = p_{atm} \cdot A \cdot x + \frac{1}{2}kx^2 + mgx$$

13. (b,c) ΔU (monatomic gas) $= \frac{f}{2} nR\Delta T$

$$= 1.5 \times 1 \times 8.3 \times 2 = 24.9 \text{ J}$$
$$\Delta W = \Delta Q - \Delta U = 42 - 24.9 = 17.1 \text{ J}$$
$$\Delta W = p\Delta V = pA \cdot x = 17.1$$

$\Rightarrow$
$$x = \frac{17.1}{100 \times 8.5 \times 10^{-4}} = 0.2 \text{ m} = 20 \text{ cm}$$

14. (a,b) $\Delta Q = \Delta U + \frac{2\Delta U}{f}$

$$\Delta U = n \cdot \frac{f}{2} \cdot R \cdot (\Delta T) = \frac{f}{2 + f} \Delta Q = \frac{140 \text{ J}}{1.4} = 100 \text{ J}$$

Work done by the gas $= \Delta Q - \Delta U$

$$= 140 \text{ J} - 100 \text{ J} = 40 \text{ J}$$

15. (a,b,c) $V_C = 2V$, $T_B = 2T_0$, $T_C = 4T_0$

$$\Delta Q = \text{Area under cycle} = \frac{1}{2} p_0 V_0 = \Delta W$$

Heat is rejected in process $CA = Q_{CA} = \Delta W_{CA} + \Delta U_{CA}$

$$= \frac{1}{2}(p_0 + 2p_0)V_0 - \frac{5}{2} nR(T_C - T_B)$$
$$= -9 p_0 V_0$$

Also, $\Delta Q_{injected} = \Delta Q_{AB} + \Delta Q_{BC} = \frac{19}{2} p_0 V_0$

In a cyclic process, $\Delta U = 0$

16. (d) When the stone is about to enter the water,

$$\Delta KE = mgh = 2943 \text{ J}$$
$$\Delta PE = -2943 \text{ J}$$
$$\Delta U = 0$$

Also, $$\Delta Q = \Delta W = 0$$

17. (d) When stone comes to rest, $\Delta Q = 0, \Delta W = 0$ and $\Delta KE = 0$

$$\Delta U = -\Delta PE = -(-2943) = 2943 \text{ J}$$

18. (c) When both stone and water comes to their original temperature,

$$\Delta W = 0,$$

Also, $$\Delta Q = -\Delta U = -2943 \text{ J}$$

19. (a) $W = \int_{V_1}^{V_2} p\,dV = \int_{V_1}^{V_2} \left(\frac{5}{V} + 1.5\right) dV$

$$= \left(5 \log_e \frac{V_2}{V_1} + 1.5 (V_2 - V_1)\right) \times 10^5$$
$$= 10^5(-5.49 - 0.15) = -564 \text{ kJ}$$

20. (c) From first law of thermodynamics,

$$\Delta Q = \Delta U + \Delta W$$
$$\Rightarrow \quad -45 = \Delta U - 564$$
$$\Rightarrow \quad \Delta U = 519 \text{ kJ}$$

21. (c) $W_{cda} = W_{cd} = p_c \Delta V_{cd} = \frac{1}{2} p_b (V_a - V_b)$

$$= -\frac{1}{2} W_{abc} = -\frac{1}{2}(-54) = 27 \text{ J}$$
$$\Delta E_{cda} = Q_{cda} - W_{cda}$$
$$\Rightarrow \quad Q_{cda} = \Delta E_{cda} + W_{cda}$$
$$= 28 + 27 = 55 \text{ J}$$
$$\Delta E_{\text{int}, a} - \Delta E_{\text{int}, c} = \Delta E_{\text{int}; ca}$$
$$= -\Delta E_{\text{int}, ac} = 28 \text{ J}$$

Also, $$\Delta E_{\text{int}, da} = Q_{da} - W_{da}$$
$$\therefore \quad Q_{da} = \Delta E_{\text{int}, da} + W_{da}$$
$$= 16 \text{ J} + 0 = 16 \text{ J}$$

Hence, (i) → (s), (ii) → (r), (iii) → (q), (iv) → (p)

22. (2)

Gas is diatomic, $\gamma = \frac{7}{5}$,

$\because$ Heat is not conducted process, it is adiabatic.

$$\therefore \quad pV^\gamma = \text{constant}$$

or $$p^{1-\gamma} \cdot T^\gamma = \text{constant}$$

$$\Rightarrow \quad \frac{d}{dh}(p^{1-\gamma} \cdot T^\gamma) = 0$$
$$\Rightarrow \quad T^\gamma (1-\gamma) p^{-\gamma} \cdot \frac{dp}{dh} + p^{1-\gamma} \cdot \gamma \cdot T^{\gamma - 1} \cdot \frac{dT}{dh} = 0$$
$$\Rightarrow \quad (1-\gamma)\frac{dp}{dh} + \gamma \frac{p}{T} \cdot \frac{dT}{dh} = 0$$

But, $$\frac{dp}{dh} = -\rho g$$

$$\Rightarrow \quad (1-\gamma)(-\rho g) + \gamma \cdot \frac{p}{T} \cdot \frac{dT}{dh} = 0$$
$$\Rightarrow \quad \frac{dT}{dh} = \frac{1-\gamma}{\gamma} \cdot \frac{T}{p} \cdot \rho g$$

From ideal gas law,

$$pV = NkT$$
$$\frac{T}{p} = \frac{V}{Nk} = \frac{V_m}{Mk} = \frac{m}{\rho k}$$
$$\therefore \quad \frac{dT}{dh} = \frac{(1-\gamma) mg}{\gamma k} = -\frac{2}{7}\left(\frac{mg}{k}\right)$$

So, $$N = 2$$

23. (3) Free expansion is an adiabatic process, $\gamma_{\text{air}} = 1.4$

$$T_{\text{out}} = T_{\text{in}} \left(\frac{p_{\text{out}}}{p_{\text{in}}}\right)^{\frac{\gamma - 1}{\gamma}}$$
$$= (273 + 150) \times \left(\frac{6.9}{13.8}\right)^{\frac{0.4}{1.4}} = 347 \text{ K}$$

Applying energy conservation to nozzle,

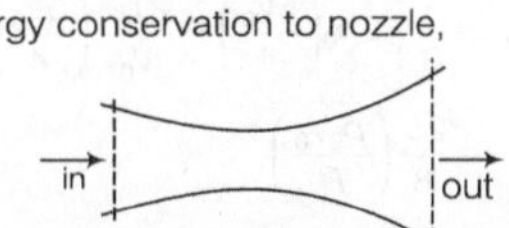

$$Q - W = m\left[C_p(T_o - T_i) + \left(\frac{V_o^2 - V_i^2}{2}\right) + g(h_2 - h_1)\right]$$

as $h_2 = h_1$ and $Q = 0$ Adiabatic expansion

Also, $$W = 0$$

$$\therefore \quad 0 = mC_p(T_o - T_i) + m\frac{(V_o^2 - V_i^2)}{2}$$
$$\Rightarrow \quad 1.005\,(347 - 423) + \left(\frac{V_o^2 - 10^2}{2000}\right) = 0$$
$$\Rightarrow \quad V_o = \sqrt{100 + 2000 \times 1.005 \times 76}$$
$$\Rightarrow \quad V_o = 390 \text{ ms}^{-1} = 3 \times 130 \text{ ms}^{-1}$$

DPP-6 **Thermodynamics-II** (Thermodynamic Processes)

1. (i) $ABCA$ is a cyclic process.

So, work done by the gas

$$= \text{area of } \Delta ABC$$
$$= \frac{1}{2} \times \text{base} \times \text{height}$$
$$= \frac{1}{2}(2V_0 - V_0)(3p_0 - p_0) = p_0V_0 \text{ J}$$

(ii) Number of moles = 1

Gas is ideal and it is monatomic

So, $$C_V = \frac{3}{2}R \text{ and } C_p = \frac{5}{2}R$$

Now, heat rejected in path $CA = nC_p\Delta T$ (process is isobaric)

$$= 1 \times C_p \times (T_f - T_i)$$
$$= C_p\left(\frac{p_fV_f}{R} - \frac{p_iV_i}{R}\right)$$
$$= \frac{5}{2}R\left(\frac{p_0V_0}{R} - \frac{p_0\,2V_0}{R}\right)$$
$$= -\frac{5}{2}p_0V_0 \text{ (negative sign shows rejection of heat)}$$

$\therefore$ Heat rejected C to $A = \frac{5}{2}p_0V_0$ J

(iii) AB process is isochoric.

$\therefore$ Heat absorbed $= nC_V\Delta T = C_V(T_f - T_i)$

$$= C_V\left(\frac{p_fV_f}{R} - \frac{p_iV_i}{R}\right) = \frac{3}{2}R\left(\frac{3p_0V_0}{R} - \frac{p_0V_0}{R}\right) = 3p_0V_0 \text{ J}$$

(iv) As process BC is not defined, we first calculate total heat absorbed.

$\therefore$ Total heat absorbed

$$= Q_{CA} + Q_{AB} + Q_{BC}$$
$$= -\frac{5}{2}p_0V_0 + 3p_0V_0 + Q_{BC} = \frac{p_0V_0}{2} + Q_{BC}$$

In cyclic process $ABCA$, change in internal energy is zero. (As, system is restored to same condition)

So, $$\Delta Q = \Delta W$$

$$\Rightarrow \quad Q_{BC} + \frac{p_0V_0}{2} = p_0V_0 \Rightarrow Q_{BC} = \frac{p_0V_0}{2}$$

(v) Maximum temperature of the gas will be between B and C. Line BC is a straight line. So, pV equation for process BC is

$$p = -mV + c$$

$$m = \text{slope} = \frac{y_2 - y_1}{x_2 - x_1} = \frac{2p_0}{V_0}$$

and intercept, $c = 5p_0$

$$\therefore \quad p = -\left(\frac{2p_0}{V_0}\right)\cdot V + 5p_0$$

$$\Rightarrow \quad pV = -\frac{2p_0}{V_0}\cdot V^2 + 5p_0V$$

or $$RT = -\frac{2p_0}{V_0}\cdot V^2 + 5p_0V$$

$$\Rightarrow \quad T = \frac{1}{R}\left(5p_0V - \frac{2p_0V^2}{V_0}\right) \quad \dots(i)$$

T is maximum, $\frac{dT}{dV} = 0$

$$\Rightarrow \quad 5p_0 - \frac{4p_0V}{V_0} = 0 \Rightarrow V = \frac{5V_0}{4}$$

Substituting this in Eq. (i), we get

$$T_{max} = \frac{1}{R}\left[5p_0\left(\frac{5V_0}{4}\right) - \frac{2p_0}{V_0}\left(\frac{5V_0}{4}\right)^2\right]$$

$$= \frac{25}{8}\left(\frac{p_0V_0}{R}\right)$$

2. 1 mole of an ideal gas at NTP has a volume of 22.4 L.

So, 0.1 mole of an ideal gas has volume of 2.24 L. At point A, $p_A = 1$ atm, $T_A = 300$ K. So by the gas law,

$$\frac{V_A}{T_A} = \left(\frac{V}{T}\right)_{NTP}$$

$$\Rightarrow \quad V_A = \frac{300 \times 2.24}{273} = 2.46 \text{ L}$$

At point B, $V_A = V_B = 2.46$ L process AB is isochoric.

$$T_B = 600 \text{ K}$$

So, $$\frac{p_B}{T_B} = \frac{p_A}{T_A}$$

$$\Rightarrow \quad p_B = \frac{T_B \times p_A}{T_A} = \frac{600 \times 1}{300} = 2 \text{ atm}$$

At point C, we have $p_A = p_C = 1$ atm (pressure constant C to A)

$$T_C = 455 \text{ K}$$

So, $$\frac{V_C}{T_C} = \frac{V_A}{T_A} \Rightarrow V_C = \frac{T_C \times V_A}{T_A}$$

$$\Rightarrow \quad V_C = \frac{455 \times 2.46}{300} = 3.73 \text{ L}$$

Now, work during AB process is zero (as $\Delta V = 0$). The work done by the gas during adiabatic expansion BC is

$$W_{BC} = \frac{nR(T_B - T_C)}{\gamma - 1}$$

$$= \frac{0.1 \times 8.31 \times (600 - 455)}{\frac{5}{3} - 1} = 181 \text{ J}$$

The work done on gas during isobaric compression CA is

$$W_{CA} = p\Delta V = nR\Delta T$$

$$= 0.1 \times 8.31 \times (300 - 455) = -129 \text{ J}$$

So, net work done by the gas is

$$W_{net} = 181 - 129 = 52 \text{ J}$$

3. (c) $pV = nRT \Rightarrow pdV = nRdT$

$$\Rightarrow \quad W = nRdT$$

$$= \frac{10}{32} \times 1.98\left(\frac{\text{cal}}{\text{mol-K}}\right) \times 100 (\text{K})$$

Note That $\Delta T = 100°\text{C} = 100$ K

$$\therefore \quad W = 62 \text{ cal}$$

As, $$C_V = 5.03 \frac{\text{cal}}{\text{mol-K}} \text{ and } R = 1.98 \frac{\text{cal}}{\text{mol-K}}$$

$$C_p = C_V + R = 5.03 + 1.98$$

$$= 7.01 \frac{\text{cal}}{\text{mol-K}}$$

So, heat supplied is

$$Q = nC_p\Delta T = \frac{10}{32} \times 7.01 \times 100 = 219 \text{ cal}$$

So, $\Delta U = \Delta Q - \Delta W = 219 - 62 = 157$ cal is used to increase the temperature of the gas.

4. (d) For an adiabatic change $TV^{\gamma - 1}$ = constant

$$\Rightarrow \quad \frac{T_f}{T_i} = \left(\frac{V_i}{V_f}\right)^{\gamma - 1}$$

$$T_i = 17 + 273 = 290 \text{ K}, \frac{V_i}{V_f} = 10$$

When the gas is monatomic, $\gamma = \frac{5}{3}$,

$$\therefore \quad T_f = 290 \times (10)^{\frac{5}{3} - 1} = 290 \times 10^{\frac{2}{3}}$$

When the gas is diatomic, $\gamma = \frac{7}{5}$

$$\therefore \quad T_f = 290 \times 10^{\frac{7}{5} - 1} = 290 \times 10^{\frac{2}{5}}$$

Hence, ratio $= \frac{290 \times 10^{2/3}}{290 \times 10^{2/5}} = 10^{\frac{2}{3} - \frac{2}{5}} = 10^{4/15}$

5. (d) For isothermal compression, pV = constant

$$V_i = V, V_f = \frac{V}{2}, p_i = p$$

$$\therefore \quad p_f = \frac{p_iV_i}{V_f} = \frac{p \times V}{\left(\frac{V}{2}\right)} = 2p$$

For adiabatic compression, pV^{γ} = constant

$$\therefore \quad p_f = 2^{\gamma} \cdot p$$

So, the ratio is

$$\frac{(p_f)_{iso}}{(p_f)_{Adi}} = \frac{2p}{2^{\gamma} \cdot p} = \frac{2}{2^{\gamma}} = 2^{1-\gamma}$$

6. (b) Work done = area of loop

$$= \frac{1}{2} \text{ base} \times \text{height}$$

$$= \frac{1}{2} \times 2 \text{ m}^3 \times 20 \frac{\text{N}}{\text{m}^2} = 20 \text{ J}$$

As, the cycle is anticlockwise, work is done on the gas.

7. (d) $W = \int_{V_1}^{V_2} pdV = \int_{V_1}^{V_2}\left(\frac{a}{V} + b\right)dV$

$$= [a \log V + bV]_{V_1}^{V_2}$$

$$= a \log V_2 + bV_2 - a \log V_1 - bV_1$$

$$= a \log\left(\frac{V_2}{V_1}\right) + b(V_2 - V_1)$$

8. (c) AB; V = constant; $W_{AB} = 0$

BC; Isothermal process, $W_{BC} = RT_2 \ln\left(\frac{V_2}{V_1}\right)$

CA; $W_{AC} = \int_{V_2}^{V_1} pdV = R(T_1 - T_2)$

9. (d) $AB: p =$ constant, V decreases

So, AB is compression at constant pressure.

BC : As slope of BC is higher, it must be an adiabatic curve.

So, BC is adiabatic compression.

CA is isothermal expansion.

10. (b) For process A to B, $pT =$ constant

$$\Rightarrow \quad p^2 \cdot V = \text{constant} = k \text{ (say)}$$

$$\Rightarrow \quad p = \frac{\sqrt{k}}{\sqrt{V}}$$

So, $$W_{A \to B} = \int_{V_A}^{V_B} p \cdot dV = \int_{V_A}^{V_B} \frac{\sqrt{k}}{\sqrt{V}} \cdot dV$$

$$= 2\sqrt{k}(\sqrt{V_B} - \sqrt{V_A})$$

$$= 2(\sqrt{p_B^2 V_B \cdot V_B} - \sqrt{p_A^2 V_A \cdot V_A})$$

$$= 2(p_B V_B - p_A V_A) = 2(nRT_B - nRT_A)$$

$$= 2nR\,(T_1 - 2T_1)$$

$$= 2 \times 2 \times R \times (300 - 600)$$

$$= -1200\, R \text{ J}$$

So, work done on the gas is 1200 R J

11. (a) $W_{AB} = p\Delta V = 10 \times (2 - 1) = 10$ J

$W_{BC} = 0$

From first law of thermodynamics,

$$\Delta Q = \Delta W + \Delta U$$

$\Delta U = 0$ as $ABCA$ is cyclic.

$$\therefore \quad \Delta Q = \Delta W_{AB} + \Delta W_{BC} + \Delta W_{CA}$$

$$\Rightarrow \quad \Delta W_{CA} = \Delta Q - \Delta W_{AB} - \Delta W_{BC}$$

$$= 5 - 10 - 0 = -5 \text{ J}$$

12. (d) Piston of A is free to move.

$\therefore$ Heat is supplied at constant pressure,

$$\Delta Q_A = nC_p(\Delta T)_A$$

When piston of B is fixed, heat is supplied at constant volume,

$$\Delta Q_B = nC_V(\Delta T)_B$$

$$\Delta Q_A = \Delta Q_B \Rightarrow nC_p\Delta T_A = nC_V\Delta T_B$$

$$\Rightarrow \quad \Delta T_B = \frac{C_p}{C_V} \cdot \Delta T_A = \gamma \Delta T_A$$

Given, $\gamma = 1.4$, $\Delta T_A = 30$ K

$$\therefore \quad \Delta T_B = 1.4 \times 30 = 42 \text{ K}$$

13. (c)

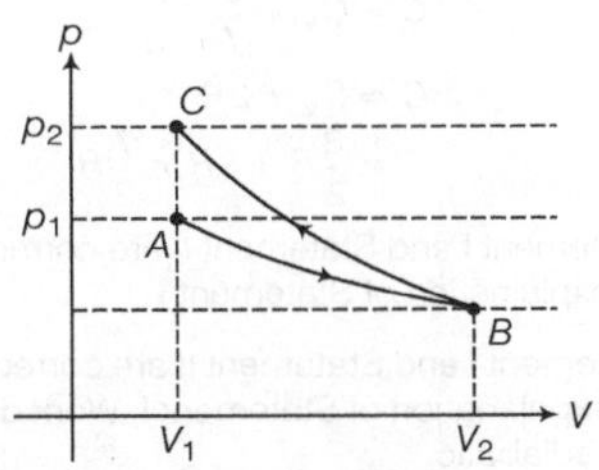

$W_{AB} =$ positive as volume is increasing

$W_{BC} =$ negative as volume is decreasing

Also, $p_2 > p_1$

14. (b)

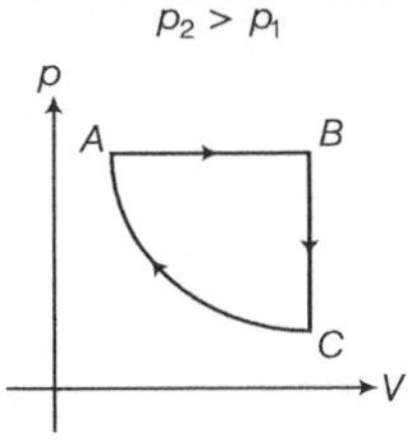

$A \to B$, $p =$ constant

$B \to C$, $V =$ constant

$C \to A$, $T =$ constant

15. (a,b,c,d) Molar specific heat for a gas, $C = \frac{1}{n} \cdot \frac{dQ}{dT}$

It is possible to obtain any set of values for ΔQ and ΔT by proper selection of process.

As, $\Delta Q = 0$ is adiabtic process

$$\therefore \quad C = 0$$

and $\Delta T = 0$ is isothermal process

$$\therefore \quad C = \infty$$

16. (a,b,d) As, $\rho \propto p$ and $\rho \propto \frac{1}{T}$

$$\Rightarrow \quad pT = \text{constant}$$

Also, $pV = nRT$

Combining both, we have

$$p^2V = \text{constant or } pV^{\frac{1}{2}} = k$$

Also, $$\Delta Q = \Delta U + \Delta W$$

$$C\Delta T = C_V T + \frac{p_f V_f - p_i V_i}{2}$$

$$\Rightarrow \quad C = C_V + 2R$$

$$\Rightarrow \quad 33.24 = \frac{R}{\gamma - 1} + 2R$$

$$\Rightarrow \quad \gamma = 1.5 \text{ and } f = 4$$

17. (b,c) Work is done by the gas (expansion results)

$$W = -\int_A^B p \cdot dV$$

$= -$ (Area under arc $\overline{AB}$)

$= -$ (area of semicircle + area of rectangular part)

$$= -\left(\frac{\pi r^2}{2} + (\text{side} \times \text{side})\right)$$

$$= -\left[\left(\frac{\pi}{2} \times 2.4 \times 10^{-3} \times 200\right) + (3 \times 10^5 \times 4.8 \times 10^{-3})\right]$$

$$= -(784 + 1440) = -2190 \text{ J}$$

$$\Delta E_{\text{int}} = Q + W$$

$$= 5790 - 2190 = 3.6 \text{ kJ}$$

18. (a,b) AB and CD are isothermal

$$\therefore \quad p_1V_A = p_2V_B \text{ and } p_2V_C = p_1V_D$$

$$\Rightarrow \quad \frac{V_B}{V_1} = \frac{p_1}{p_2} \text{ and } \frac{V_2}{V_C} = \frac{p_2}{p_1}$$

Also, $$\Sigma W = -nRT_1\left[\ln\left(\frac{p_1}{p_2}\right)\right] - nRT_2\left[\ln\left(\frac{p_2}{p_1}\right)\right]$$

$$nRT_1\left[\ln\left(\frac{p_2}{p_1}\right)\right] - nRT_2\left[\ln\left(\frac{p_2}{p_1}\right)\right] = -nR(T_2 - T_1)\ln\left(\frac{p_2}{p_1}\right)$$

$$= -p_1(V_2 - V_1)\ln\left(\frac{p_2}{p_1}\right)$$

19. (a,b,c,d) For a cycle, initial and final values of T are same.

$$\therefore \quad \Delta E_{\text{int}} = 0$$

So, $\Delta Q = -\Delta W$

Now, $$\Delta W = W_{DA} + W_{AB} + W_{BC} + W_{CD}$$

$$= -p_i(V_i - 3V_i) + 0 - 3p_i(3V_i - V_i) + 0$$

$$= -4p_iV_i$$

$$\Delta W = -4p_iV_i = -4nRTi$$

$$= -4 \times 1 \times 8.314 \times 273 = -9.08 \text{ kJ}$$

20. (a) For isothermal process,

$$T = \text{constant}, \Delta E_{\text{int}} = 0 \Rightarrow \Delta Q = \Delta W$$

For adiabatic process,

$$\Delta Q = 0 \Rightarrow \Delta E_{\text{int}} = -\Delta W$$

For free expansion,

$$\Delta Q = \Delta W = \Delta E_{\text{int}} = 0$$

Hence, (i) $\to$ (r), (ii) $\to$ (q), (iii) $\to$ (s), (iv) $\to$ p

21. (c) $|\Delta Q| = \frac{p_0V_0}{2}, |\Delta U| = \frac{5}{2}p_0V_0$

$$|\Delta W| = |\Delta U| - |\Delta Q| = \left(\frac{5}{2} - \frac{1}{2}\right)(p_0V_0) = 2p_0V_0$$

Hence, (i) → (s), (ii) → (r), (iii) → (q), (iv) → (p)

22. (b) $\Delta Q_{ABC} = \Delta U_{ABC} - \Delta W_{ABC}$

$$= 800 - (-500) = 1300 \text{ J}$$

$$\Delta W_{CDA} = W_{CD} = \frac{1}{5}p_A\Delta V_{AB}$$

$$= -\frac{1}{5}W_{AB} = 100 \text{ J}$$

ΔE_{int} or $\Delta U_{CD} = \Delta E_{\text{int }CDA} - \Delta E_{\text{int }DA}$

$$= -800 \text{ J} - 500 \text{ J} = -1300 \text{ J}$$

$$\Delta Q_{CD} = \Delta E_{\text{int }CD} - W_{CD}$$

$$= -1300 \text{ J} - 100 \text{ J} = -1400 \text{ J}$$

Hence, (i) → (s), (ii) → (r), (iii) → (q), (iv) → (p)

23. (a) The isobaric process $n = 0$, we have p = constant

The isothermal process, $n = 1$, we have pV = constant

The adibatic process $n = \gamma$, we have pV^γ = constant

Let p_2 = final pressure

For isobaric process,

$$p_2 = p_1 : \text{curve } AB$$

For adiabatic process,

$$p_2V_2^\gamma = p_1V_1^\gamma$$

$$p_2 = p_1\left(\frac{V_1}{V_2}\right)^\gamma$$

For isothermal process,

$$p_2 = p_1\left(\frac{V_1}{V_2}\right)$$

For expansion, $\frac{V_1}{V_2} < 1, \left(\frac{V_1}{V_2}\right)^\gamma < \frac{V_1}{V_2}$

∴ For adiabatic expansion final pressure is much lower so, AC is an isothermal and AD is an adiabatic.

24. (c) Work done = Area under p-V curve

As, area enclosed is largest for $AB (n = 0)$ and least for $AD (n = \gamma)$

So, work done in adiabatic process AD, $n = \gamma$ is least.

25. (c) ΔU for $n = 1$ is 0. As, in isothermal process T = a constant

ΔU is maximum for adiabatic process $n = \gamma$, as expansion is done on the expense of internal energy of the gas.

26. (b) As, $\rho \times p$ or $\rho = \frac{pM}{RT}$, M = molecular mass of the gas

AB is an isothermal process.

So, $W_{AB} = RT_A \ln\left(\frac{p_A}{p_B}\right) = RT_A \ln\left(\frac{1}{2}\right) = -\frac{p_1M}{\rho_1}\ln(2)$

$$W_{BC} = p_B(V_C - V_B) = 2p_1\left(\frac{M}{\rho_C} - \frac{M}{\rho_B}\right) = \frac{2p_1M}{2\rho_1} = \frac{p_1M}{\rho_1}$$

$W_{CA} = 0$, then ρ = constant, so V = constant and CA is isochoric process.

So, $W_{AB} : W_{BC} : W_{CA} :: -\log 2 : 1 : 0$

27. (d) Heat is supplied to gas is isobaric process BC

$$Q_{BC} = W_{BC} + \Delta U_{BC}$$

$$= \frac{p_1M}{\rho_1} + C_V\Delta T$$

$$= \frac{p_1M}{\rho_1} + \frac{3}{2}R\cdot\left(\frac{p_1M}{\rho_1R}\right)$$

$$= \frac{5p_1M}{2\rho_1}$$

And heat is rejected in processes AB and CA

$$= Q_{AB} + Q_{CA}$$

$$= \left|\left(-\frac{p_1M}{\rho_1}\log 2\right)\right| + \left|\left(\frac{-3p_1M}{2\rho_1}\right)\right|$$

$$= \frac{p_1M}{\rho_1}\left(\frac{3}{2} + \log 2\right)$$

So, $\text{ratio} = \frac{\text{Heat rejected}}{\text{Heat supplied}}$

$$= \frac{\frac{p_1M}{\rho_1}\cdot\left(\frac{3}{2} + \log 2\right)}{\frac{5}{2}\left(\frac{p_1M}{\rho_1}\right)}$$

$$= \frac{3 + 2\log 2}{5}$$

$$= \frac{3 + \log 4}{5}$$

28. (a) $p = \frac{a}{T}$ and $pV = RT$

Combining, we get

$$V = \frac{RT}{p} = \frac{RT^2}{a}$$

So, $$\frac{dV}{dT} = \frac{2RT}{a}$$

As, $$dQ = dU + dW$$

$$C = \frac{dQ}{dT} = \frac{dU}{dT} + \frac{dW}{dT}$$

$$\Rightarrow \quad C = \frac{d}{dT}(C_VdT) + p\frac{dV}{dT}$$

$$\Rightarrow \quad C = C_V + p\left(\frac{2RT}{a}\right)$$

$$\Rightarrow \quad C = C_V + \frac{a}{T}\cdot\frac{2RT}{a}$$

$$\Rightarrow \quad C = C_V + 2R$$

$$= \frac{3}{2}R + 2R = \frac{7}{2}R$$

So, both Statement I and Statement II are correct and Statement II is a correct explaination of Statement I.

29. (a) Both Statement I and Statement II are correct and Statement II is a correct expalanation of Statement I. Work done is indepedent of particular adiabatic.

30. (d) Statement I is incorrect, as in free expansion, $\Delta U = 0$ and $\Delta T = 0 \Rightarrow T$ = a constant. Statement II is correct.

DPP-7 **Thermodynamics-III** (Engines and Fridges)

1. (c) Second law states about reversibility and irreversibility by defining entropy.

2. (a) In a reversible adiabatic process, there is no entropy change ($\Delta S = 0$ reversible adiabatic).

3. (a) For any reversible process, $\Delta S_{system} + \Delta S_{surroundings} = 0$

4. (c) No spontaneous heat transfer from a cold body to a hot body is possible.

5. (a) Entropy may be expressed in terms of p and T.

6. (c) $\Delta S = \int \frac{dQ}{T}, dQ =$ Negative as heat is absorbed.

7. (c) Heat received $= 1500 \text{ kJmin}^{-1}$

$$= \frac{1500}{60} = 25 \text{ kJs}^{-1}$$

Work output $= 8.2 \text{ kW} = 8.2 \text{ kJs}^{-1}$

Heat rejected rate $= Q_{in} - W$

$$= 25 - 8.2 = 16.8 \text{ kJs}^{-1}$$

8. (d) $\Delta Q = 30$ kJ, $\Delta W = 60$ kJ for process 1-2,

$$\Delta Q_{1\text{-}2} = \Delta U_{1\text{-}2} + \Delta W_{1\text{-}2}$$

$$30 = \Delta U_{1\text{-}2} + 60$$

$$\Rightarrow \quad \Delta U_{1\text{-}2} = -30 \text{ kJ}$$

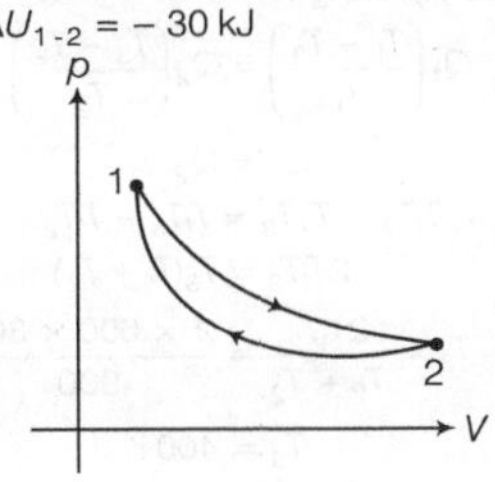

For process 2-1,

$$Q_{21} = \Delta U_{21} + W_{21}$$

$$0 = U_1 - U_2 + W_{21}$$

$$\Rightarrow \quad W_{21} = -30 \text{ kJ}$$

So, by doing a work of 30 kJ on the system, it can go back to initial state.

9. (c) Temperature of hot reservoir = Furnace temperature

$$= 2100 + 273 = 2373 \text{ K}$$

Temperature of cold reservoir $= -35.7°\text{C}$

$$= -35.7 + 273 = 237.3 \text{K}$$

Maximum possible efficiency,

$$\eta_{max.} = (\eta_{Carnot}) = 1 - \frac{T_2}{T_1}$$

$$= 1 - \frac{237.3}{2373} = 1 - \frac{1}{10} = \frac{9}{10}$$

$$= \frac{9}{10} \times 100 = 90\%$$

10. (b) $Q_{hot} = 80$ mW, $Q_{cold} = 50$ mW

Work output, $W = Q_H - Q_C = 30$ mW

$$\eta = \text{efficiency} = \frac{\text{Work output}}{\text{Heat input}}$$

$$= \frac{30}{80} = 0.375 \text{ or } 37.5\%$$

11. (a,b) In free expansion, $\Delta W = 0$

$$\Delta Q = 0$$

$\therefore \Delta U = 0$

So, temperature of gas remains at 300 K and energy unavailable.

$$= T \cdot \Delta S = nRT \ln\left(\frac{V_2}{V_1}\right)$$

$$= 5 \times 8.314 \times 300 \times \ln\left(\frac{152}{2}\right)$$

$$= 5 \times 8.314 \times 300 \times \ln(76)$$

$$= 5 \times 8.314 \times 300 \times 4.3 \approx 54 \text{ kJ}$$

In reversible isothermal process, work of 54 kJ can be recovered.

12. (c,d) Entropy is a state function. Entropy change over a closed path is zero.

$$\therefore \quad \Delta S_{12} = -\Delta S_{21}$$

13. (a,b,c,d)

$$\Delta S_{hot} = \frac{Q}{T_h} = \frac{8}{373} = 0.0214 \text{ JK}^{-1}$$

$$\Delta S_{cold} = \frac{Q}{T_c} = \frac{-8}{273} = -0.0293 \text{ JK}^{-1}$$

$$\Delta S_{hot} + \Delta S_{cold} = -0.0079 \text{ JK}^{-1}$$

This is violation of second law, so process is not possible spontaneously.

14. (a,b,c) $Q_{cold} = -Q_{hot}$

$$m_1C_1(T_f - T_c) = -m_2C_2(T_f - T_h)$$

$\Rightarrow T_f$ = final equilibrium temperature

$$= \frac{m_1C_1T_c + m_2C_2T_h}{m_1C_1 + m_2C_2}$$

$$\therefore \quad T_f = \frac{mC(27 + 273) + 2m(1.5C)(400)}{mC + 2m(1.5)C}$$

$$= \frac{300 + 400 \times 3}{4} = \frac{1500}{4}$$

$$\Rightarrow \quad T_f = \frac{1500}{4} \text{K} = 375 \text{K}$$

$$\Delta S = \int \frac{dQ_{cold}}{T} + \int \frac{dQ_{hot}}{T}$$

$$= mC \int_{T_c}^{T_f} \frac{dT}{T} + 2m(1.5C) \int_{T_h}^{T_f} \frac{dT}{T}$$

$$= mC \left[\log \frac{T_f}{T_c} + 3 \log \frac{T_f}{T_h}\right]$$

$$= mC \left[\log\left(\frac{375}{300}\right) + 3\log\left(\frac{375}{400}\right)\right]$$

$$= mC[0.22 + 3(-0.0645)]$$

$$= mC[0.22 - 0.19] = 0.03 \, mC$$

15. (a,c,d) Outcomes are listed in the table given below :

Result	Possible combination	Total
All heads	*HHHH*	1
3*H*, 1*T*	*THHH, HTHH, HHTH, HHHT*	4
2*T*, 2*H*	*TTHH, THTH, HTHT, HHTT, HTTH, THHT*	6
1*H*, 3*T*	*HTTT, THTT, TTHT, TTTH*	4
All tails	*TTTT*	1

Most probable state is 2*H* and 2*T*.

$$\therefore \quad p(2H, 2T) = \frac{6}{16} = \frac{3}{8}$$

Most probable state is also most disordered state (2*H*, 2*T*)

$$\therefore \quad p(2H, 2T) = \frac{3}{8}$$

Most stable state is least likely state (*HHHH* or *TTTT*)

$$\therefore \quad p(4H, 4T) = \frac{2}{16} = \frac{1}{8}$$

16. (d) 210 kJ/cycle is rejected.

$$\Delta S = \oint \frac{dQ}{T} = \frac{+420}{(327+273)} - \frac{210}{(27+273)} = 0$$

As, entropy change is zero, cycle is reversible and possible.

105 kJ/cycle is rejected,

$$\Delta S = \oint \frac{dQ}{T} = \frac{420}{600} - \frac{105}{300} = 0.35$$

$\Delta S > 0$ so, cycle is impossible (by Clausius inequality second law).

315 kJ/cycle is rejected,

$$\Delta S = \oint \frac{dQ}{T} = \frac{420}{600} - \frac{315}{300} = -0.35$$

Cycle is irreversible but possible (but cycle does not occur naturally).

Hence, (i) $\to$ (q), (ii) $\to$ (p), (iii) $\to$ (r)

17. (d) For isothermal process, $\frac{p_1V_1}{T_1} = \frac{p_2V_2}{T_2} \Rightarrow p_2 = 2p_1$

For adiabatic process, $p_1V_1^\gamma = p_2V_2^\gamma$

$$p_2 = 2^\gamma p_1$$

$\Rightarrow$ Ratio $\frac{p_f(\text{Isothermal})}{p_f(\text{Adiabatic})} = \frac{2p_1}{2^\gamma p_1} = 2^{1-\gamma}$

18. (d) For adiabatic process,

$$\Delta Q = 0 \quad \therefore \Delta S = 0$$

For isothermal process,

$$\Delta S = \oint \frac{dQ}{dt} = \frac{1}{T}\oint dQ$$

$$= \frac{1}{T} nRT \ln(V_2/V_1) = nR\log\left(\frac{V_2}{V_1}\right)$$

$$= -nR\log 2$$

19. (c) $\Delta S_{\text{system}} + \Delta S_{\text{surroundings}} = 0$

As, processes are reversible

$\therefore \quad \Delta S_{\text{surroundings}} = -\Delta S_{\text{system}}$

$$= nR\log 2$$

20. (3) Given situation is

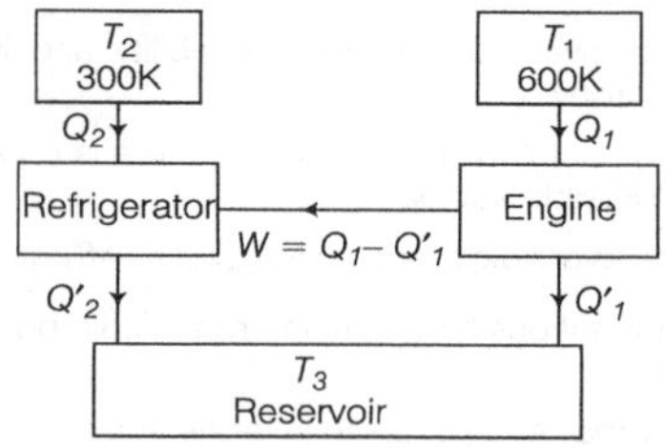

$$\frac{(\text{Work})_{\text{Carnot engine}}}{\text{Heat input to engine }(Q_1)} = \frac{T_1 - T_3}{T_1}$$

$\Rightarrow \quad W_{\text{Carnot}} = Q_1\left(\frac{T_1 - T_3}{T_1}\right) \quad \ldots\text{(i)}$

Also, COP refrigerator

$$= \frac{T_2}{T_3 - T_2} = \frac{\text{Heat absorbed by refrigerator}}{W_{\text{Carnot}}}$$

$\Rightarrow \quad \frac{T_2}{T_3 - T_2} = \frac{Q_2}{W_{\text{Carnot}}}$

$\therefore \quad W_{\text{Carnot}} = \frac{Q_2(T_3 - T_2)}{T_2} \quad \ldots\text{(ii)}$

From Eqs. (i) and (ii), we get

$$Q_1\left(\frac{T_1 - T_3}{T_1}\right) = Q_2\left(\frac{T_3 - T_2}{T_2}\right)$$

Given, $\quad Q_1 = Q_2$

$\Rightarrow \quad T_1T_2 - T_2T_3 = T_1T_3 - T_1T_2$

$$2T_1T_2 = T_3(T_1 + T_2)$$

$\Rightarrow \quad T_3 = \frac{2T_1T_2}{T_1 + T_2} = \frac{2 \times 600 \times 300}{900}$

$\Rightarrow \quad T_3 = 400\text{ K}$

So, $(\text{COP})_{\text{refrigerator}}$

$$= \frac{T_2}{T_3 - T_2} = \frac{300}{400 - 300} = 3$$

Revisal Problems (JEE Main)

1. (c) $\quad T = A \cdot e^X + B$

At $\quad T = 0°\text{C}, X = 2.0$

$\Rightarrow \quad 0 = A \cdot 2^2 + B \quad \ldots\text{(i)}$

At $\quad T = 100°\text{C}, X = 3.0$

$\Rightarrow \quad 100 = A \cdot 2^3 + B \quad \ldots\text{(ii)}$

Subtracting Eq. (i) from Eq. (ii), we get

$$100 = A(2^3 - 2^2) = 4A$$

$\Rightarrow \quad A = 25$ units

Substituting in Eq. (i), we get

$$B = -A \cdot 2^2 = 100 \text{ units}$$

So, $\quad T = 25 \times 2^X + 100$

When property has a value of -1.0, $T = 25 \times 2^{-1} + 100$

$$= \frac{25}{2} + 100 = 112.5°\text{C}$$

2. (d) $e = 0.2t - 5 \times 10^{-4} t^2$

At $t = 0°\text{C}$ (ice point),

$$e = 0$$

at $t = 100°\text{C}$ (steam point)

$$e = 0.2 \times 100 - 5 \times 10^{-4} \times (100)^2$$

$$= 15 \text{ mV}$$

Also, at $\quad t = 70°\text{C}$

$$e = 0.2 \times 70 - 5 \times 10^{-4} \times (70)^2$$

$$= 11.55 \text{ mV}$$

$\therefore$ When the gas thermometer reads 70°C, thermocouple will read t°C such that

$$\frac{t}{100 - 0} = \frac{e_t}{e_{100} - e_0}$$

$\Rightarrow \quad t = \frac{100 \times 11.5}{(15 - 0)} = 77°\text{C}$

3. (d) KE w.r.t. box is converted into heat

$\therefore \quad \frac{1}{2}mv_0^2 = nC_V\Delta T$

$\Rightarrow \quad v_0 = \sqrt{\frac{2nC_V\Delta T}{m(\text{gas})}}$

$$C_V = \frac{f}{2}R,$$

$$m_{\text{gas}} = n(32 \times 10^{-3}\text{kg})$$

$$R = 8.31 \text{ J/mol-K}, \Delta T = 1\text{K}$$

$\therefore \quad v_0 = \sqrt{\frac{2 \times n \times 8.31 \times \frac{5}{2} \times 1}{n(32 \times 10^{-3})}}$

$$= 36 \text{ ms}^{-1}$$

4. (b) $l_2 - l_1 = (l_0' - l_0) + t\,(l_0'\alpha_2 - l_0\alpha_1)$

As, $\quad l_2 - l_1 = l_0' - l_0$

$\therefore \quad t\,(l_0'\alpha_2 - l_0\alpha_1) = 0$

$$\Rightarrow \quad l_0' = \frac{l_0\alpha_1}{\alpha_2} = \frac{40 \times 6 \times 10^{-6}}{4 \times 10^{-6}} = 60 \text{ cm}$$

5. (c) $\frac{BD}{DA} = \frac{BC}{CA} = \frac{L}{\sqrt{2} \cdot L} = \frac{1}{\sqrt{2}}$

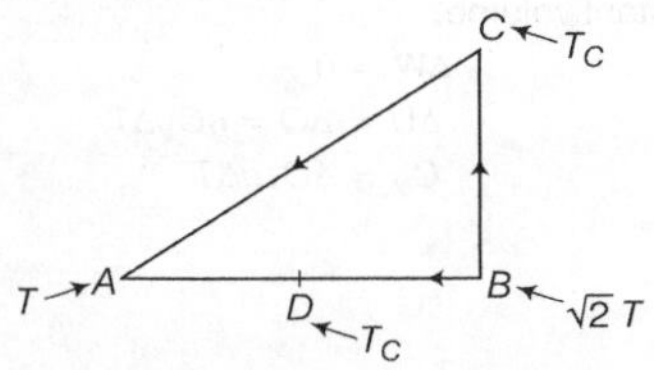

6. (b) Graph follows $y^2 = 4ax$

$$T_0^2 = 4a\left(\frac{1}{V_0}\right)$$

$\Rightarrow \quad T_0^2 V_0 = \text{constant}$

$\Rightarrow \quad T_0 V_0^{1/2} = \text{constant} \qquad (\because TV^{\gamma - 1} = \text{constant})$

$\therefore \quad \gamma - 1 = 1/2 \Rightarrow \gamma = 3/2$

$$\Rightarrow \quad \frac{v_{rms}}{v_{sound}} = \frac{\sqrt{\frac{3RT}{M}}}{\sqrt{\gamma RT/M}} = \sqrt{\frac{3}{\gamma}} = \sqrt{2}$$

7. (d) Heat flow rate through each face is same (in steady state),

$$\therefore \quad \frac{KA(200 - \theta_1)}{L} = \frac{2KA(\theta_1 - \theta_2)}{L} = \frac{1.5\,KA(\theta_2 - 18)}{L}$$

$\Rightarrow \quad 200 - \theta_1 = 2\theta_1 - 2\theta_2 = 1.5\,(\theta_2 - 18)$

$\Rightarrow \quad \theta_1 = 116°\text{C}, \theta_2 = 74°\text{C}$

8. (c) By Steafan-Boltzmann's law, $P \propto T^4 \cdot A$

9. (a) Use Newton's law of cooling.

$$\frac{Q_2 - Q_1}{\Delta t} = -K\{Q - Q_s\}$$

Case I $\quad \frac{80 - 85}{10} = -K\left\{\frac{80 + 85}{2} - 32.5\right\} \qquad \ldots\text{(i)}$

Case II $\quad \frac{80 - 85}{\Delta t} = -K\left\{\frac{165}{2} - 57.5\right\} \qquad \ldots\text{(ii)}$

On solving Eqs. (i) and (ii), we get

$\Delta t = 20$ min

10. (a) $C = \frac{\Delta Q}{\Delta T} = \frac{\Delta U + \Delta W}{\Delta T}$

ΔU = same for both paths, $\Delta W_2 > \Delta W_1$

$\Rightarrow \quad C_2 > C_1 \Rightarrow \frac{C_1}{C_2} < 1$

11. (c) For adiabatic process, $TV^{\gamma - 1}$ = constant and $U \propto T$

$\Rightarrow \quad UV^{\gamma - 1} = \text{constant}$

$\Rightarrow \quad U^{\frac{1}{\gamma - 1}} \cdot V = \text{constant}$

12. (d) $W_{12} + W_{23} + W_{31} = -300$

$\because \quad \Delta U = 0$ in an cyclic process

$W_{31} = 0$ as $\Delta V = 0$

$W_{12} = p(V_2 - V_1) = nR\,(T_2 - T_1)$

$W_{12} = 2 \times 8.31(600 - 300) = 4986$ J

13. (d) $Tp^{-2/5}$ = constant

Also, $\quad pV = nRT$

Combining both equations, we get

$pV^{5/3}$ = constant

$\therefore \quad \gamma = \frac{5}{3}$; the process is adiabatic, thus $\Delta Q = 0$

14. (c) Total internal energy of system is conserved.

$$nC_V T_1 + nC_{V_2} T_2 = (nC_{V_1} + nC_{V_2})\,T$$

$$\Rightarrow \quad \frac{3}{2}R \times 300 + \frac{5}{2}R \times 600 = \left(\frac{3}{2}R + \frac{5}{2}R\right)T$$

$$\Rightarrow \quad T = \left(\frac{9 + 30}{8}\right) \times 100$$

$= 487$ K (nearly)

15. (b) For pV^X = constant

$$C = \frac{R}{\gamma - 1} + \frac{R}{X - 1}, \gamma = \frac{5}{3}$$

C' corresponds to $X = 0$

$$\Rightarrow \quad C' = \frac{R}{\frac{5}{3} - 1} + \frac{R}{0 - 1} = \frac{5}{2}R$$

Also, X' corresponds to $C = 0$

$$\Rightarrow \quad X = \gamma = \frac{5}{3}$$

16. (a) From first law of thermodynamics

$$dQ = dU + pdV$$

For adiabatic process,

$$dQ = 0 \Rightarrow \frac{dU}{dV} = -p$$

Also, $\quad U = A + B\,(pV) \Rightarrow \frac{dU}{dV} = B\left(p + V\frac{dp}{dV}\right)$

From Eqs. (i) and (ii), we get

$$-p = B\left(p + \frac{Vdp}{dV}\right)$$

$$\Rightarrow \quad Bp + p + \frac{BVdp}{dV} = 0$$

$$\Rightarrow \quad p(BH) + BV\frac{dp}{dV} = 0$$

$$\Rightarrow \quad \left(\frac{B + 1}{B}\right)\frac{dV}{V} + \frac{dp}{p} = 0$$

$$\Rightarrow \quad \log p + \left(\frac{B + 1}{B}\right)\log V = 0$$

$$\Rightarrow \quad pV^{\frac{B + 1}{B}} = \text{constant}$$

$$\therefore \quad \gamma = \frac{B + 1}{B}$$

17. (b) Maximum thermal efficiency of engine is i.e. engine delivers an output

$$= \eta \times \text{input} = \frac{50}{100} \times 105$$

$= 52.5$ mJ $= 14.58$ kWh

As advertisor's claim (in ideal condition) is of 15 kWh which is never achieved in practice, engine is not suitable to buy.

18. (b) For vander Waals' gas,

$$T_c = \frac{8a}{27bR}, p_c = \frac{R}{27b^2}$$

$$\therefore \quad \frac{a}{b} = \frac{\frac{27R^2T_c^2}{64p_c}}{\frac{RT_c}{8p_c}} = \frac{27}{8}RT_c\left(\frac{\text{N-m}}{\text{mol}}\right)$$

19. (b) $W = p\Delta V$

$= 2 \times 1.01 \times 10^5 \times (2 \times 10^{-3} - 10 \times 10^{-3})$

$= -1.6 \times 10^3$ J $\qquad$ (isobaric compression)

As points B and A are on same isotherm, $T_A = T_B \Rightarrow \Delta U_{BDA} = 0$

So, $\quad Q = W = -1.6 \times 10^3$ J

$\therefore$ 1600 J of heat flows out of the gas in the process BDA.

20. (c) From Stefan-Boltzmann's law, $4\pi r^2 = A = \frac{\Delta Q/\Delta t}{\varepsilon\sigma T^4}$

$$\Rightarrow \quad \frac{r_B^2}{r_S^2} = \frac{(\Delta Q/\Delta t)_B}{(\Delta Q/\Delta t)_S} \times \frac{T_S^4}{T_B^4} = 10^4 \times 2^4 = 16 \times 10^4$$

$$\therefore \quad r_B = \sqrt{16 \times 10^4} \cdot r_S = 3 \times 10^{11}\,\text{m}$$

21. (a) Both Statement I and Statement II are correct and Statement II is correct explanation of Statement I.

Hot air forced down in summer increases rate of vapourisation there by increases cooling effect.

In winter fan is set to "pull up" the air, this will cause convection current to mix warm and cold air without creating a direct cold breeze.

22. (c) Statement II is incorrect. Wood absorbs more energy but its conductivity is less, so metal piece will feel hotter.

23. (a) Entropy is a state function so, its value is same along any path a to b even an isotherm.

24. (a) $V_2 = \frac{V_1}{(p_1/p_2)^{1/\gamma}}$

$\therefore$ V_2 is high, when γ is high.

25. (a) At constant volume,

$$\Delta W = 0$$

$$\therefore \quad \Delta U = \Delta Q = nC_V\Delta T$$

$$\therefore \quad C_V = \Delta Q/n\Delta T$$

and positive

Revisal Problems (JEE Advanced)

1. (a) $Q = \int_{T_i}^{T_f} mC\Delta T = \int_{25}^{125} 6 \times (0.4 + 0.004T)\,dT$

$$= 6\,[0.4T + 0.002T^2]_{25}^{125} = 420\text{ kJ}$$

2. (d) For process C, $\Delta W = -20$ J

$$\Delta U = U_f - U_i = 20\text{ J}$$

$$\therefore \quad \Delta Q = \Delta U + \Delta W = 0$$

So, process C is adiabatic.

3. (a) Work is done by the gas and

$$W_{\text{iso}} = RT_i \log_e (V_f/V_i)$$

$$= RT \log 3 \approx 0.69RT$$

Also, $$W_{\text{Adi}} = \frac{RT}{\gamma - 1}\left(1 - \left(\frac{V_i}{V_f}\right)^{\gamma-1}\right) = 0.60\,RT$$

$$\therefore \quad W_{\text{iso}} > W_{\text{Adi}}.$$

4. (d) As, rod is in steady state.

$\therefore$ It does not gain or loss heat it so,

$$dQ_{\text{rod}} = 0 \Rightarrow \Delta S_{\text{rod}} = 0$$

5. (b) $dS = \frac{dQ}{T} \Rightarrow \frac{dS}{dT} = \frac{nC}{T}$

$$\Rightarrow \quad \frac{dT}{dS} = \text{Slope on } T\text{-}S \text{ graph} = \frac{T}{nC}$$

So, $$\frac{\text{Slope of isobaric curve}}{\text{Slope of isochoric curve}} = \frac{\frac{T}{nC_p}}{\frac{T}{nC_V}} = \frac{C_V}{C_p} = \frac{1}{\gamma}$$

6. (b) Efficiencies of AB, AC and AD are

$$1 - \frac{T_B}{T_A},\ 1 - \frac{T_C}{T_A} \text{ and } 1 - \frac{T_D}{T_A}$$

As, per given condition,

$$\eta_{AC} = \frac{1}{2}(\eta_{AB} + \eta_{AD})$$

$$1 - \frac{T_C}{T_A} = \frac{1}{2}\left\{\left(1 - \frac{T_B}{T_A}\right) + \left(1 - \frac{T_D}{T_A}\right)\right\}$$

$$1 - \frac{T_C}{T_A} = 1 - \frac{T_B + T_D}{2T_A}$$

$$\Rightarrow \quad T_C = \frac{T_B + T_D}{2}$$

7. (c) $\Delta W = p\Delta V = 1.013 \times 10^5\,(1671 \times 10^{-6} - 1 \times 10^6)$

$$= 169.2\text{ J} \approx 170\text{ J}$$

$$\Delta Q = 539 \times 4.18 = 2253\text{ J}$$

$$\Delta U = \Delta Q - \Delta W$$

$$= 2253 - 170$$

$$= 2083\text{ J}$$

8. (a,b,c) As, $\frac{Q_1}{Q_2} = \frac{T_1}{T_2}$ and $Q_2 = Q_1 + W$

$$Q_1 = Q_2 - W = Q_1 \cdot \frac{T_2}{T_1} - W$$

$$W = Q_1\left(\frac{T_2}{T_1} - 1\right)$$

Rearranging and dividing by time, we have

$$\frac{Q_1}{t} = \text{rate of heat removal} = P\left(\frac{T_1}{T_2 - T_1}\right)$$

Rate at which heat passes into room is $R = K\Delta T$

$\therefore$ $K\Delta T = P\left(\frac{T_1}{T_2 - T_1}\right)$, for stable room temperature

$$\Rightarrow \quad K(\Delta T)^2 = P\,(T_2 - \Delta T)$$

Above equation is a quadratic equation in ΔT.

solving, $$\Delta T = \frac{-P \pm \sqrt{P^2 + 4P\,KT_2}}{2k}$$

Only positive root has physical significance so,

$$\Delta T = \frac{x}{2}\left(\sqrt{\left(1 + \frac{4T_2}{x}\right)} - 1\right)$$

where, $$x = \frac{p}{K}$$

So, $$T_1 = \text{room temperature}$$

$$= T_2 - \frac{x}{2}\left(\sqrt{1 + \left(\frac{4T_2}{x}\right)} - 1\right)$$

Also, $$P = \frac{K(\Delta T)^2}{T_1} = 130\text{ W}$$

9. (b,d) Heat absorbed by (either of) balls is

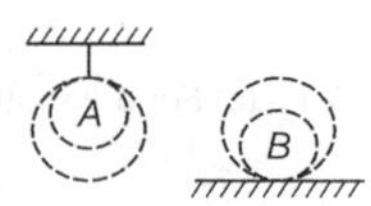

$$\Delta Q = mC(\Delta T)$$

Change in potential energy of balls is :

$$\Delta U = mg(r\,\alpha\,\Delta T)$$

Clearly for both balls, centre of mass moves through a distance of $r\,\alpha\,\Delta T$.

When COM of A is shifted downwards, then work is done by gravity and in case of B, work is done against gravity by the expanding ball.

$\therefore$ For A, heat used up in raising temperature is

$$\Delta Q - \Delta U = (mC - mgr\,\alpha)\Delta T$$

and in case of B, it is

$$\Delta Q + \Delta U = (mC + mgr\,\alpha) \cdot \Delta T$$

So, temperature of A is higher than that of B.

Also, $$T_{fA} - T_{fB} = \frac{2gr \cdot \alpha}{\{C^2 - (gr\alpha)^2\}m}$$

Neglecting term containing α^2, being very small,

$$T_{fA} - T_{fB} = \left(\frac{2gr}{mC^2}\right) \cdot \alpha$$

$\therefore$ Temperature difference of A and B is proportional to coefficient of expansion of the material of ball.

10. (a,c) $W_{BC} = \int p dV = p(V_C - V_B)$

$$= 100\ (10 - 2) = 800 \text{ kJ}$$

$$W_{CA} = mRT \ln\left(\frac{V_A}{V_C}\right)$$

where, $T_C = \dfrac{p_C V_C}{mR}$

$\therefore$ $W_{CA} = mR \dfrac{(100 \times 10)}{mR} \cdot \ln \dfrac{2}{10}$

$$= -1609 \text{ kJ} \qquad \left(\because \ln\frac{1}{5} = -1.609 \text{ J}\right)$$

So, total work done during the complete cycle is

$$W_{net} = W_{AB} + W_{BC} + W_{CA}$$
$$= 0 + 800 - 1609$$
$$= -809 \text{ kJ}$$

11. (a,b,c) Process AB is isothermal process,

$$T_{AB} = \text{constant}, \ p_B > p_A$$

p-V graph, $p \propto \dfrac{1}{V}$ so, pV graph is a hyperbola with $V_B < V_A$

and V-T graph is a straight line with $V_B < V_A$

p-T graph; $\rho \propto p, \rho$-T graph is a line parallel to ρ-axis with $\rho_B > \rho_A$

S-T graph; $\Delta S = \oint \dfrac{dQ}{T}$, as heat must be rejected during process AB

$\therefore S_B < S_A$, S-T graph is not correct.

12. (b,c) $p \propto V \Rightarrow pV^{-1}$ = constant, process is polytropic with $x = -1$

C = specific heat for the process

$$= \frac{R}{\gamma - 1} + \frac{R}{1 - x} = C_V + \frac{R}{1 - x}$$
$$= \frac{5}{2}R + \frac{R}{2} = 3R$$
$$\Delta Q = nC\Delta T = nC(T_f - T_i)$$
$$= nC(4T_i - T_i) = 3n\,CT_i$$
$$= 9n\,RT_i = 9p_iV_i$$

As, $v_{rms} \propto \sqrt{T}$, rms speed is doubled.

$\therefore$ Temperature becomes four times.

13. (c) In a C-R circuit, during discharging,

$$q = Q \cdot e^{-t/CR}$$

$\therefore$ Charge on the plates of capacitor at $t = 2.5 \ln 4$ is

$$q = Q_0\, e^{\frac{-2.5 \times 60 \ln 4}{180}}$$

where, $R \times C = 2000 \times 75 \times 10^{-3} = 150$ (s)

$\therefore$ q (at $t = 2.5 \ln 4$) $= Q_0\, e^{-\ln 4}$

$$= \frac{Q_0}{4}$$

Loss of PE of capacitor = heat produced

$$= \frac{Q_0^2}{2C} - \frac{Q_0^2/16}{2C} = \frac{15}{32}\frac{Q_0^2}{C}$$
$$= \frac{15}{32} \times \frac{(C \times V_0)^2}{C} = \frac{15 \times 16 \times 16}{32 \times 75 \times 10^{-3}}$$
$$= 1600 \text{ J} = 1.6 \text{ kJ}$$

14. (a) Gas expands at constant pressure

$\therefore$ $\Delta W = nR\Delta T = 1 \times 8.3 \times 72$

$$= 0.596 \text{ kJ} \approx 0.6 \text{ kJ}$$
$$\Delta U = \Delta Q - \Delta W = 1.6 - 0.6 = 1 \text{ kJ}$$

15. (b) When engines produce same work, i.e. $W_1 = W_2$

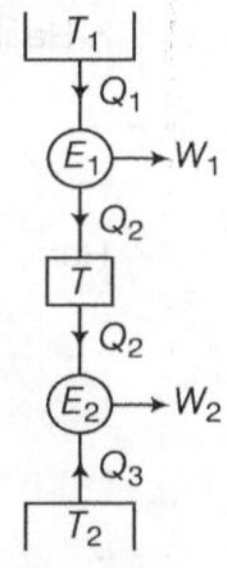

$$\eta_1 Q_1 = \eta_2 Q_2$$

$$\left(1 - \frac{T}{T_1}\right)\left(\frac{T_1}{T}\right)Q_2 = \left(1 - \frac{T_2}{T}\right)Q_2$$

As, $\dfrac{Q_1}{T_1} = \dfrac{Q_2}{T} \Rightarrow Q_1 = \dfrac{T_1}{T} \cdot Q_2$

or $\dfrac{T_1}{T} - 1 = 1 - \dfrac{T_2}{T}$

$\Rightarrow$ $T = \dfrac{T_1 + T_2}{2}$

16. (a) With same efficiency, $\eta_1 = \eta_2$

$\Rightarrow$ $1 - \dfrac{T}{T_1} = 1 - \dfrac{T_2}{T}$

$\Rightarrow$ $T = \sqrt{T_1 T_2}$

17. (a) Starting with Clausius definition,

$$\Delta S = \frac{dQ}{T}$$

We have, entropy of a system in a given state can be written as

$$S = k_B \log_e W$$

where, K_B = Botzmann's constant and W is number of microstates corresponding to that state.

Now, number of microstates for $A = W_A = \dfrac{10!}{10!\ 0!} = 1$, as $0! = 1$

Note *Number of ways in arranging N atoms such that N_2 atoms are in LHS and N_R atoms in RHS is* $= \dfrac{N!}{N_L!\,N_R!}$; *a case of repeating permutations.*

Also, number of microstates for $B = W_B = \dfrac{10!}{5!\ 5!} = 252$

So, when system goes from state A to state B,

$$\Delta S = S_f - S_i = S_B - S_A$$
$$= k_B \log_e W_B - k_B \log_e W_A$$
$$= k_B \log_e \frac{W_B}{W_A} = k_B \log 252$$

As $\Delta S > 0$, this process can occur naturally (free expansion).

18. (b) $\Delta S = S_A - S_B$

$$= k_B \log_e \left(\frac{W_A}{W_B}\right) = k_B \log_e \left(\frac{1}{252}\right)$$
$$= -k_B \log_e 252$$

As, $\Delta S < 0$ process cannot occurs naturally.

19. (b) Total thermal resistance is a combination of conduction and convection resistance.

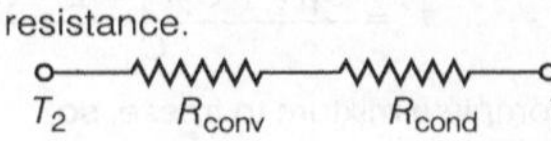

$$R_{cond} = \frac{1}{4\pi k}\left(\frac{1}{r_1} + \frac{1}{r_2}\right)$$

Note *For a sphere,* $q = -kA\dfrac{dT}{dr}$

$$= k4\pi r^2 \frac{dT}{dr}$$

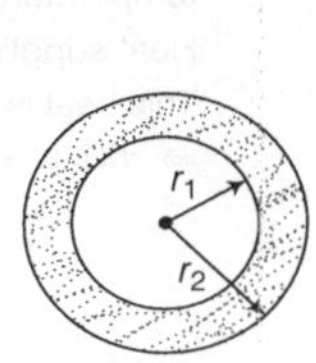

$$\frac{q}{4\pi}\int_{r_1}^{r_2} \frac{dr}{r^2} = -\int_{T_1}^{T_2} k \cdot dT$$

$\Rightarrow$ $q = \dfrac{4\pi k(T_1 - T_2)}{\left(\dfrac{1}{r_1} - \dfrac{1}{r_2}\right)}$

$\Rightarrow$ $R_{(cond)} = \dfrac{1}{4\pi k}\left(\dfrac{1}{r_1} - \dfrac{1}{r_2}\right)$

and $R_{conv} = \dfrac{1}{h4\pi r^2}$

$\therefore$ Heat transfer rate will be

$$q = \frac{T_2 - T_1}{\frac{1}{4\pi k}\left(\frac{1}{r_1} - \frac{1}{r_2}\right) + \frac{1}{h4\pi r_2^2}}$$

Here, $T_2 = 300\text{ K}, T_1 = 77\text{ K}$

$k = 0.0017\text{ W/m-K}$

$h = \frac{20\text{ W}}{\text{m}^2\text{-K}}$

$r_1 = 0.25\text{ m}, r_2 = 0.275\text{m}$

$\therefore \quad q = \left(\frac{22.3}{17.02 + 0.05}\right)$

$= 13.06\text{ W} \approx 13\text{W}$

20. (d) Boil off rate of nitrogen is

$$\frac{dm}{dt} = \frac{q}{L} = \frac{13.06}{2 \times 10^5} \qquad [\because mL = Q]$$

$= 6.53 \times 10^{-5}\text{ kgs}^{-1}$

So, loss per day $= 6.53 \times 10^{-5} \times 3600 \times 24$

$= 5.64\text{ kg/day}$

Hence, volume evaporated per day

$$= \frac{dV}{dt} = \frac{dm/dt}{\rho}$$

$$= \frac{5.64\text{ kg/day}}{804\text{ kg/m}^3}$$

$= 0.007\text{ m}^3\text{/day}$

$= 7\text{ litres/day}$

21. (c) To make complete mixture, ice should not melt untill the temperature of water and calorimeter reached at 0°C. Heat given by calorimeter and water in it.

$$mC_1t_2 + mC_2t_2$$

This heat is used by ice to increase its temperature

$$= mC_3(t + t_3)$$

$\Rightarrow \quad mC_3(t + t_3) = mC_1t_2 + mC_2t_2$

$\Rightarrow \quad t = \frac{C_1t_2 + C_2t_2 - C_3t_3}{C_3}$

If $t < 0$, means ice is still below 0°C

$\Rightarrow \quad \frac{C_1t_2 + C_2t_2 - C_3t_3}{C_3} < 0$

Now, the water supplies heat to freeze so;

$$mC_3(t' + t_3) = mC_1t_2 + mC_2t_2 + mL$$

$\Rightarrow \quad t' = \frac{C_1t_2 + C_2t_2 + L - C_3t_3}{C_3}$

As, $t' < 0$, for complete mixture to freeze, so,

$$C_1t_2 + C_2t_2 + L - C_3t_3 < 0$$

22. (b) For making complete mixture water, ice has to melt completely and then its temperature must rise to reach equilibrium temperature.

Heat supplied by ice $= m \cdot C_3 \cdot t_3 + mL$

This heat is gained by calorimeter and water.

$\Rightarrow mC_1(t - t_2) + mC_2(t - t_2)$

$= mC_3t_3 + mL$

$\Rightarrow \quad t = \frac{C_3t_3 + L + C_1t_2 + C_2t_2}{C_1 + C_2}$

and $\quad t > 0$

$\Rightarrow \quad C_3t_3 + L + C_1t_2 + C_2t_2 > 0$

23. (b) For AB, U = constant and ρ increases

$\therefore T$ = constant and p increases as

$$p = \rho\frac{RT}{M}$$

Also, as ρ increases, V decreases

$$\rho = \frac{M}{V}$$

For BC, U increases at constant density so, T increases.

24. (a) AB and CD are isothermal processes. In AB, volume increases and in CD volume decreases. BC and DA are constant volume process.

$\therefore$ p-T graph is a straight line that passes through origin.

25. (c) $E_{\text{int }a} - E_{\text{int }c} = \Delta E_{CA}$

$= -\Delta E_{\text{int, }ac} = -(Q_{ac} - W_{ac})$

$= (-85\text{ J} - (-55\text{ J})) = 30\text{ J}$

$\Delta E_{\text{int, }cda} = Q_{cda} - W_{cda}$

$\Rightarrow \quad Q_{cda} = \Delta E_{\text{int, }cda} + W_{cda}$

$= \Delta E_{\text{int, }ca} + W_{cda}$

$= 30\text{ J} + 38\text{ J} = 68\text{ J}$

26. (d) $W_{abc} = W_{ab} = p_a\,\Delta V_{ab}$

$= p_a(V_b - V_a) = 2.2\,p_d(V_c - V_d)$

$= -2.2W_{cda} = -2.2 \times 38 = -84\text{ J}$

$\Delta E_{\text{int, }abc} = Q_{abc} - W_{abc}$

$\Rightarrow \quad Q_{abc} = \Delta E_{\text{int, }abc} + W_{abc}$

$= -30\text{ J} - 84\text{ J} = -114\text{ J}$

27. (b) Energy delivered by water

$$= gh = 9.81 \times 50 \times \frac{1}{1000}$$

$= 0.491\text{ kJ/kg}$

Rate at which mechanical energy is supplied to turbine

$$= 5000\,\frac{\text{kg}}{\text{s}} \times 0.491\frac{\text{kJ}}{\text{kg}}$$

$= 2455\text{ kW}$

$\therefore \quad \eta_{\text{overall}} = \frac{\text{Work output}}{\text{Work input}} = \frac{\text{Electrical output}}{\text{Work input}}$

$= \frac{1862}{2455} = 0.76\text{ or }76\%$

28. (a) $_{(\text{turbine - generator})} = \frac{0.76}{0.95} = 0.80$

$\therefore$ Shaft power output of turbine = 80% of 2455

$= 1964\text{ kW}$

Hence, power loss (turbine – generator)

$= 1964 - 1862$

$= 102\text{ kW}$

29. (d) Heat produced by coil in time t is

$$dQ = I^2r\,dt$$

Also, $\quad dQ = dU + dW$

$\Rightarrow \quad I^2r\,dt = C_V dT + p dV$

$\Rightarrow \quad I^2r = C_V\left(\frac{dT}{dt}\right) + p\left(\frac{dV}{dt}\right)$

$$= C_V\left(\frac{dT}{dt}\right) + R\left(\frac{dT}{dt}\right) \quad \text{...(i)}$$

As, $\quad T = T_0 + at + bt^2$

$$\frac{dT}{dt} = a + 2bt \quad \text{...(ii)}$$

From Eqs. (i) and (ii), we get

$$I = \left(\frac{5R}{2r}(2bt + a)\right)^{1/2}$$

Also, $pAdx = RdT$

$$v = \text{Velocity}, \frac{dx}{dt}$$

$$= \frac{R}{pA} \cdot \frac{dT}{dt} = \frac{R}{pA} \cdot (2bt + a)$$

and acceleration

$$a = \frac{dv}{dt} = \frac{2Rb}{pA}$$

Hence, (i) → (q), (ii) → (p), (iii) → (r), (iv) → (s)

30. (c) In an isothermal expansion, volume increases and pressure decreases.

$$p_1V_1 = p_2V_2$$

In an adiabatic expansion, volume increases, pressure decreases and as work is done as the expense of internal energy, $\Delta U < 0$

∴ Temperature decreases $\Delta T < 0$.

In an isobaric process , p = constant

so, $\Delta U > 0$

and volume increases on heating.

In an isochoric cooling, $\Delta V = 0$

⇒ p and T both decreases.

Hence, (i) → (r, s), (ii) → (q,r,s), (iii) → (p,r), (iv) → (q,s)

31. (d) For reversible adiabatic $\Delta Q = 0$

$$\Rightarrow \quad \Delta S = 0$$

For all other processes, $\Delta S = \int \frac{dQ}{T}$

Hence, (i) → (r), (ii) → (q), (iii) → (p), (iv) → (s), (v) → (t)

32. (a) For a polytropic process, pV^n = a constant.

Work done $= \int_{V_1}^{V_2} pdV$

$= \frac{p_2V_2 - p_1V_1}{1 - n}$ (equation is valid except for $n = 1$)

Each process is associated with a particular value of n.

$n = 1$, isothermal

$n = \infty$, constant volume

$n = 0$, constant pressure

$n = \gamma$, adiabatic

Hence, (i) → (p), (ii) → (q), (iii) → (r), (iv) → (s) (v) → (t)

33. (1) Heat lost by room per second.

= heat generated by heater in 1 second

$$\therefore \quad \frac{V^2}{R} = \frac{20 - 15.24}{l/KA}$$

$$\Rightarrow \quad \frac{200 \times 200}{20} = \frac{4.76}{\dfrac{0.2 \times 10^{-2}}{0.2 \times A \times 4.2}}$$

$$\Rightarrow \quad 2000 \times 10^{-2} = 4.76 \times A \times 4.2$$

$$\Rightarrow \quad A = 1.0004 \text{ m}^2$$

$$\Rightarrow \quad A \approx 1 \text{ m}^2$$

34. (7) By Newton's law of cooling,

$$\frac{T_1 - T_2}{t} = K\left[\frac{T_1 + T_2}{2} - T_0\right]$$

$$\frac{50 - 40}{5 \times 60} = K\left(\frac{50 + 40}{2} - 20\right) \quad \ldots(i)$$

$$\frac{40 - T}{5 \times 60} = K\left(\frac{40 + T}{2} - 20\right) \quad \ldots(ii)$$

From Eqs. (i) and (ii), $T = 33.33°C$

So, $\Delta T = 6.67°C = 7°C$

35. (9) In first stage,

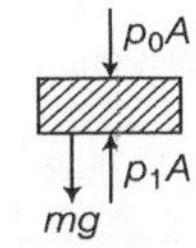

In second stage,

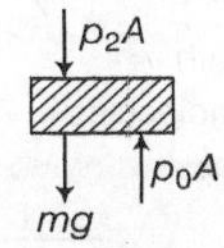

So, $p_0A + mg = p_1A \quad \ldots(i)$

and $p_2A + mg = p_0A \quad \ldots(ii)$

$$\Rightarrow \quad p_1 = p_0 + \frac{mg}{A}$$

$$p_2 = p_0 - \frac{mg}{A}$$

For isothermal change, $p_1V_1 = p_2V_2$

$$\Rightarrow \quad \left(p_0 + \frac{mg}{A}\right)(AL_1) = \left(p_0 - \frac{mg}{A}\right)(AL_2)$$

$$\Rightarrow \quad \frac{L_2}{L_1} = \left(\frac{p_0 + mg/A}{p_0 - mg/A}\right) = \frac{10^5 + \dfrac{0.002 \times 10}{20 \times 10^{-6}}}{10^5 - \dfrac{0.002 \times 10}{20 \times 10^{-6}}}$$

$$= \frac{101}{99} = \frac{101}{9 \times 11}$$

$$\therefore \quad K = 9$$

36. (7)

$$\eta = \frac{\text{Total work done}}{\text{Total heat supplied}}$$

$W_{AB} = p\Delta V = nR\Delta T$ (isochoric process)

$$= 2 \times \frac{25}{3} \times 300 = 5000 \text{ J}$$

$$W_{BC} = nRT \ln\left(\frac{T_2}{T_1}\right) = nRT \ln\left(\frac{p_1}{p_2}\right)$$

$$= 2 \times R \times 600 \ln\left(\frac{2 \times 10^5}{1 \times 10^5}\right)$$

$$= 1200 \, R \ln 2$$

$$W_{CA} = \frac{nR\Delta T}{1 - \gamma} = -1200 \, R$$

$$Q_{AB} = nC_p\Delta T = 2 \times \frac{5}{2} R \times 300 = 1500 \, R$$

$$Q_{BC} = W_{BC} = 1200 \, R \ln 2$$

$$\eta = \frac{600 \, R \, (2 \ln 2 - 1)}{1800 \, R + 1200 R \ln 2}$$

$$= 1 - \frac{21}{15 + 12 \ln 2}$$

So, $x = 7$

37. (5) Work needed to raise the piston and to compress the spring.

$$W = mgh + \frac{1}{2}kx^2 + p_{atm} Ah$$

$$= 60 \times 9.81 \times 0.05 + \frac{1}{2} \times 50000 \times (0.05)^2 + 100000 \times \frac{\pi \times (0.2)^2}{4} \times 0.05$$

$$= 250 \text{ J}$$

From first law of thermodynamics

$$\Delta Q - \Delta W = \Delta U$$

$$\Rightarrow \quad \Delta U = 200 - 250 = -50 \text{ J}$$

$$= -10 \times 5 \text{ J}$$

$$\therefore \quad K = 5$$

JEE Main & AIEEE Archive

1. (c) Average time taken collision,

$$f = \frac{\lambda}{V_{runs}},$$

$$\lambda = \frac{1}{\sqrt{2}\frac{N}{V}\pi d^2}$$

where, λ = Mean free path
N = Number of molecules
V = Volume of gas

$$\therefore \quad t = \frac{\frac{1}{\sqrt{2}\frac{N}{V}\pi d^2}}{\sqrt{\frac{3RT}{M}}}$$

$$= \frac{\sqrt{M}\cdot V}{\sqrt{2}\, N\pi d^2\, \sqrt{3R}\, \sqrt{T}}$$

$$t = k\frac{V}{\sqrt{T}}$$

where, k is a constant $= \frac{\sqrt{M}}{\sqrt{2}\, N\pi d^2 \sqrt{3R}}$

$$\therefore \quad T \propto \frac{V^2}{t^2}$$

For adiabatic process,

$$TV^{\gamma-1} = \text{constant}$$

$$\frac{V^2}{t^2}V^{\gamma-1} = \text{constant}$$

$$\frac{V^{\gamma+1}}{t^2} = \text{constant}$$

$$\therefore \quad t \propto V^{\frac{\gamma+1}{2}}$$

$$\therefore \quad q = \frac{\gamma+1}{2}$$

2. (a) In this question, the system is accelerating horizontally, hence the pressure in the vertical direction will remain unaffected.

i.e. $\quad p_1 = p_0 + \rho g h$

Again, we have to use the concept that the pressure in the same level will be same.

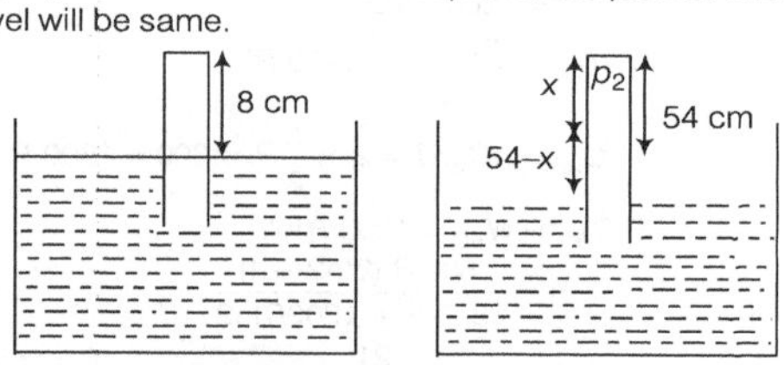

For air trapped in tube, $p_1V_1 = p_2V_2$

$$p_1 = p_{atm} = \rho g \times 76$$

$$V_1 = A \times 8$$

[$\because A$ = area of cross-section]

$$p_2 = p_{atm} - \rho g\,(54 - x) = \rho g\,(22 + x)$$

$$V_2 = A \cdot x$$

At constant temprature,

$$p_1V_1 = p_2V_2$$

$$\rho g \times 76 \cdot A \times 8 = \rho g\,(22 + x)\,Ax$$

$$x^2 + 22x - 78 \times 8 = 0$$

$$\Rightarrow \quad x = 16 \text{ cm}$$

3. (d) According to first law of thermodynamics, we get

(i) Change in internal energy from A to B i.e. ΔU_{AB}

$$\Delta U_{AB} = nC_V\,(T_B - T_A)$$

$$= 1 \times \frac{5R}{2}(800 - 400)$$

$$= 1000\,R$$

(ii) Change in internal energy from B to C i.e. ΔU_{BC}

$$\Delta U_{BC} = nC_V(T_C - T_B)$$

$$= 1 \times \frac{5R}{2}(600 - 800) = -500\,R$$

(iii) $\Delta U_{total} = 0$

(iv) Change in internal energy from C to A i.e. ΔU_{CA}

$$\Delta U_{CA} = nC_V\,(T_A - T_C)$$

$$= 1 \times \frac{5R}{2}(400 - 600)$$

$$= -500\,R$$

4. (b) Heat supplied $= nC_V\,(2T_0 - T_0) + nC_p + (2T_0 - T_0)$

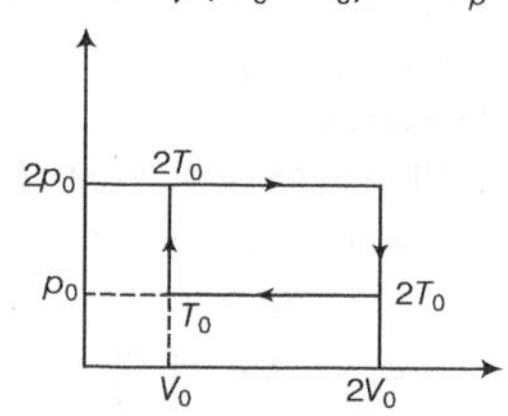

$$= \frac{n3RT_0}{2} + \frac{n5R}{2}2T_0$$

$$= \frac{13}{2}nRT_0 = \frac{13}{2}p_0V_0$$

5. (a) Efficiency of a process is defined as the ratio of work done to energy supplied. Here,

$$\eta = \frac{\Delta W}{\Delta Q} = \frac{\text{Area under } p\text{-}V \text{ diagram}}{\Delta Q_{AB} + \Delta Q_{BC}}$$

$$\therefore \quad \eta = \frac{p_0V_0}{nC_V\Delta T_1 + nC_p\Delta T_2}$$

$$= \frac{p_0V_0}{\frac{3}{2}nR(T_B - T_A) + \frac{5}{2}nR(T_C - T_D)}$$

$$= \frac{p_0V_0}{\frac{3}{2}(2p_0V_0 - p_0V_0) + \frac{5}{4}(4p_0V_0 - 2p_0V_0)}$$

[as $pV = nRT$]

$$= \frac{p_0V_0}{\frac{3}{2}p_0V_0 + \frac{5}{2}\cdot 2p_0V_0}$$

$$= \frac{1}{6.5} = 15.4\%$$

6. (c) Efficiency, $\eta = 1 - \frac{T_{sink}}{T_{source}}$

Now, $\quad 0.4 = 1 - \frac{T_{sink}}{500\text{ K}}$

$\Rightarrow \quad T_{sink} = 0.6 \times 500\text{ K} = 300\text{ K}$

Thus, $\quad \eta = 1 - \frac{300\text{ K}}{T'_{source}}$

$\Rightarrow \quad T'_{source} = \frac{300\text{ K}}{0.4} = 750\text{ K}$

7. (c) Heat required to change the temperature of vessel by a small amount dT,

$$-dQ = mC_p dT$$

Total heat required,

$$-Q = m\int_{20}^{4} 32\left(\frac{T}{400}\right)^3 dT$$

$$= \frac{100 \times 10^{-3} \times 32}{(400)^3}\left[\frac{T^4}{4}\right]_{20}^{4}$$

$\Rightarrow$ $$Q = 0.001996 \text{ kJ}$$

Work done required to maintain the temperature of sink to T_2,

$$W = Q_1 - Q_2 = \frac{Q_1 - Q_2}{Q_2} Q_2 = \left(\frac{T_1}{T_2} - 1\right) Q_2$$

$\Rightarrow$ $$W = \left(\frac{T_1 - T_2}{T_2}\right) Q_2$$

For $T_2 = 20$ K,

$$W_1 = \frac{300 - 20}{20} \times 0.001996$$

$$= 0.028 \text{ kJ}$$

For $T_2 = 4$ K,

$$W_2 = \frac{300 - 4}{4} \times 0.001996 = 0.148 \text{ kJ}$$

As, temperature is changing from 20 K to 4 K, work done required will be more than W_1 but less than W_2.

8. (a) Internal energy of the gas remains constant, hence

$$T_2 = T$$

Using at constant temperature, $p_1 V_1 = p_2 V_2$

$$p_2 = \frac{p}{2}$$

9. (c) As no heat is lost,

Loss of kinetic energy = Gain of internal energy of gas

$$\frac{1}{2}mv^2 = nC_V \Delta T$$

$\Rightarrow$ $$\frac{1}{2}mv^2 = \frac{m}{M} \cdot \frac{R}{\gamma - 1}\Delta T$$

$\Rightarrow$ $$\Delta T = \frac{Mv^2(\gamma - 1)}{2R} \text{ K}$$

10. (d) As efficiency, $\eta_1 = 1 - \dfrac{T_2}{T_1}$

$\Rightarrow$ $$\frac{1}{6} = 1 - \frac{T_2}{T_1}$$

$\Rightarrow$ $$\frac{T_2}{T_1} = \frac{5}{6} \quad \text{...(i)}$$

$$\eta_2 = 1 - \frac{T_2 - 62}{T_1}$$

$\Rightarrow$ $$\frac{1}{3} = 1 - \frac{T_2 - 62}{T_1} \quad \text{...(ii)}$$

On solving Eqs. (i) and (ii), we get

$$T_1 = 372 \text{ K}$$

and $$T_2 = 310 \text{ K}$$

11. (a) $\dfrac{f}{2}n_1kT_1 + \dfrac{f}{2}n_2kT_2 + \dfrac{f}{2}n_3kT_3 = \dfrac{f}{2}(n_1 + n_2 + n_3)kT$

($\because f$ is degree of freedom)

Temperature of mixture,

$$T = \frac{n_1T_1 + n_2T_2 + n_3T_3}{n_1 + n_2 + n_3}$$

12. (c) $U(x) = \dfrac{a}{x^{12}} - \dfrac{b}{x^6}$

$$U(x = \infty) = 0$$

As, $$F = -\frac{dU}{dx} = -\left[\frac{12a}{x^{13}} + \frac{6b}{x^7}\right]$$

At equilibrium, $F = 0$

$\therefore$ $$x^6 = \frac{2a}{b}$$

$\therefore$ $$U_{\text{at equilibrium}} = \frac{a}{\left(\frac{2a}{b}\right)^2} - \frac{b}{\left(\frac{2a}{b}\right)} = \frac{-b^2}{4a}$$

$\therefore$ $$D = [U(x = \infty) - U_{\text{at equilibrium}}] = \frac{b^2}{4a}$$

Explanations for Q. Nos. (13-15)

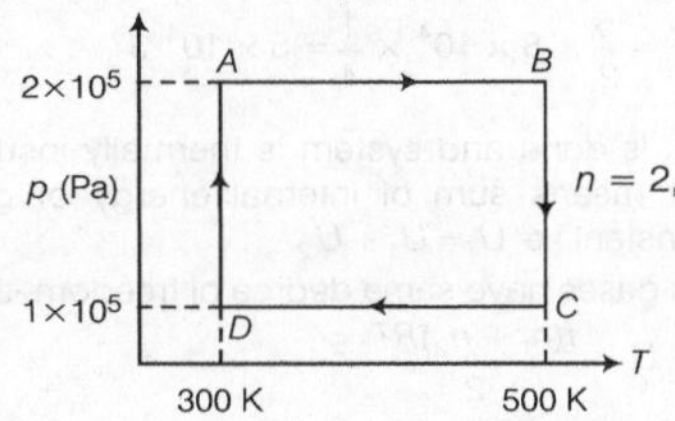

$n = 2, \gamma = 1.67$

$$W_{AB} = \Delta Q - \Delta U = nC_p dT - nC_V dT$$

(at constant pressure)

$$= n(C_p - C_V)dT = nR\,dT$$

$$= 2 \times R \times (500 - 300)$$

$$= 400R$$

At constant temperature (isothermal process),

$$W_{DA} = nRT\ln\left(\frac{p_1}{p_2}\right) = 2.303\,nRT\log\left(\frac{p_1}{p_2}\right)$$

$$= 2.303 \times 2R \times 300\log\left(\frac{10^5}{2 \times 10^5}\right)$$

$$= 2.303 \times 600R\log\left(\frac{1}{2}\right)$$

$$= 0.693 \times 600R = -414R$$

So, work done on the gas is $+414R$.

$$W_{AB} = nR\Delta T = 2R \times 200 = 400R$$

$$W_{BC} = nRT\log\left(\frac{p_1}{p_2}\right)$$

$$= 2.303 \times R \times 500\log\left(\frac{2 \times 10^5}{10^5}\right)$$

$$= 2 \times 346R = 692R$$

$$W_{CD} = nR\Delta T = 2R \times 200 = 400R$$

$$W_{DA} = 2.303 \times R \times 300\log\left(\frac{10^5}{2 \times 10^5}\right)$$

$$= -415.963R$$

$\therefore$ Net work done in a cycle

$$= W_{AB} + W_{BC} + W_{CD} + W_{DA}$$

$$= 400R + 692R + 400R - 415.963R$$

$$= 1076R$$

13. (c)

14. (b)

15. (c)

16. (b) Thermal energy corresponds to internal energy

$$\text{Mass} = 1 \text{ kg}$$

$$\text{Density} = 4 \text{ kg m}^{-3}$$

$$\Rightarrow \quad \text{Volume} = \frac{\text{Mass}}{\text{Density}} = \frac{1}{4} \text{ m}^3$$

$$\text{Pressure} = 8 \times 10^4 \text{ Nm}^{-2}$$

As $\quad \text{internal energy} = \frac{f}{2} nRT = \frac{f}{2} pV$

Degree of freedom for diatomic = 5

$\therefore \quad \text{Internal energy} = \frac{5}{2} p \times V$

Putting the values of $p = 8 \times 10^4 \text{ Nm}^{-2}$

and $\quad V = \frac{1}{4} \text{m}^3$, we get

$$\text{Internal energy} = \frac{5}{2} \times 8 \times 10^4 \times \frac{1}{4} = 5 \times 10^4 \text{ J}$$

17. (a) As no work is done and system is thermally insulated from surrounding, it means sum of internal energy of gas in two partitions is constant i.e. $U = U_1 + U_2$.
Assuming both gases have same degree of freedom, then

$$\text{Internal energy, } U = \frac{f(n_1 + n_2)RT}{2}$$

and $\quad U_1 = \frac{fn_1RT_1}{2}, \; U_2 = \frac{fn_2RT_2}{2}$

Solving, we get

$$\text{Common temperature, } T = \frac{(p_1V_1 + p_2V_2)\, T_1T_2}{p_1V_1T_2 + p_2V_2T_1}$$

18. (a) According to Mayer's relation,

$$C_p - C_V = \frac{R}{m}$$

For nitrogen, $m = 28$

$$\therefore \quad C_p - C_V = \frac{R}{28}$$

19. (a) From first law of thermodynamics,

$$\Delta Q = \Delta U + W$$

For path *iaf*,

$$50 = \Delta U + 20$$

$$\therefore \quad \Delta U = U_f - U_i = 30 \text{ cal}$$

For path *ibf*,

$$\Delta Q = \Delta U + W$$

or $\quad W = \Delta Q - \Delta U$

$$= 36 - 30 = 6 \text{ cal}$$

20. (b) For Carnot engine using as refrigerator,

$$W = Q_2 \left(\frac{T_1}{T_2} - 1 \right)$$

and $\quad \eta = \frac{W_1}{Q_2} = \left(\frac{T_1}{T_2} - 1 \right)$

It is given that, $\eta = \frac{1}{10}$

$$\Rightarrow \quad \eta = 1 - \frac{T_2}{T_1}$$

or $\quad \frac{T_2}{T_1} = \frac{9}{10}$

So, $\quad Q_2 = 90 \text{ J} \qquad (\text{as } W = 10 \text{ J})$

21. (c) Here, change in internal energy of the system is zero, i.e. increase in internal energy of one is equal to decrease in internal energy of other.

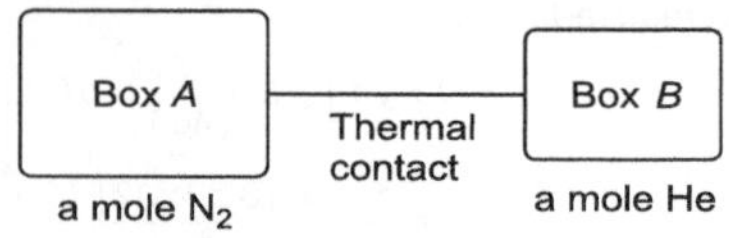

Change in internal energy in box A,

$$\Delta U_A = 1 \times \frac{5}{2} R\, (T_f - T_0)$$

Change in internal energy in box B,

$$\Delta U_B = 1 \times \frac{3R}{2} \left(T_f - \frac{7}{3} T_0 \right)$$

Now, $\quad \Delta U_A + \Delta U_B = 0$

$$\Rightarrow \quad \frac{5R}{2}(T_f - T_0) + \frac{3R}{2}\left(T_f - \frac{7T_0}{3} \right) = 0$$

or $\quad 5T_f - 5T_0 + 3T_f - 7T_0 = 0$

or $\quad 8T_f = 12T_0$

or $\quad T_f = \frac{12}{8} T_0 = \frac{3}{2} T_0$

22. (a) For adiabatic process, no exchange of heat.

$\therefore \quad dQ = 0$

So, $\quad dU = -\Delta W$

$$\Rightarrow \quad nC_V\, dT = +146 \times 10^3 \text{ J}$$

$$\Rightarrow \quad \frac{nfR}{2} \times 7 = 146 \times 10^3$$

(where, f is degree of freedom)

$$\Rightarrow \quad \frac{10^3 \times f \times 8.3 \times 7}{2} = 146 \times 10^3$$

$$\therefore \quad f = 5.02 \approx 5$$

So, it is a diatomic gas.

23. (a) The change in internal energy does not depend upon path followed by the process. It only depends on initial and final states.

Hence, $\quad \Delta U_1 = \Delta U_2$

24. (c) According to the figure,

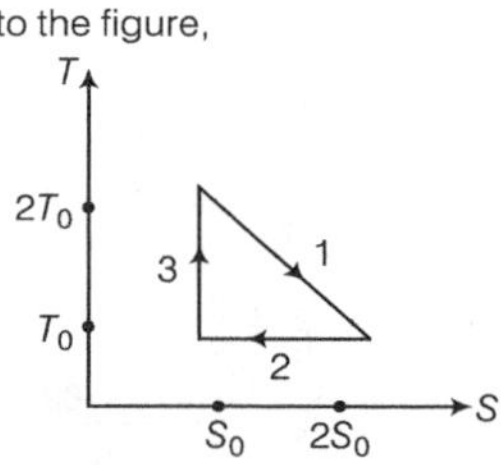

For process 1, $\quad Q_1 = T_0S_0 + \frac{1}{2} T_0S_0 = \frac{3}{2} T_0S_0$

For process 2, $\quad Q_2 = T_0(2S_0 - S_0) = T_0S_0$

For process 3, $\quad Q_3 = 0$

$$\therefore \quad \eta = \frac{W}{Q_1} = \frac{Q_1 - Q_2}{Q_1} = 1 - \frac{Q_2}{Q_1}$$

$$= 1 - \frac{2}{3} = \frac{1}{3}$$

25. (d) Statements (a) and (c) are wrong. Concept of entropy is associated with second law of thermodynamics.

26. (b) For mixture, $C_V = \dfrac{n_1 C_{V_1} + n_2 C_{V_2}}{n_1 + n_2}$

For helium,

Number of moles, $n_1 = \dfrac{16}{4} = 4$ and $\gamma_1 = \dfrac{5}{3}$

For oxygen,

Number of moles, $n_2 = \dfrac{16}{32} = \dfrac{1}{2}$ and $\gamma_2 = \dfrac{7}{5}$

Then, $$C_{V_1} = \frac{R}{\gamma_1 - 1} = \frac{R}{\frac{5}{3} - 1} = \frac{3}{2}R$$

and $$C_{V_2} = \frac{R}{\gamma_2 - 1} = \frac{R}{\frac{7}{5} - 1} = \frac{5}{2}R$$

∴ For mixture,

$$C_V = \frac{4 \times \frac{3}{2}R + \frac{1}{2} \times \frac{5}{2}R}{4 + \frac{1}{2}} = \frac{6R + \frac{5}{4}R}{\frac{9}{2}}$$

$$= \frac{29R \times 2}{9 \times 4} = \frac{29R}{18}$$

Now, $$C_V = \frac{R}{\gamma - 1}$$

$\Rightarrow$ $$\gamma - 1 = \frac{R}{C_V}$$

or $$\gamma = \frac{R}{C_V} + 1 = \frac{R}{\frac{29}{18}R} + 1$$

$\Rightarrow$ $$\frac{C_p}{C_V} = \frac{18}{29} + 1 = \frac{18 + 29}{29} = 1.62$$

27. (a) Mayer's formula is

$$C_p - C_V = R \text{ and } \gamma = \frac{C_p}{C_V}$$

Therefore, using above two relations, we find

$$C_V = \frac{R}{\gamma - 1}$$

For a mole of monatomic gas, $\gamma = \dfrac{5}{3}$

∴ $$C_V = \frac{R}{\left(\frac{5}{3}\right) - 1} = \frac{3}{2}R$$

For a mole of diatomic gas, $\gamma = \dfrac{7}{5}$

∴ $$C_V = \frac{R}{\left(\frac{7}{5}\right) - 1} = \frac{5}{2}R$$

When these two moles are mixed, heat required to raise the temperature to 1°C is

$$C_V = \frac{3}{2}R + \frac{5}{2}R = 4R$$

Hence, for one mole, heat required $= \dfrac{4R}{2} = 2R$

∴ $$C_V = 2R$$

$\Rightarrow$ $$\frac{R}{\gamma - 1} = 2R \text{ or } \gamma = \frac{3}{2}$$

Alternate Method

$$\frac{n_1 + n_2}{\gamma - 1} = \frac{n_1}{\gamma_1 - 1} + \frac{n_2}{\gamma_2 - 1}$$

Here, $n_1 = 1, n_2 = 1, \gamma_1 = \dfrac{5}{3}, \gamma_2 = \dfrac{7}{5}$

∴ $$\frac{1 + 1}{\gamma - 1} = \frac{1}{\left(\frac{5}{3}\right) - 1} + \frac{1}{\left(\frac{7}{5}\right) - 1}$$

$\Rightarrow$ $$\frac{2}{\gamma - 1} = \frac{3}{2} + \frac{5}{2} \text{ or } \frac{2}{\gamma - 1} = \frac{8}{2}$$

or $$\frac{2}{\gamma - 1} = 4 \Rightarrow \gamma = \frac{2}{4} + 1$$

Hence, $$\gamma = \frac{3}{2}$$

28. (b) Internal energy does not change in isothermal process. ΔS can be zero for adiabatic process. Work done in adiabatic process may be non-zero.

29. (a) Heat cannot flow itself from a body at lower temperature to a body at higher temperature.

This corresponds to second law of thermodynamics.

30. (d) Given, $$p \propto T^3 \quad \text{...(i)}$$

In an adiabatic process,

$$T^\gamma p^{1-\gamma} = \text{constant} \quad \left(\text{as } \gamma = \frac{C_p}{C_V}\right)$$

$$T \propto \frac{1}{p^{(1-\gamma)/\gamma}}$$

$$T^{(\gamma/\gamma - 1)} \propto p \quad \text{...(ii)}$$

Comparing Eqs. (i) and (ii), we get

$$\frac{\gamma}{\gamma - 1} = 3$$

$\Rightarrow$ $$3\gamma - 3 = \gamma \Rightarrow 2\gamma = 3$$

or $$\frac{C_p}{C_V} = \gamma = \frac{3}{2}$$

31. (c) Work does not characterise the thermodynamic state of matter, it is a path function giving only relationship between two quantities.

32. (b) $T_1 = 627 + 273 = 900\text{ K}$

$$Q_1 = 3 \times 10^6 \text{ cal}$$

$$T_2 = 27 + 273 = 300\text{ K}$$

Now, from Carnot theorem,

$$\frac{Q_1}{T_1} = \frac{Q_2}{T_2}$$

or $$Q_2 = \frac{T_2}{T_1} \times Q_1$$

$\Rightarrow$ $$Q_2 = \frac{300}{900} \times 3 \times 10^6 = 1 \times 10^6 \text{ cal}$$

∴ $$\text{Work done} = Q_1 - Q_2$$

$$= 3 \times 10^6 - 1 \times 10^6$$

$$= 2 \times 10^6 \text{ cal}$$

$$= 2 \times 4.2 \times 10^6 \text{ J}$$

$$= 8.4 \times 10^6 \text{ J}$$

33. (c) The efficiency of Carnot engine is

$$\eta = 1 - (T_2 / T_1)$$

where, T_1 is the temperature of the source and T_2 that of sink.

Since, $$\frac{T_2}{T_1} = \frac{Q_2}{Q_1}$$

So, $$\eta = 1 - \frac{Q_2}{Q_1}$$

To obtain 100% efficiency (i.e. $\eta = 1$), Q_2 must be zero i.e. if a sink at absolute zero would be available, all the heat taken from the source would have been converted into work.

The temperature of sink means a negative temperature on the absolute scale at which the efficiency of engine is greater than unity. This would be a violation of the second law of thermodynamics. Hence, a negative temperature on the absolute scale is impossible. Hence, we cannot reach absolute zero temperature.

34. (a) Efficiency of all reversible cycles depends upon temperature of source and sink which will be different. Rest of the statements are correct.

35. (c) Using the relation,

$$\frac{n_1 + n_2}{\gamma - 1} = \frac{n_1}{\gamma_1 - 1} + \frac{n_2}{\gamma_2 - 1}$$

$$\Rightarrow \quad \frac{1+1}{\gamma - 1} = \frac{1}{(5/3 - 1)} + \frac{1}{(7/5 - 1)}$$

or $$\frac{2}{\gamma - 1} = \frac{3}{2} + \frac{5}{2} = 4 \Rightarrow \gamma = \frac{3}{2} = \frac{24}{16}$$

36. (d) The rms velocity of the molecule of a gas of molecular weight M at temperature T is given by

$$c_{rms} = \sqrt{\left(\frac{3RT}{M}\right)}$$

Let M_O and M_H be molecular weights of oxygen and hydrogen and T_O and T_H be the corresponding kelvin temperatures at which c_{rms} is same for both gases.

i.e. $$c_{rms\,(O)} = c_{rms\,(H)}$$

$$\sqrt{\left(\frac{3RT_O}{M_O}\right)} = \sqrt{\left(\frac{3RT_H}{M_H}\right)}$$

Hence, $$\frac{T_O}{M_O} = \frac{T_H}{M_H}$$

Given, $$T_O = 273 + 47 = 320\text{ K}$$

and $$M_O = 32,\ M_H = 2$$

$\therefore$ $$T_H = \frac{2}{32} \times 320 = 20\text{ K}$$

JEE Advanced & IIT JEE Archive

1. (b) We have, $$\frac{p_1V_1}{T_1} = \frac{p_2V_2}{T_2}$$

(a) $$\frac{p_1V_1}{T_1} = p_2\frac{2V_1}{3T_1} \Rightarrow p_2 = \frac{3}{2}p_1 \qquad (\because S = \text{cross-section area})$$

$$p_2S = kx$$

$$\Rightarrow \quad p_2Sx = kx^2$$

$$\Rightarrow \quad \frac{1}{2}kx^2 = \frac{1}{2}p_2Sx = kx^2 - \frac{1}{2}p_2(V_2 - V_1)$$

$$\Rightarrow \text{Energy stored in spring} = \frac{1}{2} \times \frac{3}{2}p_1(2V_1 - V_1) = \frac{3}{4}p_1V_1$$

(b) $$\Delta U = \frac{3}{2}nR(T_2 - T_1) = \frac{3}{2}(p_1V_2 - p_2V_1)$$

$$\frac{p_1V_1}{T_1} = \frac{p_2 2V_1}{3T_1}$$

$$\Rightarrow \quad p_2 = \frac{3}{2}p_1$$

$$\Delta U = \frac{3}{2}\left[\frac{3}{2}p_1 \cdot 2V_1 - p_1V_1\right] = 3p_1V_1$$

(c) $$\frac{p_1V_1}{T_1} = p_2\frac{3V_1}{4T_1} \Rightarrow p_2 = \frac{4}{3}p_1$$

$$p_2S = Kx \Rightarrow p_2\frac{Sx}{2} = \frac{1}{2}kx^2 = W$$

$\because$ $$W = \frac{1}{2}\cdot\frac{4}{3}p_1(V_2 - V_1)$$

$$\frac{2}{3}p_1(3V_1 - V_1) = \frac{4}{3}p_1V_1$$

(d) $$\frac{4}{3}p_1V_1 + \frac{3}{2}(p_2V_2 - p_1V_1) = \frac{4}{3}p_1V_1 + \frac{3}{2}\left[\frac{4}{3}p_1 3V_1 - p_1V_1\right]$$

$$= \frac{4}{3}p_1V_1 + \frac{9}{2}p_1V_1$$

$$= \frac{8p_1V_1 + 27p_1V_1}{6} = \frac{35}{6}p_1V_1$$

2. (d) $$U = \frac{n_1}{2}f_1\,RT + \frac{n_2}{2}f_2\,RT$$

$$= \frac{1}{2}5RT + \frac{1}{2}3RT$$

$$= 4RT$$

$\therefore$ Average energy per mole of the mixture

$$= \frac{4RT}{2} = 2RT$$

$\because$ $$v_{rms} = \sqrt{\frac{3RT}{M}}$$

$\therefore$ $$v_{rms} \propto \frac{1}{\sqrt{M}}$$

So, $$\frac{v_{rms\,He}}{v_{rms\,H_2}} = \frac{\sqrt{\frac{1}{4}}}{\sqrt{\frac{1}{2}}} = \frac{1}{\sqrt{2}}$$

3. (d) Let final equilibrium temperature of gases is T

Heat rejected by gas by lower compartment

$$= nC_V\Delta T = 2 \times \frac{3}{2}R\,(700 - T)$$

Heat received by the gas in above compartment

$$= nC_p\Delta T = 2 \times \frac{7}{2}R\,(T - 400)$$

Equating the two, we get

$$2100 - 3T = 7T - 2800 \Rightarrow T = 490\text{ K}$$

4. (d)

$$\Delta W_1 + \Delta U_1 = \Delta Q_1 \qquad \text{...(i)}$$

$$\Delta W_2 + \Delta U_2 = \Delta Q_2 \qquad \text{...(ii)}$$

$$\Delta Q_1 + \Delta Q_2 = 0$$

$\therefore$ $$(nC_p\Delta T)_1 + (nC_p\Delta T)_2 = 0$$

But $$n_1 = n_2 = 2$$

$$\therefore \frac{5}{2}R\,(T - 700) + \frac{7}{2}R\,(T - 400) = 0$$

Solving, we get $T = 525$ K

Now, from Eqs. (i) and (ii), we get

$$\Delta W_1 + \Delta W_2 = -\Delta U_1 - \Delta U_2$$

As $\Delta Q_1 + \Delta Q_2 = 0$

$$\therefore \ \Delta W_1 + \Delta W_2 = -[(nC_V\Delta T)_1 + (nC_V\Delta T)_2]$$

$$= -\left[2 \times \frac{3}{2}R \times (525 - 700) + 2 \times \frac{5}{2}R \times (525 - 400)\right]$$

$$= -100\,R$$

5. (d) Minimum length $= \frac{\lambda}{4} \Rightarrow \lambda = 4l$

Now, $\quad v = f\lambda = (244) \times 4 \times l$

as $\quad l = 0.350 \pm 0.005$

$\Rightarrow v$ lies between 336.7 m/s to 346.5 ms^{-1}

Now, $v = \sqrt{\frac{\gamma RT}{M \times 10^{-3}}}$, here M is molecular mass in gram.

$$= \sqrt{100\gamma RT} \times \sqrt{\frac{10}{M}}$$

For monatomic gas, $\gamma = 1.67 \Rightarrow v = 640 \times \sqrt{\frac{10}{M}}$

For diatomic gas, $\gamma = 1.4 \Rightarrow v = 590 \times \sqrt{\frac{10}{M}}$

$\therefore\ v_{Ne} = 640 \times \frac{7}{10} = 448\ ms^{-1}$

$v_{Ar} = 640 \times \frac{17}{32} = 340\ ms^{-1}$

$v_{O_2} = 590 \times \frac{9}{16} = 331.8\ ms^{-1}$

$v_{N_2} = 590 \times \frac{3}{5} = 354\ ms^{-1}$

$\therefore$ Only possible answer is argon.

6. (2) $W_{ibf} = W_{ib} + W_{bf} = 50\ J + 100\ J = 150\ J$

$W_{iaf} = W_{ia} + W_{af} = 0 + 200\ J = 200\ J$

$Q_{iaf} = 500\ J$

So, $\quad \Delta U_{iaf} = Q_{iaf} - W_{iaf}$

$= 500\ J - 200\ J = 300\ J = U_f - U_i$

So, $\quad U_f = U_{iaf} + U_i$

$= 300\ J + 100\ J = 400\ J$

$\because \quad \Delta U_{ib} = U_b - U_i$

$= 200\ J - 100\ J = 100\ J$

$Q_{ib} = \Delta U_{ib} + W_{ib} = 100\ J + 50\ J = 150\ J$

$Q_{ibf} = \Delta U_{ibf} + W_{ibf}$

$= \Delta U_{iaf} + W_{ibf}$

$= 300\ J + 150\ J = 450\ J$

So, the required ratio

$$\frac{Q_{bf}}{Q_{ib}} = \frac{Q_{ibf} - Q_{ib}}{Q_{ib}} = \frac{450 - 150}{150} = 2.$$

7. (d) $\rho = \frac{pM}{RT}$

$\therefore \quad \rho \propto pM \Rightarrow \frac{\rho_1}{\rho_2} = \left(\frac{p_1}{p_2}\right)\left(\frac{M_1}{M_2}\right) = \left(\frac{2}{3}\right)\left(\frac{4}{3}\right) = \frac{8}{9}$

8. (a)

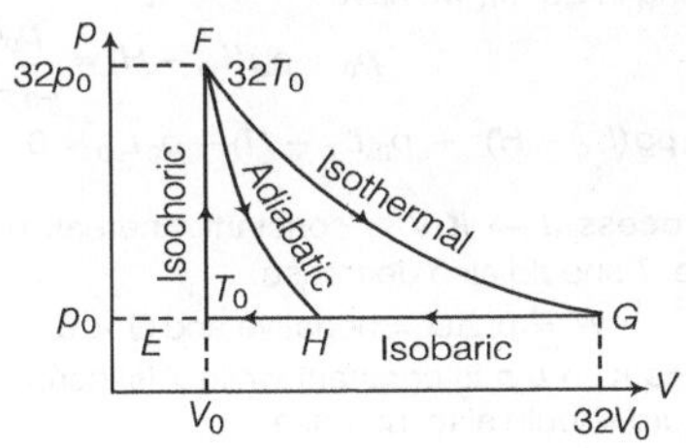

In $F \to G$, work done in isothermal process is,

$$\Delta W = nRT \ln\left(\frac{V_f}{V_i}\right) = 32\ p_0V_0 \ln\left(\frac{32V_0}{V_0}\right)$$

$$= 32\ p_0V_0 \ln 2^5 = 160p_0\ V_0 \ln 2$$

In $G \to E$, $\Delta W = p_0\Delta V = p_0(31\ V_0) = 31\ p_0V_0$

In $G \to H$, work done is less than $31\ p_0V_0$ i.e. $24\ p_0V_0$

In $F \to H$, work done is $36\ p_0V_0$.

Hence, (i) $\to$ (s), (ii) $\to$ (r), (iii) $\to$ (q), (iv) $\to$ (p)

9. (b,c,d) $Q = mCT$

$$\frac{dQ}{dt} = mc\frac{dT}{dt}$$

R = rate of absorption of heat $= \frac{dQ}{dt} \propto C$

(i) In 0-100 K C increases, So, R increases but not linearly.

(ii) $\Delta Q = mC\Delta T$ as C is more in (400 K - 500 K) then (0-100 K), so heat is increasing.

(iii) C remains constant, so there are no change in R from (400 K-500 K).

(iv) C is increase, so R increases in range (200 K-300K).

10. (d) $v_{rms} = \sqrt{\frac{3RT}{M}} \Rightarrow v_{rms} \propto \frac{1}{\sqrt{M}}$

$\therefore \quad \frac{(v_{rms})_{He}}{(v_{rms})_{Ar}} = \sqrt{\frac{M_{Ar}}{M_{He}}} = \sqrt{\frac{40}{4}} = \sqrt{10} \approx 3.16$

11. (a) At STP, 22.4 L of any gas is 1 mole.

$5.6\ L = \frac{5.6}{22.4} = \frac{1}{4}$ mole $= n$

In adiabatic process, $TV^{\gamma-1}$ = constant

$\therefore \quad T_2V_2^{\gamma-1} = T_1V_1^{\gamma-1} \Rightarrow T_2 = T_1\left(\frac{V_1}{V_2}\right)^{\gamma-1}$

$\gamma = \frac{C_p}{C_V} = \frac{5}{3}$ for monatomic He gas.

$\therefore \quad T_2 = T_1\left(\frac{5.6}{0.7}\right)^{\frac{5}{3}-1} = 4T_1$

Further in adiabatic process, $Q = 0$

$\therefore \quad W + \Delta U = 0$

$\Rightarrow \quad W = -\Delta U = -nC_V\Delta T = -n\left(\frac{R}{\gamma - 1}\right)(T_2 - T_1)$

$$= -\frac{1}{4}\left(\frac{R}{\frac{5}{3} - 1}\right)(4T_1 - T_1) = -\frac{9}{8}RT_1$$

12. (d) Internal energy $\propto T \propto pV$

This is because $\quad U = \frac{nf}{2}RT = \frac{f}{2}(pV)$

Here, n = number of moles, f = degree of freedom.

$\therefore$ If the product pV increases, then internal energy will increase and if product decreases, then internal energy will decrease.

Further, work is done on the gas, if volume of gas decreases.

For heat exchange,

$$Q = W + \Delta U$$

Work done is area under p - V graph. If volume increases, work done by gas is positive and if volume decreases, work done by gas is negative. Further ΔU is positive if product of pV is increasing and ΔU is negative if product of pV is decreasing.

If heat is taken by the gas, Q is positive and if heat is lost by the gas, Q is negative.

Keeping the above points in mind the answer to this question is as under.

Hence, (i) $\to$ (p,r, t), (ii) $\to$ (p,r), (iii) $\to$ (q,s), (iv) $\to$ (r,t)

13. (a,b) $T_A = T_B$

$\therefore \quad U_A = U_B$

$$W_{AB} = (1)(R)T_0 \ln\left(\frac{V_f}{V_i}\right)$$

$$= RT_0 \ln\left(\frac{4V_0}{V_0}\right) = p_0V_0 \ln(4)$$

Information regarding p and T at C can not be obtained from the given graph. Unless it is mentioned that line BC passes through origin or not.

Hence, the correct options are (a) and (b).

14. (4) In adiabatic process,

$TV^{\gamma - 1}$ = constant

$\therefore \quad T_i V_i^{0.4} = T_f V_f^{0.4}$ (as $\gamma = 1.4$ for diatomic gas)

$\Rightarrow \quad T_i V_i^{0.4} = (a\, T_i)\left(\frac{V_i}{32}\right)^{0.4} \Rightarrow \quad a = 4$

15. (d) A real gas behaves like an ideal gas at low pressure and high temperature.

16. (b,d) For monatomic gas, $C_p = \frac{5}{2}R$ and $C_V = \frac{3}{2}R$

For diatomic gas, $\quad C_p = \frac{7}{2}R$

and $\quad C_V = \frac{5}{2}R$

17. (b,d) (a) *p-V* graph is not rectangular hyperbola. Therefore, process *AB* is not isothermal.

(b) In process *BCD*, product of *pV* (therefore temperature and internal energy) is decreasing. Further, volume is decreasing. Hence, work done is also negative. Hence, *Q* will be negative or heat will flow out of the gas.

(c) W_{ABC} = positive

(d) For clockwise cycle on *p-V* diagram with *p* on *Y*-axis, net work done is positive.

18. (c) pT^2 = constant

$\therefore \quad \left(\frac{nRT}{V}\right)T^2 = \text{constant} \Rightarrow T^3V^{-1} = \text{constant}$

Differentiating the equation, we get

$$\frac{3T^2}{V}.dT - \frac{T^3}{V^2}dV = 0$$

or $\quad 3dT = \frac{T}{V}.dV \quad$ ... (i)

From the equation, $\quad dV = V\gamma dT$

(where, γ = coefficient of volume)

Expansion of gas $= \frac{dV}{V.dT}$

From Eq. (i), $\gamma = \frac{dV}{V.dT} = \frac{3}{T}$

19. (a) (i) In case of free expansion under adiabatic conditions, change in internal energy $\Delta U = 0$.

$\therefore$ Internal energy and temperature will remain constant.

(ii) $p \propto \frac{1}{V^2}$

$\therefore \quad pV^2$ = constant ... (i)

or $\quad \left(\frac{nRT}{V}\right).V^2$ = constant

$\therefore \quad T \propto \frac{1}{V}$ ... (ii)

If volume is doubled, temperature will decrease as per Eq. (ii). Further, molar heat capacity in process pV^x = constant is

$$C = C_V + \frac{R}{1-x}$$

From Eq. (i), $x = 2$

$\therefore \quad C = \frac{3}{2}R + \frac{R}{1-2} = +\frac{R}{2}$

Since, molar heat capacity is positive, according to $Q = nC\Delta T$, *Q* will be negative, if ΔT is negative or gas loses heat, if temperature is decreasing.

(iii) $p \propto \frac{1}{V^{4/3}}, \; pV^{4/3}$ = constant

$\therefore \quad \left(\frac{nRT}{V}\right)V^{4/3}$ = constant

$\therefore \quad T \propto \frac{1}{V^{1/3}}$

Further, with increase in volume temperature will decrease.

Here, $\quad x = \frac{4}{3}$

$\therefore \quad C = \frac{3}{2}R + \frac{R}{1-\frac{4}{3}} = -1.5R$

As molar heat capacity is negative, *Q* will be positive if ΔT is negative or gas gains heat with decrease in temperature.

(iv) $T \propto pV$

In expansion from V_1 to $2\,V_1$, product of *pV* is increasing. Therefore, temperature will increase, i.e. ΔU = positive. Further in expansion, work done is also positive.

Hence, $Q = W + \Delta U$ = positive or gas gains heat.

Hence, (i) → (q), (ii) → (p,r), (iii) → (p,s), (iv) → (q,s)

20. (b) Total translation kinetic energy $= \frac{3}{2}nRT = \frac{3}{2}pV = 1.5\,pV$

21. (a) Since it is open from top, pressure will be p_0.

$\therefore$ Correct option is (a).

22. (d) Let *p* be the pressure in equilibrium.

Then, $\quad pA = p_0A - Mg$

$$p = p_0 - \frac{Mg}{A} = p_0 - \frac{Mg}{\pi R^2}$$

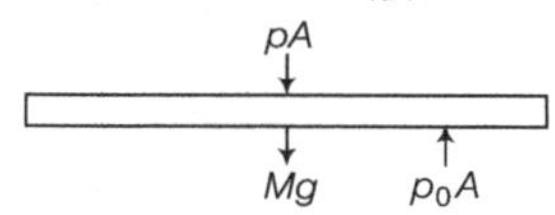

Applying, $p_1V_1 = p_2V_2$

$\therefore \quad p_0(2AL) = (p)(AL')$

$$\therefore \quad L' = \frac{2p_0L}{p} = \left(\frac{p_0}{p_0 - \frac{Mg}{\pi R^2}}\right)(2L) = \left(\frac{p_0\pi R^2}{\pi R^2 p_0 - Mg}\right)(2L)$$

23. (c) $p_1 = p_2$

$$p_0 + \rho g(L_0 - H) = p \quad \text{...(i)}$$

Now, applying $p_1V_1 = p_2V_2$ for the air inside the cylinder, we have

$$p_0(L_0) = p(L_0 - H)$$

$$\therefore \quad p = \frac{p_0L_0}{L_0 - H}$$

Substituting in Eq. (i), we have

$$p_0 + \rho g(L_0 - H) = \frac{p_0L_0}{L_0 - H}$$

or $\quad \rho g(L_0 - H)^2 + p_0(L_0 - H) - p_0L_0 = 0$

24. (b) **In process $J \to K$** *V* is constant whereas *p* is decreasing. Therefore, *T* should also decrease.

$\therefore \quad W = 0, \Delta U$ = negative and $Q < 0$

In process $K \to L$ *p* is constant while *V* is increasing. Therefore, temperature should also increase.

$\therefore \quad W > 0, \Delta U > 0$ and $Q > 0$

In process $L \to M$ This is inverse of process $J \to K$.

$\therefore \quad W = 0, \Delta U > 0$ and $Q > 0$

In process $M \to J$ *V* is decreasing. Therefore, $W < 0$

$(pV)_J < (pV)_M$

$\therefore \quad T_J < T_M$ or $\Delta U < 0$

Therefore, $\quad Q < 0$.

Hence, (i) → (s), (ii) → (p,r), (iii) → (p), (iv) → (q,s)

Last 3 Years' Questions
JEE Main & Advanced
(2016-2018)

JEE Main

2016

1. An ideal gas undergoes a quasi static, reversible process in which its molar heat capacity C remains constant. If during this process the relation of pressure p and volume V is given by pV^n = constant, then n is given by (here, C_p and C_V are molar specific heat at constant pressure and constant volume, respectively)

[Kinetic Theory of Gases and Thermodynamics]

(a) $n = \frac{C_p}{C_V}$ (b) $n = \frac{C - C_p}{C - C_V}$

(c) $n = \frac{C_p - C}{C - C_V}$ (d) $n = \frac{C - C_V}{C - C_p}$

2. n moles of an ideal gas undergoes a process A and B as shown in the figure. The maximum temperature of the gas during the process will be

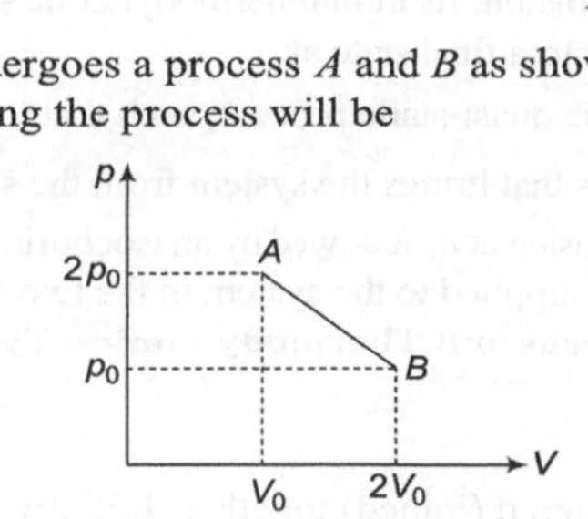

[Kinetic Theory of Gases and Thermodynamics]

(a) $\frac{9}{4}\frac{p_0V_0}{nR}$ (b) $\frac{3}{2}\frac{p_0V_0}{nR}$

(c) $\frac{9}{2}\frac{p_0V_0}{nR}$ (d) $\frac{9p_0V_0}{nR}$

2017

3. C_p and C_V are specific heats at constant pressure and constant volume, respectively. It is observed that $C_p - C_V = a$ for hydrogen gas $C_p - C_V = b$ for nitrogen gas. The correct relation between a and b is **[Kinetic Theory of Gases and Thermodynamics]**

(a) $a = b$ (b) $a = 14\,b$

(c) $a = 28\,b$ (d) $a = \frac{1}{14}\,b$

4. A copper ball of mass 100 g is at a temperature T. It is dropped in a copper calorimeter of mass 100 g, filled with 170 g of water at room temperature. Subsequently, the temperature of the system is found to be 75°C. T is (Given, room temperature = 30°C, specific heat of copper = 0.1 cal/g°C) **[Heat, Temperature and Heat Transfer]**

(a) 885°C (b) 1250°C

(c) 825°C (d) 800°C

5. The temperature of an open room of volume 30 m^3 increases from 17°C to 27°C due to the sunshine. The atmospheric pressure in the room remains 1×10^5 Pa. If n_i and n_f are the number of molecules in the room before and after heating, then $n_f - n_i$ will be **[Kinetic Theory of Gases and Thermodynamics]**

(a) 1.38×10^{23}
(b) 2.5×10^{25}
(c) -2.5×10^{25}
(d) -1.61×10^{23}

2018

6. Two moles of an ideal monoatomic gas occupies a volume V at 27°C. The gas expands adiabatically to a volume $2V$. Calculate (i) the final temperature of the gas and (ii) change in its internal energy. **[Kinetic Theory of Gases and Thermodynamic]**

(a) (i) 189 K (ii) 2.7 kJ
(b) (i) 195 K (ii) −2.7 kJ
(c) (i) 189 K (ii) −2.7 kJ
(d) (i) 195 K (ii) 2.7 kJ

JEE Advanced

2016

1. A metal is heated in a furnace where a sensor is kept above the metal surface to read the power radiated (P) by the metal. The sensor has a scale that displays $\log_2(P/P_0)$, where P_0 is a constant. When the metal surface is at a temperature of 487°C, the sensor shows a value 1. Assume that the emissivity of the metallic surface remains constant. What is the value displayed by the sensor when the temperature of the metal surface is raised to 2767°C? **[Heat, Temperature and Heat Transfer, Type 3]**

2. A gas is enclosed in a cylinder with a movable frictionless piston. Its initial thermodynamic state at pressure $P_i = 10^5$ Pa and volume $V_i = 10^{-3}$ m^3 changes to a final state at $P_f = (1/32) \times 10^5$ Pa and $V_f = 8 \times 10^{-3}$ m^3 in an adiabatic quasi-static process, such that P^3V^5 = constant. Consider another thermodynamic process that brings the system from the same initial state to the same final state in two steps : an isobaric expansion at P_i followed by an isochoric (isovolumetric) process at volume V_f. The amount of heat supplied to the system in the two-step process is approximately **[Kinetic Theory of Gases and Thermodynamics, Type 1]**

(a) 112 J
(b) 294 J
(c) 588 J
(d) 813 J

3. The ends Q and R of two thin wires, PQ and RS, are soldered (joined) together. Initially, each of the wire has a length of 1 m 10°C. Now, the end P is maintained at 10°C, while the end S is heated and maintained at 400°C. The system is thermally insulated from its surroundings. If the thermal conductivity of wire PQ is twice that of the wire RS and the coefficient of linear thermal expansion of PQ is 1.2×10^{-5} K^{-1}, the change in length of the wire PQ is **[Heat, Temperature and Heat Transfer, Type 1]**

(a) 0.78 mm
(b) 0.90 mm
(c) 1.56 mm
(d) 2.34 mm

4. A water cooler of storage capacity 120 litres can cool water at a constant rate of P watts. In a closed circulation system (as shown schematically in the figure), the water from the cooler is used to cool an external device that generates constantly 3 kW of heat (thermal load).

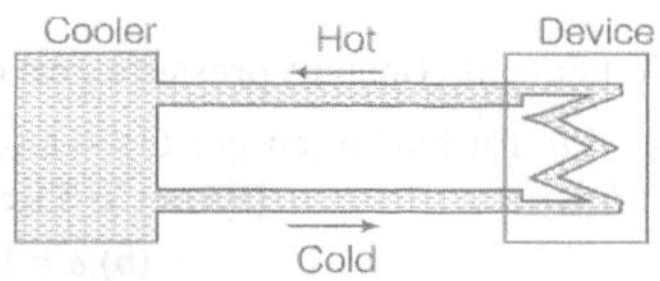

The temperature of water fed into the device cannot exceed 30°C and the entire stored 120 litres of water is initially cooled to 10°C. The entire system is thermally insulated. The minimum value of P (in watts) for which the device can be operated for 3 hours is **[Heat, Temperature and Heat Transfer, Type 2]**

(Specific heat of water is 4.2 kJ $kg^{-1}K^{-1}$ and the density of water is 1000 kg m^{-3})

(a) 1600
(b) 2067
(c) 2533
(d) 3933

Note ***Type-1*** *Only one correct option,* ***Type-2*** *More then one correct option,* ***Type-3*** *Single digit Answer,* ***Type-4*** *Numerical value based*

2017

5. A human body has a surface area of approximately 1 m^2. The normal body temperature is 10K above the surrounding room temperature T_0. Take the room temperature to be $T_0 = 300$ K. For $T_0 = 300$ K, the value of $\sigma T_0^4 = 460\ Wm^{-2}$ (where σ is the Stefan Boltzmann constant). Which of the following options is/are correct?

[Kinetic Theory of Gases and Thermodynamics, Type 2]

(a) If the body temperature rises significantly, then the peak in the spectrum of electromagnetic radiation emitted by the body would shift to longer wavelengths

(b) If the surrounding temperature reduces by a small amount $\Delta T_0 << T_0$, then to maintain the same body temperature the same (living) human being needs to radiate $\Delta W = 4\sigma T_0^3 \Delta T_0$ more energy per unit time

(c) The amount of energy radiated by the body in 1s is close to 60 J

(d) Reducing the exposed surface area of the body (e.g. by curling up) allows humans to maintain the same body temperature while reducing the energy lost by radiation

Directions (Q.Nos. 6-8) Matching the information given in the three columns of the following table.

An ideal gas is undergoing a cyclic thermodynamic process in different ways as shown in the corresponding *p-V* diagrams in column 3 of the table. Consider only the path from state 1 to state 2. *W* denotes the corresponding work done on the system. The equations and plots in the table have standards notations and used in thermodynamic processes. Here γ is the ratio of heat capacities at constant pressure and constant volume. The number of moles in the gas is *n*.

Column 1	Column 2	Column 3
(I) $W_{1\to 2} = \frac{1}{\gamma - 1}(p_2V_2 - p_1V_1)$	(i) Isothermal	(P) 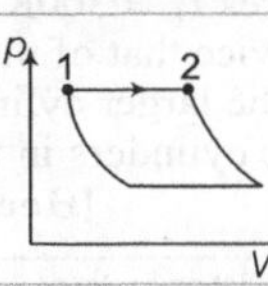
(II) $W_{1\to 2} = -pV_2 + pV_1$	(ii) Isochoric	(Q) 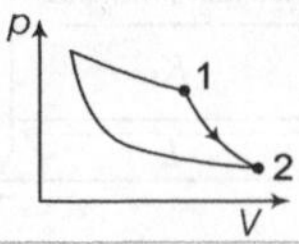
(III) $W_{1\to 2} = 0$	(iii) Isobaric	(R) 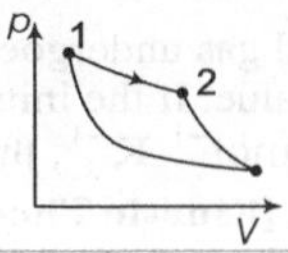
(IV) $W_{1\to 2} = -nRT \ln\left(\frac{V_2}{V_1}\right)$	(iv) Adiabatic	(S) 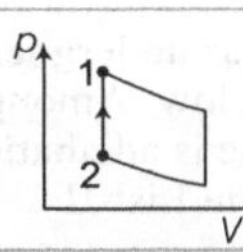

[Heat, Temperature and Heat Transfer]

6. Which one of the following options correctly represents a thermodynamic process that is used as a correction in the determination of the speed of sound in an ideal gas?

(a) (IV) (ii) (R) (b) (I) (ii) (Q)

(c) (I), (iv) (Q) (d) (III) (iv) (R)

7. Which of the following options is the only correct representation of a process in which $\Delta U = \Delta Q - p\Delta V$?

(a) (II) (iii) (S) (b) (II) (iii) (P)

(c) (III) (iii) (P) (d) (II) (iv) (R)

8. Which one of the following options is the correct combination?

(a) (II) (iv) (P) (b) (III) (ii) (S)

(c) (II) (iv) (R) (d) (IV) (ii) (S)

9. A flat plane is moving normal to its plane through a gas under the action of a constant force F. The gas is kept at a very low pressure. The speed of the plate v is much less than the average speed u of the gas molecules. Which of the following options is/are true?

[Heat, Temperature and Heat Transfer, Type 2]

(a) At a later time the external force F balances the resistive force

(b) The plate will continue to move with constant non-zero acceleration, at all times

(c) The resistive force experienced by the plate is proportional to v

(d) The pressure difference between the leading and trailing faces of the plate is proportional to uv

2018

10. One mole of a monoatomic ideal gas undergoes a cyclic process as shown in the figure (where, V is the volume and T is the temperature). Which of the statements below is (are) true?

[Kinetic Theory of Gases and Thermodynamics, Type 2]

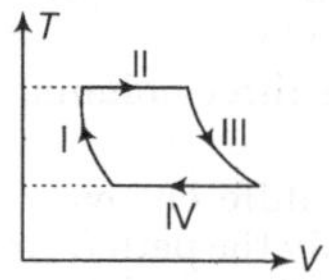

(a) Process I is an isochoric process

(b) In process II, gas absorbs heat

(c) In process IV, gas releases heat

(d) Processes I and III are not isobaric

11. Two conducting cylinders of equal length but different radii are connected in series between two heat baths kept at temperatures $T_1 = 300\text{K}$ and $T_2 = 100\text{K}$, as shown in the figure. The radius of the bigger cylinder is twice that of the smaller one and the thermal conductivities of the materials of the smaller and the larger cylinders are K_1 and K_2, respectively. If the temperature at the junction of the cylinders in the steady state is 200K, then K_1/K_2 =.........

[Heat, Temperature and Heat Transfer, Type 4]

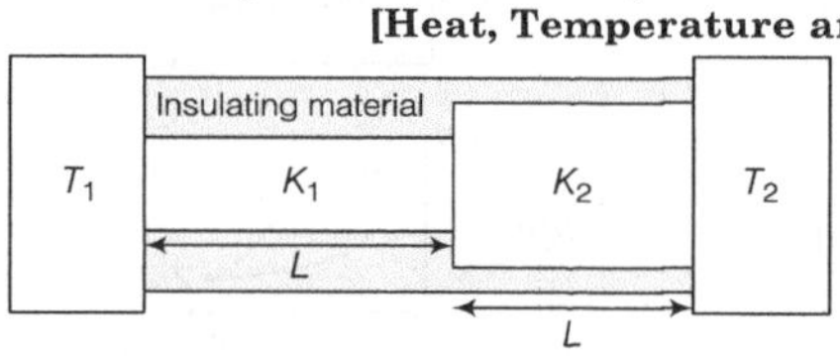

12. One mole of a monoatomic ideal gas undergoes an adiabatic expansion in which its volume becomes eight times its initial value. If the initial temperature of the gas is 100 K and the universal gas constant $R = 8.0 \text{ j mol}^{-1} \text{ K}^{-1}$, the decrease in its internal energy in joule, is ....

[Kinetic Theory of Gases and Thermodynamics, Type 4]

13. One mole of a monoatomic ideal gas undergoes four thermodynamic processes as shown schematically in the pV-diagram below. Among these four processes, one is isobaric, one is isochoric, one is isothermal and one is adiabatic. Match the processes mentioned in List-I with the corresponding statements in List-II.

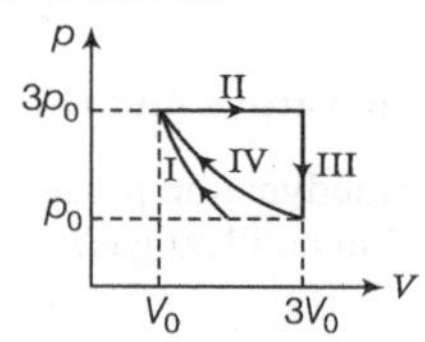

	List-I		List-II
P.	In process I	1.	Work done by the gas is zero
Q.	In process II	2.	Temperature of the gas remains unchanged
R.	In process III	3.	No heat is exchanged between the gas and its surroundings
S.	In process IV	4.	Work done by the gas is $6p_0V_0$

[Kinetic Theory of Gases and Thermodynamics]

(a) $P \to 4$; $Q \to 3$; $R \to 1$; $S \to 2$

(b) $P \to 1$; $Q \to 3$; $R \to 2$; $S \to 4$

(c) $P \to 3$; $Q \to 4$; $R \to 1$; $S \to 2$

(d) $P \to 3$; $Q \to 4$; $R \to 2$; $S \to 1$

Answer *with* Explanations

JEE Main

1. *(b)* $\Delta Q = \Delta U + \Delta W$

In the process

pV^n = constant, molar heat capacity is given by

$$C = \frac{R}{\gamma - 1} + \frac{R}{1-n} = C_V + \frac{R}{1-n}$$

$$C - C_V = \frac{R}{1-n} \Rightarrow 1 - n = \frac{C_p - C_V}{C - C_V}$$

$$\therefore \quad n = 1 - \left(\frac{C_p - C_V}{C - C_V}\right) = \frac{(C - C_V) - (C_p - C_V)}{C - C_V}$$

$$= \frac{C - C_p}{C - C_V}$$

2. *(a)* *p-V* equation for path *AB*

$$p = -\left(\frac{p_0}{V_0}\right)V + 3p_0$$

$$\Rightarrow \quad pV = 3p_0V - \frac{p_0}{V_0}V^2$$

$$\text{or} \quad T = \frac{pV}{nR} = \frac{1}{nR}\left(3p_0V - \frac{p_0}{V_0}V^2\right)$$

For maximum temperature,

$$\frac{dT}{dV} = 0 \quad \Rightarrow 3p_0 - \frac{2p_0V}{V_0} = 0$$

$$\Rightarrow \quad V = \frac{3}{2}V_0 \text{ and } p = 3p_0 - \frac{p_0}{V_0} = \frac{3p_0}{2}$$

Therefore, at these values

$$T_{\max} = \frac{\left(\frac{3p_0}{2}\right)\left(\frac{3V_0}{2}\right)}{nR} = \frac{9p_0V_0}{4nR}$$

3. *(b)* By Mayor's relation, for 1 g mole of a gas,

$$C_p - C_V = R$$

So, when *n* gram moles are given,

$$C_p - C_V = \frac{R}{n}$$

As per given question,

$$a = C_p - C_V = \frac{R}{2}; \text{ for } H_2 \quad \ldots \text{(i)}$$

$$b = C_p - C_V = \frac{R}{28}; \text{ for } N_2 \quad \ldots \text{(ii)}$$

From Eqs. (i) and (ii), we get

$$a = 14b$$

4. *(a)* Heat gained (water + calorimeter)

= Heat lost by copper ball

$$\Rightarrow \quad m_w s_w \Delta T + m_c s_c \Delta T = m_B s_B \Delta T$$

$$\Rightarrow \quad 170 \times 1 \times (75 - 30) + 100 \times 0.1 \times (75 - 30) = 100 \times 0.1 \times (T - 75)$$

$$\therefore \quad T = 885°\text{C}$$

5. *(c)* From $pV = nRT = \frac{N}{N_A}RT$

We have, $n_f - n_i = \frac{pVN_A}{RT_f} - \frac{pVN_A}{RT_i}$

$$\Rightarrow \quad n_f - n_i = \frac{10^5 \times 30}{8.3} \times 6.02 \times 10^{23} \cdot \left(\frac{1}{300} - \frac{1}{290}\right)$$

$$= -2.5 \times 10^{25}$$

$$\therefore \quad \Delta n = -2.5 \times 10^{25}$$

6. *(c)* For adiabatic process relation of temperature and volume is,

$$T_2V_2^{\gamma-1} = T_1V_1^{\gamma-1}$$

$$\Rightarrow \quad T_2(2V)^{2/3} = 300(V)^{2/3}$$

$$\left[\gamma = \frac{5}{3} \text{ for monoatomic gases}\right]$$

$$\Rightarrow \quad T_2 = \frac{300}{2^{2/3}} \approx 189\text{K}$$

Also, in adiabatic process,

$$\Delta Q = 0, \Delta U = -\Delta W$$

$$\text{or} \quad \Delta U = \frac{-nR(\Delta T)}{\gamma - 1}$$

$$= -2 \times \frac{3}{2} \times \frac{25}{3}(300 - 189)$$

$$\approx -2.7 \text{ kJ}$$

$$T_2 \approx 189K, \Delta U \approx 2.7\text{kJ}$$

JEE Advanced

1. *(9)* $\log_2 \frac{p_1}{p_0} = 1$

Therefore, $\frac{p_1}{p_0} = 2$

According to Stefan's law, $p \propto T^4$

$$\Rightarrow \quad \frac{p_2}{p_1} = \left(\frac{T_2}{T_1}\right)^4 = \left(\frac{2767 + 273}{487 + 273}\right)^4 = 4^4$$

$$\frac{p_2}{p_1} = \frac{p_2}{2p_0} = 4^4$$

$$\Rightarrow \quad \frac{p_2}{p_0} = 2 \times 4^4$$

$$\log_2 \frac{p_2}{p_0} = \log_2[2 \times 4^4]$$

$$= \log_2 2 + \log_2 4^4$$

$$= 1 + \log_2 2^8 = 1 + 8 = 9$$

2. *(c)* In the first process : $p_iV_i^\gamma = p_fV_f^\gamma$

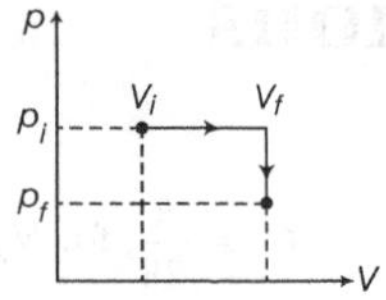

$$\Rightarrow \quad \frac{p_i}{p_f} = \left(\frac{V_f}{V_i}\right)^\gamma \Rightarrow 32 = 8^\gamma$$

$$\gamma = \frac{5}{3} \quad \ldots\text{(i)}$$

For the two step process,.

$$W = p_i(V_f - V_i) = 10^5(7\times10^{-3})$$

$$W = 7\times10^2\,\text{J}$$

$$\Delta U = \frac{f}{2}(p_fV_f - p_iV_i)$$

$$= \frac{1}{\gamma-1}\left(\frac{1}{4}\times10^2 - 10^2\right)$$

$$\Delta U = -\frac{3}{2}\cdot\frac{3}{4}\times10^2 = -\frac{9}{8}\times10^2\,\text{J}$$

$$Q - W = \Delta U$$

$$\Rightarrow \quad Q = 7\times10^2 - \frac{9}{8}\times10^2$$

$$= \frac{47}{8}\times10^2\,\text{J} = 588\ \text{J}$$

3. *(a)*

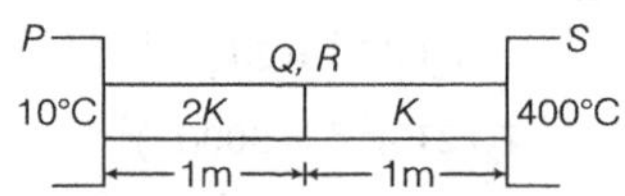

Rate of heat flow from P to Q,

$$\frac{dQ}{dt} = \frac{2KA\,(T-10)}{1}$$

Rate of heat flow from Q to S,

$$\frac{dQ}{dt} = \frac{KA\,(4000-T)}{1}$$

At steady state, state rate of heat flow is same

$$\therefore \quad \frac{2KA\,(T-10)}{1} = KA\,(400-T)$$

or $2T - 20 = 400 - T$

or $3T = 420$

$\therefore$ $T = 140°$

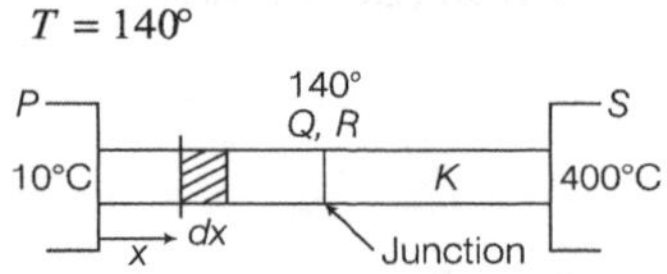

Temperature of junction is 140°C.

Temperature at a distance x from end P is

$$T_x = (130x + 10°)$$

Change in length dx is suppose dy.

Then, $\quad dy = \alpha dx\,(T_x - 10)$

$$\int_0^{\Delta y} dy = \int_0^1 \alpha dx\,(130x + 10 - 10)$$

$$\Delta y = \left[\frac{\alpha x^2}{2}\times130\right]_0^1$$

$$\Delta y = 1.2\times10^{-5}\times65$$

$$\Delta y = 78.0\times10^{-5}\,\text{m}$$

$$= 0.78\ \text{mm}$$

4. *(b)* Heat generated in device in 3 h

$$= \text{Time}\times\text{power}$$

$$= 3\times3600\times3\times10^3$$

$$= 324\times10^5\,\text{J}$$

Heat used to heat water $= ms\Delta\theta$

$$= 120\times1\times4.2\times10^3\times20\text{J}$$

Heat absorbed by coolant

$$Pt = 324\times10^5 - 120\times1\times4.2\times10^3\times20\text{J}$$

$$Pt = (325-100.8)\times10^5\,\text{J} = 223.2\times10^5\,\text{J}$$

$$P = \frac{223.2\times10^5}{3600} = 2067\ \text{W}$$

5. *(b,c,d)* Assumption $e = 1$ [black body radiation]

$$P = \sigma A(T^4 - T_0^4)$$

(c) $P_{\text{rad}} = \sigma AT^4 = \sigma\cdot1\cdot(T_0+10)^4$

$$= \sigma\cdot T_0^4\left(1+\frac{10}{T_0}\right)^4 \qquad [T_0 = 300\text{K given}]$$

$$= \sigma\cdot(300)^4\cdot\left(1+\frac{40}{300}\right) \approx 460\times\frac{17}{15} \approx 520\ \text{J}$$

$$P_{\text{net}} = 520 - 460 \approx 60\ \text{W}$$

$\Rightarrow$ Energy radiated in 1 s = 60 J

(b) $P = \sigma A(T^4 - T_0^4)$

$$dP = \sigma A(0 - 4T_0^3\cdot dT)$$

and $\quad dT = -\Delta T$

$\Rightarrow \quad dP = 4\sigma AT_0^3\Delta T$

(d) If surface area decreases, then energy radiation also decreases.

Note *While giving answer (b) and (c) it is assumed that energy radiated refers the net radiation. If energy radiated is taken as only emission, then (b) and (c) will not be included in answer.*

6. *(c)* $V_{\text{sound}} = \sqrt{\dfrac{\gamma RT}{M}}$. As the sound wave propagates, the air in a chamber undergoes compression and rarefaction very fastly, hence undergo a adiabatic process. So, curves are steeper than isothermal.

$$\left(\frac{dp}{dV}\right)_{\text{Adi}} = -\gamma\left(\frac{p}{V}\right) \quad \ldots\text{(i)}$$

$$\left(\frac{dp}{dV}\right)_{\text{Iso}} = -\left(\frac{P}{V}\right) \quad \ldots\text{(ii)}$$

Graph 'Q' satisfies Eq. (i)

7. *(b)* $\Delta U = \Delta Q - p\Delta V, \Delta U + p\Delta V = \Delta Q$

As $\Delta U \neq 0, W \neq 0, \Delta Q \neq 0$.

The process represents, isobaric process

$$W_{gas} = -p(\Delta V)$$
$$= -p(V_2 - V_1)$$
$$= -pV_2 + pV_1$$

Graph 'P' satisfies isobaric process.

8. *(b)* Work done in isochoric process is zero.

$$W_{12} = 0 \text{ as } \Delta V = 0$$

Graph 'S' represents isochoric process.

9. *(a,c,d)*

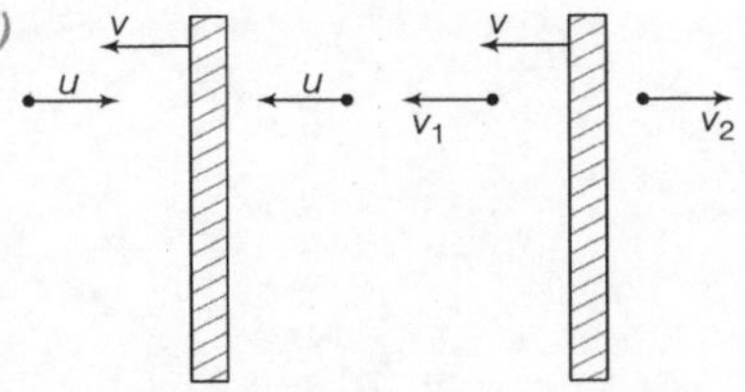

Just before the collision Just after the collision

$$v_1 = u + 2v$$
$$\Delta v_1 = (2u + 2v)$$
$$F_1 = \frac{dp_1}{dt} = \rho A(u+v)(2u+2v)$$
$$= 2\rho A(u+v)^2$$
$$v_2 = (u - 2v)$$
$$\Delta v_2 = (2u - 2v)$$
$$F_2 = \frac{dp_2}{dt} = \rho A(u-v)(2u-2v)$$
$$= 2\rho A(u-v)^2$$

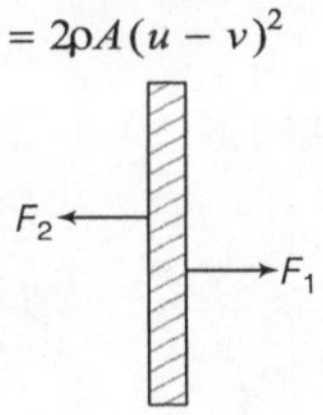

[ΔF is the net force due to the air molecules on the plate]

$$\Delta F = 2\rho A(4uv) = 8\rho Auv$$
$$P = \frac{\Delta F}{A} = 8\rho(uv)$$
$$F_{net} = (F - \Delta F) = ma$$

[m is mass of the plate]

$$F - (8\rho Au)v = ma$$

10. *(b,c,d)* (b) Process-II is isothermal expansion,

$$\Delta U = 0, W > 0$$
$$\Delta Q = W > 0$$

(c) Process-IV is isothermal compression,

$$\Delta U = 0, W < 0$$
$$\Delta Q = W < 0$$

(d) Process-I and III are not isobaric because in isobaric process $T \propto V$ hence, T-V graph should be a straight line passing through origin.

11. **(4)** Rate of heat flow will be same,

$$\therefore \quad \frac{300-200}{R_1} = \frac{200-100}{R_2} \left(\text{as } H = \frac{dQ}{dt} = \frac{T \cdot D}{R}\right)$$
$$\therefore \quad R_1 = R_2 \Rightarrow \frac{L_1}{K_1A_1} = \frac{L_2}{K_2A_2}$$
$$\Rightarrow \quad \frac{K_1}{K_2} = \frac{A_2}{A_1} = 4$$

12. *(900)* Given, $n = 1, \gamma = \frac{5}{3}$

T-V equation in adiabatic process is

$TV^{\gamma-1}$ = constant

$$\therefore \quad T_1V_1^{\gamma-1} = T_2V_2^{\gamma-1}$$
$$\Rightarrow \quad T_2 = T_1\left(\frac{V_1}{V_2}\right)^{\gamma-1} = 100 \times \left(\frac{1}{8}\right)^{\frac{2}{3}}$$
$$\Rightarrow \quad T_2 = 25\text{ K}$$

$C_V = \frac{3}{2}R$ for monoatomic gas

$$\therefore \quad \Delta U = nC_V\Delta T = n \times \left(\frac{3R}{2}\right)(T_2 - T_1)$$
$$= 1 \times \frac{3}{2} \times 8 \times (25 - 100)$$
$$= -900\text{ J}$$

$\therefore$ Decrease in internal energy = 900 J

13. *(c)* $\left(\frac{dp}{dV}\right)_{adiabatic} = \gamma\left(\frac{dp}{dV}\right)_{isothermal}$

List-I

(P) Process I $\Rightarrow$ Adiabatic $\Rightarrow Q = 0$

(Q) Process II $\Rightarrow$ Isobaric

$\therefore \quad W = p\Delta V = 3p_0[3V_0 - V_0] = 6p_0V_0$

(R) Process III $\Rightarrow$ Isochoric $\Rightarrow W = 0$

(S) Process (IV) $\Rightarrow$ Isothermal

$\Rightarrow$ Temperature = Constant